S0-CKG-381

Molecules and Models

Molecules and Models

The molecular structures of main group element compounds

ARNE HAALAND
Department of Chemistry
University of Oslo
May 2007

OXFORD
UNIVERSITY PRESS

Great Clarendon Street, Oxford OX2 6DP

Oxford University Press is a department of the University of Oxford.
It furthers the University's objective of excellence in research, scholarship,
and education by publishing worldwide in

Oxford New York

Auckland Cape Town Dar es Salaam Hong Kong Karachi
Kuala Lumpur Madrid Melbourne Mexico City Nairobi
New Delhi Shanghai Taipei Toronto

With offices in

Argentina Austria Brazil Chile Czech Republic France Greece
Guatemala Hungary Italy Japan Poland Portugal Singapore
South Korea Switzerland Thailand Turkey Ukraine Vietnam

Published in the United States
by Oxford University Press Inc., New York

First published 2008

British Library Cataloguing in Publication Data
Data available

Library of Congress Cataloging in Publication Data
Data available

Typeset by Newgen Imaging Systems (P) Ltd., Chennai, India
Printed in Great Britain
on acid-free paper by
Biddles Ltd., King's Lynn, Norfolk

ISBN 978–0–19–923535–3

1 3 5 7 9 10 8 6 4 2

Preface

In this book we provide a survey of the molecular structures of some 300 representative compounds of the main group elements, i.e. the elements in Groups 1 and 2, and 12 through 18. The emphasis is on their equilibrium structures, but some thermochemical parameters, particularly bond energies, are included when available. Most of the quoted structures have been determined in the gas phase, but crystal structures are included when gas phase structures are lacking, or when the change of phase is accompanied by a significant change of structure. The aim has been to organize the survey in a manner that brings out similarities and trends, and at the same time give adequate attention to those molecules that break well-established patterns.

Chemists use models to interpret and systematize information and to extrapolate from the known to the unknown. A good model is simple, has a wide validity range, and precise predictive power. However, since they all represent simplifications of a more complex reality, they will all fail in some instances. In this book we describe these models in parallel with the molecular structures for which they are most successful. The aim has been to provide a clear description of their physical basis, and to delineate the limits of their applicability.

During four decades of research on molecular structure and bonding, I have benefited from fruitful collaboration and innumerable discussions with fellow scientists in Norway and abroad. I wish, however, to express my particular gratitude to my colleagues at the University of Oslo, Professors Knut Fægri, Grete Gundersen, Trygve Helgaker, Svein Samdal, Tor Strand and Einar Uggerud; and to some of my students and postdocs, Richard Blom, Paul Kiprof, Kjell-Gunnar Martinsen, Kristin Rypdal, Wolfgang Scherer, Dmitry J. Shorokhov, Vasily I. Sokolov, Tatyana Strenalyuk, Andrey V. Tutukin, Natalya V. Tverdova, Hans Peter Verne, and Kari-Anne Østby.

The information contained in this book has been collected by hundreds of scientists and published in hundreds, perhaps a thousand, articles. We do hope that the reader will accept that we have found it too cumbersome to give a complete set of references to the primary research literature. Instead we refer to widely available tabulations of structure data, thermodynamic data and atomic energy levels. Only when the original work is considered to be of exceptional importance or originality, when it is relatively recent, or when it may be particularly difficult to locate, have we included a full reference.

Finally: this is a book that I have wanted to write, and I should be delighted to learn that somebody has wanted to read it!

Oslo,
June 2007

Arne Haaland

Contents

List of tables

Chapter 1

Atomic properties: one-electron atoms

Introduction

What would a molecule look like if it was large enough to be seen? In Fig. 1.1 we show a picture of a water molecule as many chemists would imagine it. You can recognize the three spherical atoms, the O atom larger than the H atoms. The atoms do not only touch, they penetrate each other, much like burs sticking together. The structure of the molecule is described by the two O–H bond distances, defined as the distances between the O and H nuclei, and the valence angle, defined as the angle between the lines connecting the O nucleus to the two H nuclei. *To determine the structure of a molecule is to determine the relative positions of the atoms.*

One thing is to know the structure of a molecule, another is to have the feeling that you understand it. It is an experimental fact that the equilibrium bond distances are equal (95.8 pm) and that the valence angle is 104.5°, but why are the bond distances equal, and why is the valence angle not tetrahedral (109.5°) or even 180°?

The answer to such questions is sought in the nature of the atoms involved: how large and how electronegative are they, what is the number of valence electrons on each of them? We therefore begin by describing the properties of separate atoms. As you know, such a description must be based on the laws of physics, in particular those of quantum mechanics.

1.1 A dash of physics: a particle in motion

The position of a particle that is free to move in three directions may be described by giving its coordinates in a Cartesian coordinate system: x, y and z. Alternatively the position may be indicated by drawing a vector $\mathbf{r}$ from the origin of the coordinate system to the

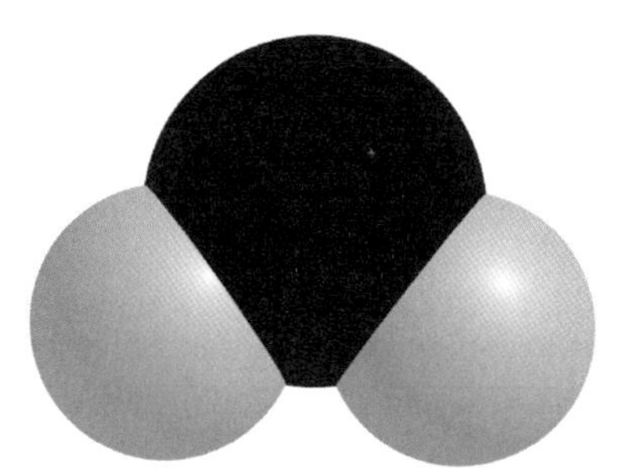

Fig. 1.1. A so-called "space-filling" representation of the H_2O molecule. The atoms have been drawn as spheres with radii corresponding to their van der Waals radii. See Chapter 9.

particle position:

$$\mathbf{r} = x\mathbf{i} + y\mathbf{j} + z\mathbf{k}$$

The magnitude, or length, of the vector is denoted by r and is given by:

$$r^2 = |\mathbf{r}|^2 = x^2 + y^2 + z^2$$

The velocity of the particle in the x-, y-, or z-direction is given by:

$$v_x = \mathrm{d}x/\mathrm{d}t, \quad v_y = \mathrm{d}y/\mathrm{d}t, \quad \text{or} \quad v_z = \mathrm{d}z/\mathrm{d}t$$

where t denotes the time. Like the position, the velocity may be described by a vector:

$$\mathbf{v} = v_x\mathbf{i} + v_y\mathbf{j} + v_z\mathbf{k}$$

The magnitude of the velocity, v, is given by:

$$v^2 = |\mathbf{v}|^2 = v_x^2 + v_y^2 + v_z^2$$

Finally we define the linear momentum of a particle by the relationship:

$$\mathbf{p} = m\mathbf{v}$$

where m is the mass of the particle. The components of the linear momentum vectors are:

$$p_x = mv_x, \quad p_y = mv_y, \quad \text{and} \quad p_z = mv_z$$

We now turn our attention to the *kinetic energy* of the particle. According to classical, that is non-relativistic and non-quantum, mechanics it is given by:

$$\mathrm{KE} = (1/2)mv^2 = (1/2)m\,(v_x^2 + v_y^2 + v_z^2)$$

Or, in terms of the linear momentum as:

$$\mathrm{KE} = p^2/(2m) = (p_x^2 + p_y^2 + p_z^2)/(2m)$$

It follows from this expression that *the kinetic energy cannot be negative*. It is zero if the particle is at rest and positive if it is in motion relative to the origin of the coordinate system.

1.2 A dash of physics: the forces acting between particles carrying electric charge

The electric charge on a particle may be positive or negative, and is measured in coulombs. Thus the charge on an electron is

$$-e = -1.602 \times 10^{-19}\,\mathrm{C}$$

Chemists often use the elementary charge (e) as an "atomic unit" (a. u.) of charge.

The force acting between two particles with electric charges Q_1 and Q_2 is given by:

$$F = \frac{Q_1 Q_2}{4\pi \varepsilon_0 r^2}$$

where r is the distance between the two particles and ε_0 the vacuum permittivity:

$$\varepsilon_0 = 8.854 \times 10^{-12}\,\mathrm{C^2\,J^{-1}\,m^{-1}}$$

If Q_1 and Q_2 are given in coulombs, and the distance in meters, the force will be given in newtons (1 newton = 1 J $\mathrm{m^{-1}}$). If Q_1 and Q_2 have the same sign, F will be positive, indicating that the particles repel each other and that the direction of the force is such as to tend to increase r. If Q_1 and Q_2 have opposite signs, F will be negative, the particles attract one another and the direction of the force tends to decrease the distance between them.

Let us now assume that Q_1 and Q_2 have opposite signs, and that the particles initially are separated by a distance r_0. If we wish to increase the distance between them by an infinitesimal amount dr, we must apply an external force equal to (or infinitesimally greater) than F, but with the opposite direction; $F_{ex} = -F$. The work involved is

$$\mathrm{d}W = F_{ex}\,\mathrm{d}r = -F\mathrm{d}r = \left[\frac{-Q_1 Q_2}{4\pi \varepsilon_0 r^2}\right]\mathrm{d}r$$

Since the product $Q_1 Q_2$ is negative, dW is *positive*.

The work required to separate the particles completely is given by the integral

$$W = \int \mathrm{d}W = \int \left[\frac{-Q_1 Q_2}{4\pi \varepsilon_0 r^2}\right]\mathrm{d}r$$

from $r = r_0$ to infinity. Integration yields

$$W = \frac{-Q_1 Q_2}{4\pi \varepsilon_0 r_0}$$

which again is a positive quantity.

This work is equal to the increase of the *potential energy* of the particles:

$$W = \Delta\mathrm{PE} = \mathrm{PE}(r = \infty) - \mathrm{PE}(r = r_0) = \frac{-Q_1 Q_2}{4\pi \varepsilon_0 r_0} \qquad (1.1)$$

In these notes we shall adopt the usual convention that *the potential energy of the two particles is defined as zero when they are infinitely far apart*, i.e. that

$$\mathrm{PE}(r = \infty) = 0$$

Insertion of this definition in (1.1) yields $\mathrm{PE}(r_0) = (Q_1 Q_2)/(4\pi \varepsilon_0 r_0)$. Dropping the subscript to r_0, we finally obtain the general relation

$$\mathrm{PE}(r) = \frac{Q_1 Q_2}{4\pi \varepsilon_0 r} \qquad (1.2)$$

Since the product Q_1Q_2 is negative, it follows that the potential energy will be negative for all finite values of r. The maximum value for the potential energy is zero, which is only reached as the distance between the particles approaches infinity.

Since the potential energy will be zero or negative, it may not always be clear whether the terms "small *PE*" or "large *PE*" are meant to describe the potential energy itself or its absolute value. In order to minimize confusion, we shall refer to "high" or "low" potential energies.

Problem 1.1 Assume that Q_1 and Q_2 have the same sign and that the initial distance between the particles is very large (infinite). Calculate the work required to move them to a finite separation r_0. Define the potential energy as zero when r is infinite, and show that the potential energy corresponding to a given distance r is given by equation (1.2).

1.3 The energies of one-electron atoms

The term "one-electron atom" is intended to cover the hydrogen atom as well as all ions which consist of *one* nucleus and *one* electron, such as for example He^+, Li^{+2} or Be^{+3}. The total energy of such an atom may, as a good first approximation, be described as the sum of the kinetic energy of the nucleus KE_n, the kinetic energy of the electron KE_e and the electrostatic potential energy $PE(r)$:

$$\begin{aligned} E_{tot} &= KE_n + KE_e + PE(r) \\ &= \frac{1}{2}M_n\left[\left(\frac{dx_n}{dt}\right)^2 + \left(\frac{dy_n}{dt}\right)^2 + \left(\frac{dz_n}{dt}\right)^2\right] \\ &\quad + \frac{1}{2}m_e\left[\left(\frac{dx_e}{dt}\right)^2 + \left(\frac{dy_e}{dt}\right)^2 + \left(\frac{dz_e}{dt}\right)^2\right] - \frac{(Ze^2)}{(4\pi\varepsilon_0 r)} \end{aligned} \tag{1.3}$$

Here M_n, x_n, y_n and z_n denote the mass and Cartesian coordinates of the nucleus, while m_e, x_e, y_e and z_e denote the mass and Cartesian coordinates of the electron. The nuclear charge is $+Ze$ and r the distance from the nucleus to the electron. This expression implies that the kinetic energy is zero when the two particles are at rest, and the potential energy is zero when the electron and nucleus are infinitely far apart.

Equation (1.3) is not exact: relativistic effects (such as the variation of mass with velocity or the "electron spin") have been neglected. We shall have to return to this point later.

In order to separate the motion of the electron relative to the nucleus from the motion of the atom as a whole, we introduce the Cartesian coordinates of the atom's center of mass, X, Y and Z:

$$(M_n + m_e)X = M_nx_n + m_ex_e, \quad (M_n + m_e)Y = M_ny_n + m_ey_e, \quad \text{and}$$

$$(M_n + m_e)Z = M_nz_n + m_ez_e$$

and the coordinates of the electron in a Cartesian coordinate system with the origin at the nucleus:

$$x = x_e - x_n, \quad y = y_e - y_n, \quad \text{and} \quad z = z_e - z_n$$

The two kinetic energy terms in equation (1.3) may then be rearranged to yield

$$\frac{1}{2}(M_n + m_e)\left[\left(\frac{dX}{dt}\right)^2 + \left(\frac{dY}{dt}\right)^2 + \left(\frac{dZ}{dt}\right)^2\right] + \frac{1}{2}\mu\left[\left(\frac{dx}{dt}\right)^2 + \left(\frac{dy}{dt}\right)^2 + \left(\frac{dz}{dt}\right)^2\right] \tag{1.4}$$

where the parameter μ is defined by

$$\mu_e = \frac{M_n m_e}{M_n + m_e} \tag{1.5}$$

The dimension of μ_e is mass. The magnitude depends on the mass of the nucleus, but it is always slightly smaller than m_e. μ_e is therefore referred to as the *reduced mass* of the electron.

The first term in the expression (1.4) represents the kinetic energy associated with motion of the atom as a whole and the second represents the kinetic energy associated with motion of the electron relative to the nucleus.

The sum of the second term and the electrostatic interaction energy is often referred to as the *electronic energy* of the atom, E:

$$E = \frac{1}{2}\mu_e\left[\left(\frac{dx}{dt}\right)^2 + \left(\frac{dy}{dt}\right)^2 + \left(\frac{dz}{dt}\right)^2\right] - \frac{Ze^2}{4\pi\varepsilon_0 r} \tag{1.6}$$

If the electronic energy of the atom is positive, the electron may move infinitely far away from the nucleus. The potential energy would then be zero, and the kinetic energy would be equal to the total electronic energy. If the electronic energy is exactly equal to zero, the electron may move infinitely far away from the nucleus, but it will lose all its kinetic energy in doing so. If the electronic energy is negative, the electron may move away from the nucleus until the kinetic energy becomes zero. At this distance the potential energy becomes equal to the electronic energy

$$E = -\frac{Ze^2}{4\pi\varepsilon_0 r_{max}} \tag{1.7}$$

Solution of equation (1.7) with respect to r_{max} yields the maximum distance between electron and nucleus according to classical physics. Since the electron cannot escape from the nucleus, we say that it is bound to it, and *refer to states with $E < 0$ as bound states of the atom.*

Problem 1.2 Show that μ_e always will be smaller than the true electron mass m_e. In which one-electron atom is the difference between μ_e and m_e the largest? How large is the difference in per cent?

1.4 The introduction of quantum mechanics: atomic orbitals and orbital energies

By following the rules of quantum mechanics we may convert the classical expression for the internal energy (1.6) into the Hamiltonian operator of the atom:

$$\mathcal{H} = -\left(\frac{h^2}{8\pi^2\mu_e}\right)\left(\frac{\partial^2}{\partial x^2} + \frac{\partial^2}{\partial y^2} + \frac{\partial^2}{\partial z^2}\right) - \frac{Ze^2}{4\pi\varepsilon_0 r}$$

where h is the Planck constant:

$$h = 6.626 \times 10^{-36}\ \mathrm{J\,s}$$

The Hamiltonian may be written more compactly by introducing the Laplace operator

$$\nabla^2 = \frac{\partial^2}{\partial x^2} + \frac{\partial^2}{\partial y^2} + \frac{\partial^2}{\partial z^2};$$

$$\mathcal{H} = -\left(\frac{h^2}{8\pi^2\mu_e}\right)\nabla^2 - \frac{Ze^2}{4\pi\varepsilon_0 r} \tag{1.8}$$

According to quantum mechanics, all the information *that it is possible to have* about the atom is contained in its wavefunction $\Psi(x, y, z)$. If we limit ourselves to the consideration of atoms in stationary states, that is to atoms with a fixed, unchanging energy, then all possible wavefunctions are found by solving the *time-independent Schrödinger equation*

$$\mathcal{H}\Psi(x, y, z) = E\,\Psi(x, y, z)$$

where E denotes the electronic energy of the atom.

We now simplify the Hamiltonian operator, equation (1.8), by replacing μ_e by the real mass of the electron, m_e, and the Schrödinger equation becomes

$$\left(\frac{h^2}{8\pi^2 m_e}\right)\left(\frac{\partial^2\Psi(x, y, z)}{\partial x^2} + \frac{\partial^2\Psi(x, y, z)}{\partial y^2} + \frac{\partial^2\Psi(x, y, z)}{\partial z^2}\right)$$
$$-\left(\frac{Ze^2}{4\pi\varepsilon_0 r}\right)\Psi(x, y, z) = E\,\Psi(x, y, z) \tag{1.9}$$

This Schrödinger equation can be solved for any value of E that is positive or zero, that is if the energy is large enough for the electron to move infinitely far away from the nucleus, *but only for selected values of E less than zero*. These selected values represent the only energies of bound states of the atom that are allowed by quantum mechanics. The corresponding wavefunctions describe an electron which is most probably found in the vicinity of the nucleus. Such a one-electron wavefunction is referred to as an *atomic orbital*, and abbreviated AO. The corresponding energy is, in turn, referred to as the orbital energy.

1.5 Polar coordinates, the Bohr radius and the Rydberg energy unit

In these notes we shall not work out the solutions to equation (1.9), but list the allowed energies and wavefunctions and survey some of the features which are particularly relevant for discussions of chemical bonding and molecular structure.

The wavefunctions (atomic orbitals) take on a simpler mathematical form if we describe the position of the electron relative to the nucleus in terms of polar rather than Cartesian coordinates:

$$r = \sqrt{(x^2 + y^2 + z^2)}, \quad \theta = \cos^{-1}(z/r), \quad \text{and} \quad \phi = \tan^{-1}(y/x)$$

See Fig. 1.2. The radius r varies from zero to infinity, the angle θ from zero to π and ϕ from zero to 2π.

In the following we shall often refer to integrals of the type

$$\int F(x, y, z)\mathrm{d}x\ \mathrm{d}y\ \mathrm{d}z$$

If F is a function of the polar coordinates the volume element $\mathrm{d}\tau = \mathrm{d}x\ \mathrm{d}y\ \mathrm{d}z$ must be replaced by

$$\mathrm{d}\tau = r^2 \sin\theta \mathrm{d}r\, \mathrm{d}\theta\, \mathrm{d}\phi$$

See Fig. 1.2.

The mathematical form of the wavefunctions is further simplified by introduction of the "Bohr radius" which is defined by:

$$a_0 = \frac{\varepsilon_0 h^2}{\pi m_e e^2} = 52.92 \times 10^{-12}\,\mathrm{m} = 52.92\,\mathrm{pm} \tag{1.10}$$

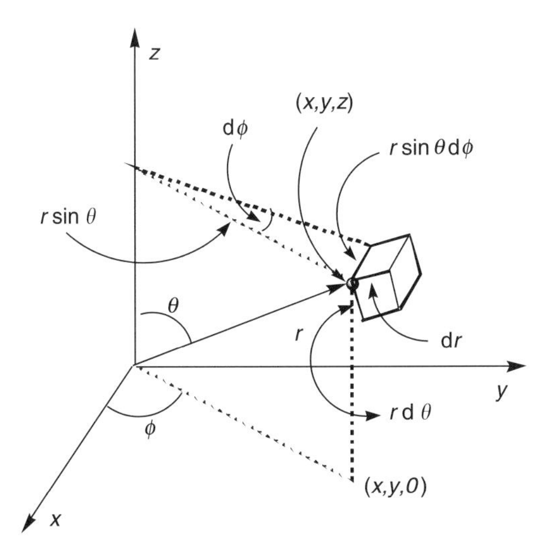

Fig. 1.2. Polar coordinates.

(1 pm = 1 picometer = 1×10^{-12} m). The Bohr radius is often used as an "atomic unit" (a.u.) of length.

Similarly the mathematical form of the energies is simplified by introduction of the "Rydberg" which is defined by

$$\mathcal{R} = \frac{m_e e^4}{8\varepsilon_0^2 h^2} = \frac{e^2}{8\pi\varepsilon_0 a_0} = 2.180 \times 10^{-18}\,\text{J} = 2.180\,\text{aJ} \tag{1.11}$$

(1 aJ = 1 attoJoule = 1×10^{-18} J). Note that the magnitude of the Rydberg corresponds to half the electrostatic potential energy of an electron and a proton at a distance of a_0.

Multiplication with Avogadro's number yields the Rydberg as a unit of energy per mole atoms or molecules:

$$\mathcal{R} = 1313\ \text{kJ mol}^{-1}$$

Sometimes the "Hartree" which is defined as $2\mathcal{R}$ is used as an atomic unit for energy.

1.6 The energy levels of one-electron atoms

The negative values of the energy for which acceptable solutions of the Schrödinger equation can be found, are given by the relation

$$E_n = \frac{-Z^2\mathcal{R}}{n^2} \tag{1.12}$$

where n is a positive integer. We refer to n as the *principal quantum number.* The lowest energy, corresponding to the *ground state* of the atom, is obtained when $n = 1$. Increasing values of n yield increasing energies. Atoms with higher energy than the ground state are said to be excited.

Problem 1.3 Draw a perpendicular energy axis and indicate the three lowest allowed energies of H and He^+ by horizontal lines. Such figures are referred to as orbital energy diagrams. Calculate the energies in kJ mol^{-1}.

1.7 The orbital energy diagrams of one-electron atoms: comparison with experiment

The accuracy of the energy level diagram of a one-electron atom obtained by calculation may be checked through comparison with spectroscopic measurements, i.e. by measurement of the energies of the photons that are emitted or absorbed when the atom goes from one energy level to another. There is in fact good, but not perfect, agreement between the experimental energy level diagram of hydrogen and that calculated from equation (1.12): According to (1.12) the first excited state ($n = 2$) of the hydrogen atom should be found $\Delta E = (1 - 0.25)\mathcal{R}$ above the ground state. The experimental energy level diagram does indeed contain a level 984.06 kJ mol^{-1}above the ground state, but on closer observation one finds that this is not a single level, but (at least) *three*. The second of the three levels is found only 0.0004 kJ mol^{-1}

above the first, and the third level is found only 0.0040 kJ mol^{-1} above the second. The second level is assigned to an atom with the electron in a 2*s* orbital, while the first and the third are assigned to atoms with the electron in a 2*p* orbital, but differing in the direction of the electron spin. The energies predicted by equation (1.12) are obviously not completely accurate, even if we base our calculations on the reduced mass rather than the real mass of the electron.

The expression given for the kinetic energy in (1.6) did not include relativistic effects, neither did the Hamiltonian operator (1.8) derived from it. The Schrödinger equation (1.9) can be solved exactly, but due to the neglect of relativistic effects the solutions are not completely accurate. When relativistic effects are included in the Schrödinger equation, the energies of 2*s* and 2*p* orbitals are no longer equal, and the degeneracy of the *p* orbitals is removed by so-called spin–orbit coupling. While the effect is very small for hydrogen atoms, it becomes increasingly important with increasing nuclear charge, and is particularly large in the atoms of sixth period elements like gold, mercury, thallium or lead.

1.8 Ionization energies

The ionization energy, IE, of a one-electron atom A^{+k} is defined as the energy that is required to remove an electron

$$A^{+k}(g) \rightarrow A^{+(k+1)}(g) + e^-(g)$$

where the reactant, $A^{+k}(g)$, and the products, $A^{+(k+1)}(g)$ and $e^-(g)$, are at rest and the latter are infinitely far apart. The ionization energy of an atom in the ground state is easier to measure than the ionization energy of an excited state, and this is the ionization energy which is meant unless there is a statement to the contrary.

The ionization energy of the H atom is the energy of the process

$$H(g) \rightarrow H^+(g) + e^-(g)$$
$$\text{IE} = E(H^+(g)) + E(e^-(g)) - E(H(g))$$

or, since the proton and electron have zero kinetic energy,

$$\text{IE} = 0 - 0 - E(H(g)) = \frac{\mathcal{R}}{n^2}$$

The highest ionization energy is found when the principal quantum number is equal to one, i.e. when the atom is in the ground state.

For one-electron atoms in general one obtains

$$\text{IE} = \frac{Z^2\mathcal{R}}{n^2} \tag{1.13}$$

Comparison with equation (1.12) shows that *the ionization energy of a one-electron atom is equal to minus the orbital energy.*

1.9 The size of one-electron atoms

If the expression "the size of an atom" is to have any meaning, it must have something to do with the distance between the electron and the nucleus. Several definitions are possible; at this stage we choose to identify the atomic radius with the maximum distance allowed by classical physics, equation (1.7):

$$E_{\mathrm{n}} = \frac{-Z^2\mathcal{R}}{n^2} = -Z^2\left(\frac{e^2}{8\pi\varepsilon_0 a_0}\right)\left(\frac{1}{n^2}\right) = \frac{-Ze^2}{4\pi\varepsilon_0 r_{\max}}$$

Solution with respect to $r_{\max}$ yields

$$r_{\max} = \frac{2n^2 a_0}{Z} \tag{1.14}$$

The information in equations (1.13) and (1.14) may be summarized as follows: *As the principal quantum number n increases, the radius of the atom increases by n^2, and the ionization energy decreases by the factor $1/n^2$.*

Comparing one-electron atoms with the same value for the principal quantum number, we see that *when the nuclear charge Z increases, the size of the atom decreases by a factor of 1/Z while the ionization energy increases by a factor of Z^2.*

Problem 1.4 What is the radius ($r_{\max}$) of a hydrogen atom in the ground state? What is the size of a H atom with principal quantum number $n = 10\,000$? The circumference of the Earth is 40 000 km. What would be the principal quantum number of a hydrogen atom with the same size? (It would obviously be difficult to find a container for such an atom!)

Problem 1.5 Calculate the ratio between the volumes and between the mass densities of H and He^+ atoms in the ground state.

1.10 The ground state wavefunctions of one-electron atoms

The atomic orbitals of one-electron atoms with principal quantum number $n = 1, 2$ or 3 are listed in Table 1.1. They all have the form

$$\Psi_{n,l,m}(r,\theta,\phi) = R_{n,l}(r)Y_{l,m}(\theta,\phi) = R_{n,l}(r)\Theta_{n,l}(\theta)\Phi_m(\phi) \tag{1.15}$$

where the functions $R_{n,l}(r)$ and $Y_{l,m}(\theta,\phi)$ are referred to as the *radial* and *angular* parts of the wavefunction respectively. The orbitals are classified by three quantum numbers, the principal quantum number n, which determines the energy according to equation (1.12), the angular quantum number l, and the magnetic quantum number m. We shall return to the significance of l and m later in this chapter.

Only one wavefunction is listed for the ground state, $n = 1$. This wavefunction is referred to as the $1s$ orbital:

$$\Psi_{1s} = 2\left(\frac{Z}{a_0}\right)^{3/2} \exp\left(\frac{-Zr}{a_0}\right)\left(\frac{1}{4\pi}\right)^{1/2}$$

Table 1.1. One-electron atoms: atomic orbitals with principal quantum numbers $n = 1, 2$ or 3. All functions are normalized and orthogonal.

AO	$n, l, \|m\|$	$R_{n,l}(r)$	$Y_{l,\|m\|}(\theta, \phi)$
$1s$	1, 0, 0	$2(Z/a_0)^{3/2} \exp[-Zr/a_0]$	$(1/4\pi)^{1/2}$
$2s$	2, 0, 0	$(2)^{-3/2}(Z/z_0)^{3/2}(2 - Zr/a_0) \exp[-Zr/(2a_0)]$	$(1/4\pi)^{1/2}$
$2p_z$	2, 1, 0	$(24)^{-1/2}(Z/a_0)^{3/2}(Zr/a_0) \exp[-Zr/(2a_0)]$	$(3/4\pi)^{1/2} \cos\theta$
$2p_x$	2, 1, 1	$(24)^{-1/2}(Z/a_0)^{3/2}(Zr/a_0) \exp[-Zr/(2a_0)]$	$(3/4\pi)^{1/2} \sin\theta \cos\phi$
$2p_y$	2, 1, 1	$(24)^{-1/2}(Z/a_0)^{3/2}(Zr/a_0) \exp[-Zr/(2a_0)]$	$(3/4\pi)^{1/2} \sin\theta \sin\phi$
$3s$	3, 0, 0	$2(3)^{-9/2}(Z/a_0)^{3/2}[27 - 18Zr/a_0 + 2(Zr/a_0)^2] \exp[-Zr/(3a_0)]$	$(1/4\pi)^{1/2}$
$3p_z$	3, 1, 0	$(2)^{-3/2}(3)^{-9/2}(Z/a_0)^{3/2}[6Zr/a_0 - (Zr/a_0)^2] \exp[-Zr/(3a_0)]$	$(3/4\pi)^{1/2} \cos\theta$
$3p_x$	3, 1, 1	$(2)^{-3/2}(3)^{-9/2}(Z/a_0)^{3/2}[6Zr/a_0 - (Zr/a_0)^2] \exp[-Zr/(3a_0)]$	$(3/4\pi)^{1/2} \sin\theta \cos\phi$
$3p_y$	3, 1, 1	$(2)^{-3/2}(3)^{-9/2}(Z/a_0)^{3/2}[6Zr/a_0 - (Zr/a_0)^2] \exp[-Zr/(3a_0)]$	$(3/4\pi)^{1/2} \sin\theta \sin\phi$
$3d_z^2$	3, 2, 0	$(3)^{-9/2}(5)^{-1/2}(Z/a_0)^{3/2}(2Zr/a_0)^2 \exp[-Zr/(3a_0)]$	$(5/16\pi)^{1/2}(3\cos^2\theta - 1)$
$3d_{xz}$	3, 2, 1	$(3)^{-9/2}(5)^{-1/2}(Z/a_0)^{3/2}(2Zr/a_0)^2 \exp[-Zr/(3a_0)]$	$(15/4\pi)^{1/2} \cos\theta \sin\theta \cos\phi$
$3d_{yz}$	3, 2, 1	$(3)^{-9/2}(5)^{-1/2}(Z/a_0)^{3/2}(2Zr/a_0)^2 \exp[-Zr/(3a_0)]$	$(15/4\pi)^{1/2} \cos\theta \sin\theta \sin\phi$
$3d_{xy}$	3, 2, 2	$(3)^{-9/2}(5)^{-1/2}(Z/a_0)^{3/2}(2Zr/a_0)^2 \exp[-Zr/(3a_0)]$	$(15/4\pi))^{1/2} \sin^2\theta \cos\phi \sin\phi$
$3d_{x^2-y^2}$	3, 2, 2	$(3)^{-9/2}(5)^{-1/2}(Z/a_0)^{3/2}(2Zr/a_0)^2 \exp[-Zr/(3a_0)]$	$(15/16\pi)^{1/2} \sin^2\theta(\cos^2\phi - \sin^2\phi)$

The variable part of the function is given by the term $\exp(-Zr/a_0)$, while the collection of constants have been introduced to *normalize* the wavefunction, i.e. to ensure that

$$\int \Psi^2(x,y,z)\mathrm{d}x\,\mathrm{d}y\,\mathrm{d}z = 1 \tag{1.16}$$

when integration is carried out over all space, i.e. from minus to plus infinity for each of the three Cartesian coordinates.

When polar coordinates are used, the normalization equation (1.16) becomes:

$$\int \Psi^2(x,y,z)r^2\sin\theta\mathrm{d}r\,\mathrm{d}\theta\,\mathrm{d}\phi$$
$$= \int R_{n,l}^2(r)Y_{l,m}^2(\theta,\phi)r^2\sin\theta\mathrm{d}r\,\mathrm{d}\theta\,\mathrm{d}\phi = 1$$

or

$$\int R_{n,l}^2(r)r^2\mathrm{d}r\int Y_{l,m}^2(\theta,\phi)\sin\theta\mathrm{d}\theta\,\mathrm{d}\phi = 1 \tag{1.17}$$

when the integration is carried out over all space.

One way to assure that equation (1.17) is satisfied is to define the two functions R and Y in such a manner that

$$\int R_{n,l}^2(r)r^2\mathrm{d}r = 1 \tag{1.18}$$

when the integration is carried out from zero to infinity, and

$$\int Y_{l,m}^2(\theta,\phi)\sin\theta\mathrm{d}\theta\,\mathrm{d}\phi = 1 \tag{1.19}$$

when θ is integrated from zero to π and ϕ from zero to 2π.

Note that the $1s$ function given above is not the only function which satisfies the Schrödinger equation for $E = E_1$: multiplication of the $1s$ wavefunction by any number c, positive or negative, real, imaginary or complex, yields a new function which is also a solution of the time-independent Schrödinger equation, equation (1.9), with the same energy:

$$\mathcal{H}(c\Psi(r,\theta\phi)) = c(\mathcal{H}\Psi(r,\theta\phi)) = cE\Psi(r,\theta,\phi) = E(c\Psi(r,\theta\phi))$$

Thus, while the energy of the ground state is unequivocally fixed, the corresponding wavefunction is not! All these wavefunctions will, however, give the same predictions as to the possible outcomes of any measurement. Clearly they describe the same state of the atom, and it is enough to list one of them.

In these notes we shall prefer to use wavefunctions which are normalized. We shall also prefer wavefunctions which are real unless there is a particular reason to use complex functions. If the wavefunction of the *ground* state of a one-electron atom ($n = 1$) is assumed to be real and normalized, it is unequivocally determined except for a factor of ± 1. In Table 1.1 we list the positive form.

Problem 1.6 Show that the $1s$ AO is normalized.

1.11 Probability densities, electron densities and the shape of 1s atomic orbitals

An electron in a given atomic orbital is moving in space, and it is not possible to predict where it will be found if we try to measure its position at a given time. We can, however, use the wavefunction to calculate the *probability* of finding the electron within a certain region: the probability of finding the electron in a small volume element $\mathrm{d}\tau$ with a corner at the point with coordinates x, y, z is given by

$$\text{Probability} = \Psi^2(x, y, z)\,\mathrm{d}x\,\mathrm{d}y\,\mathrm{d}z$$

Alternatively, if we use polar coordinates, the probability of finding the electron in the cube $\mathrm{d}\tau = r^2 \sin\theta \mathrm{d}r\, \mathrm{d}\theta\, \mathrm{d}\phi$ is given by

$$\text{Probability} = \Psi^2(r, \theta, \phi)\, r^2 \sin\theta \mathrm{d}r \mathrm{d}\theta\, \mathrm{d}\phi \tag{1.20}$$

See Fig. 1.2. This probability is a dimensionless number between zero (which means impossible) and unity (which means absolutely certain). The dimension of $\mathrm{d}\tau$ is volume; not surprisingly the probability of finding the electron in the volume element $\mathrm{d}\tau$ is proportional to its size. Since the probability is dimensionless, $\Psi^2(r, \theta, \phi)$ must have the dimension $(\text{volume})^{-1}$ and is therefore referred to as the *probability density*.

Sometimes we imagine the electron to be divided into infinitesimal parts which are distributed around the nucleus in such a way that the sum of the parts contained in $\mathrm{d}\tau$ adds up to

$$\Psi^2(r, \theta, \phi)\,\mathrm{d}\tau \text{ electrons}$$

This means that we have replaced the one solid, but moving, electron by a stationary cloud of minute electron fragments: an "electron cloud." Instead of referring to $\Psi^2(r, \theta, \phi)$ as the probability density, we refer to it as the "electron density." The electron cloud represents a *model*of the atom. Here the word "model" is not meant to denote a copy of the true atom at a smaller scale, but a *simplified representation* of the atom.

Finally, multiplying the electron density with the electronic charge we obtain the charge density:

$$\rho_{-}(r, \theta, \phi) = -e\Psi^2(r, \theta, \phi)$$

The probability density for an electron in the 1s orbital is given by

$$\Psi_{1s}^2 = \left(\frac{1}{\pi}\right)\left(\frac{Z}{a_0}\right)^3 \exp\left(\frac{-2Zr}{a_0}\right) \tag{1.21}$$

This means that the probability density of an electron in a 1s orbital of a one-electron atom falls to zero only as r approaches infinity. As we have seen above, classical physics leads to the conclusion that an electron with $E < 0$ will be unable to move further away from the nucleus than the distance at which the total energy of the electron is equal to the potential energy. According to quantum mechanics there is a definite probability of finding the electron at beyond this limit. This extension of the probability density beyond the classical limit is a "quantum effect" and is often referred to as "tunneling."

Since the probability density of a $1s$ orbital is constant for all points at the same distance from the nucleus, a one-electron atom in the ground state is often represented as a sphere. The radius of the sphere might be chosen as the classical radius or as the radius at which the probability density has fallen to some predetermined value.

Problem 1.7 Plot the probability densities of the $1s$ orbitals of H and He^+ as a function of r from $r = 0$ to $3a_0$. Where are the densities largest? Compare the probability densities at the nuclei. At what value of r do the curves cross?

Problem 1.8 Find the distances from the nuclei in H, He^+ and Li^{+2} at which the $1s$ probability density has fallen to 1/2, 1/4 and 1/8 of its maximum value.

Problem 1.9 Calculate the ratio between the probability density at the classical radius of a one-electron atom, equation (1.14), and the density at the nucleus. Calculate the probability of finding an electron in a $1s$ orbital outside the classical boundary of the atom. (The probability is higher than 20%).

1.12 The wavefunctions, probability densities and shape of 2*s* atomic orbitals

Table 1.1 lists four possible wavefunctions for the first excited state ($n = 2$). Each of these functions is real and normalized. The functions have moreover been chosen in such a way that the integral of the product of any two over all space is zero:

$$\int \Psi_i(x, y, z)\ \Psi_j(x, y, z)\, \mathrm{d}x\, \mathrm{d}y\, \mathrm{d}z = 0 \tag{1.22}$$

unless the two wavefunctions are identical, $i = j$. Such functions are described as *orthogonal*.

Problem 1.10 Show that the AO denoted $2p_z$ is orthogonal to the $2s$, $2p_x$ and $2p_y$ orbitals.

The four AOs listed for $n = 2$, correspond to the same energy. Such orbitals are referred to as *degenerate*, and the number of such orbitals is referred to as the degeneracy of the energy level. According to our calculations, the degeneracy of the first excited state of a one-electron atom is four. (As we have seen in Section 1.7 this statement is not strictly true.)

While the requirement that the $1s$ orbital be real and normalized was sufficient to determine the form except for a factor of ± 1, these requirements added to the orthogonality requirement still leave us a wide choice of form for degenerate atomic orbitals. See Problem 1.11.

Problem 1.11 Use the Schrödinger equation on the form $\mathcal{H}\Psi(x, y, z) = E_n\Psi(x, y, z)$. Assume that the level E_n is degenerate, i.e. that there are two orthogonal wavefunctions, $\Psi_a(x, y, z)$ and $\Psi_b(x, y, z)$, that satisfy the Schrödinger equation with the same energy E_n. Prove that a linear combination of the two,

$$\Psi_c(x, y, z) = c_1\Psi_a(x, y, z) + c_2\Psi_b(x, y, z)$$

where c_1 and c_2 are arbitrary numbers, also is a solution of the Schrödinger equation with the same energy.

Problem 1.12 Assume that $\Psi_a(x, y, z)$ and $\Psi_b(x, y, z)$ are normalized. Show that $\Psi_c(x, y, z)$ becomes normalized if we multiply it with the constant $1/\sqrt{D}$ where $D = c_1^2 + c_1^2$.

The results derived in Problems 1.11 and 1.12 are quite general, and we shall make frequent use of them in the following.

The angular part of the $2s$ AO is identical to the angular part of the $1s$ AO. In fact all s orbitals have the same angular parts. See Table 1.1. This means that all s orbitals yield probability densities that are independent of θ and ϕ, and that constant probability density contours are spherical.

Investigation of the radial part of the $2s$ AO shows that the radial function is zero when $r = 2a_0$. This distance is much smaller than the classical limit of $r_{max} = 8a_0$. The spherical surface where the probability density is zero is therefore regarded as being too far from the exterior surface of the atom to have chemical consequences. Again we choose to represent the shape of the AO simply by a sphere.

Surfaces where the values of an atomic orbital and hence the probability densities are exactly zero are referred to as *nodal surfaces*. The nodal surface of the $2s$ orbital is spherical.

All ns atomic orbitals have spherical shape. Each time the principal quantum number is increased by one, the size of the atom measured by the classical radius increases, the atom becomes "more diffuse," and another spherical nodal surface is introduced.

1.13 The wavefunctions, probability densities and shape of 2*p* atomic orbitals

The three AOs $2p_z$, $2p_x$ and $2p_y$ have the same radial, but different angular functions. See Table 1.1. The Cartesian coordinates of the electron may be written as functions of the polar coordinates r, θ and ϕ:

$$z = r\cos\theta, \quad x = r\sin\theta\cos\phi, \quad \text{and} \quad y = r\sin\theta\sin\phi \tag{1.23}$$

Using this result, we see that the three functions $2p_z$, $2p_x$ and $2p_y$ may be rewritten as

$$2p_q = (32\pi)^{-1/2}\left(\frac{Z}{a_0}\right)^{3/2}\left(\frac{Zq}{a_0}\right)\exp\left(-\frac{Zr}{2a_0}\right), \quad \text{where } q = x, y, \text{ or } z \tag{1.24}$$

The three functions have the same shape; they differ only in their orientation. Each of the three functions has a planar nodal surface which is defined by the three equations $x = 0$, $y = 0$, or $z = 0$ respectively. The $2p_x$ orbital is positive where x is positive, the $2p_y$ orbital is positive where y is positive, and the $2p_z$ is positive where z is positive.

Constant probability density contours of a $2p_z$ AO are shown to the left in Fig. 1.3. The points of maximum density are indicated by filled squares on the positive and negative z-axes. The density is zero in the xy plane. Since the wavefunction is independent of ϕ, the density has cylinder symmetry around the z-axis. Rotation of the figure around this axis would sweep out two three-dimensional surfaces with constant probability density.

The $3p_x$, $3p_y$ and $3p_z$ AOs have identical radial functions. The angular functions are the same as for the $2p_x$, $2p_y$ and $2p_z$ orbitals respectively. Again the AOs have the same shape, but differ in their orientation. Constant density contours for a $3p_z$ orbital are shown at the

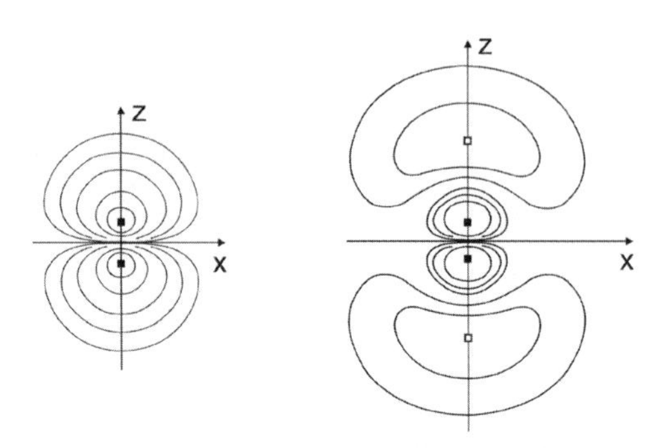

Fig. 1.3. Constant probability density contours of atomic orbitals. Left: a $2p_Z$ orbital. The points of maximum probability density are indicated by filled squares. Contours have been drawn at 66, 20, 8, 2, and 0.5% of the maximum density. Right: a $3p_Z$ orbital. The points of maximum probability density are indicated by filled squares, local maxima with lower densities are indicated by open squares. Contours have been drawn at 24, 6, and 1.5% of the maximum density. (Reproduced with permission from B. Perlmutter-Hayman, *J. Chem. Ed.*, 46 (1969) 427.) Copyright (1969) Division of Chemical Education, Inc.

right in Fig. 1.3. Like the $2s$ orbital the $3p_Z$ orbital has zero density on a sphere fairly close to the nucleus, but the shape of the probability density in the outer regions of the atom is similar to the $2p_Z$ orbital.

1.14 The wavefunctions, probability densities and shape of 3*d* atomic orbitals

The form of the d orbitals listed in Table 1.1 has again been chosen in such a way that all AOs with the principal quantum number $n = 3$ are orthogonal.

Problem 1.13 Show that the wavefunction of each of the five $3d$ orbitals may be written as

$$3d = \mathrm{K}\left(\frac{Z}{a_0}\right)^{7/2} \exp\left(\frac{-Zr}{3a_0}\right) f(x, y, z) \tag{1.25}$$

where K is a constant and $f(x, y, z)$ is one of the five functions

$$f(x, y, z) = 3z^2 - r^2, \quad f(x, y, z) = xz, \quad f(x, y, z) = yz, \quad f(x, y, z) = xy, \quad \text{or}$$
$$f(x, y, z) = x^2 - y^2 \tag{1.26}$$

(The constant K does not have the same value for all the d orbitals, but this need not concern us here.)

The atomic orbital

$$3d_z^2 = K\left(\frac{Z}{a_0}\right)^{7/2} \exp\left(-\frac{Zr}{3a_0}\right)(3z^2 - r^2)$$

will be zero when the factor $(3z^2 - r^2) = 0$, i.e. when $z = \pm r/\sqrt{3}$ or $\theta = 55°$ or $125°$. Constant probability density contours are shown in Fig. 1.4. The two points with maximum density are indicated at the positive and negative z-axes. Since the wavefunction is

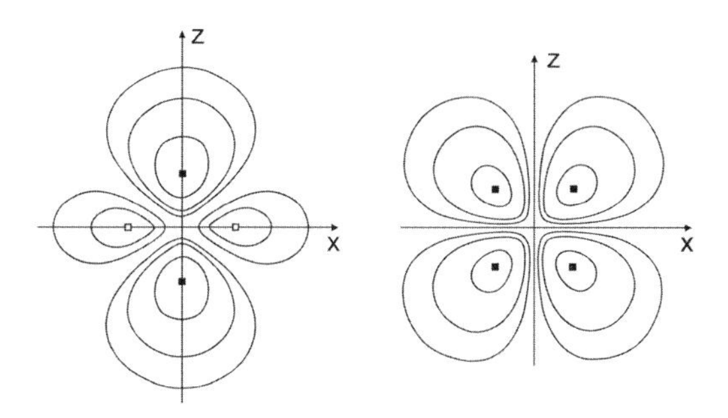

Fig. 1.4. Constant probability density contours of $3d$ atomic orbitals in the xz plane. Left: a $3d_z^2$ orbital. The points of maximum probability density are indicated by filled squares, local maxima with lower densities are indicated by open squares. Contours have been drawn at 52, 13, and 3% of the maximum density. Right: a $3d_{xz}$ orbital. Contours have been drawn at 68, 16, and 3 % of the maximum density. (Reproduced with permission from B. Perlmutter-Hayman, *J. Chem. Ed.*, 46 (1969) 427.) Copyright (1969) Division of Chemical Education, Inc.

independent of ϕ, the probability density will have cylinder symmetry about the z-axes. Rotation of the figure about this axis will sweep out three three-dimensional surfaces with constant density; two lobes at the positive and negative z-axes and a torus (doughnut) with maximum density on a circle in the xy plane.

The atomic orbital

$$3d_{xz} = K\left(\frac{Z}{a_0}\right)^{7/2} \exp\left(\frac{-Zr}{3a_0}\right) xz$$

is zero in the xy and yz planes. Constant density contours in the xz plane are shown in Fig. 1.4. Four points of maximum density are indicated by filled squares. Each lobe is somewhat more extended in the directions perpendicular to the xy plane than they are in the plane. The cross-sections of the lobes are thus oval.

Note that the $3d_z^2$ and $3d_{xz}$ orbitals have different shapes.

Though it may not be obvious from equations (1.25) and (1.26), each of the three orbitals $3d_{yz}$, $3d_{xy}$, and $3d_{x^2-y^2}$ have the same shape as the $3d_{xz}$ orbital, but have different orientations in space.

Problem 1.14 Mark the lobes of the d orbitals in Fig. 1.4 with plus or minus to indicate the sign of the wavefunction.

Problem 1.15 Make a rough sketch of the constant probability density contours of the $3d_{xy}$ and $3d_{x^2-y^2}$ orbitals in the xy plane and indicate nodes by stippled lines.

Problem 1.16 Make a rough sketch of constant probability density contours of the $3d_z^2$ orbital in the xy plane.

We have stressed that the form of the atomic orbitals in Table 1.1 has been chosen in such a way that all functions corresponding to the same principal quantum number n are orthogonal. What about the AOs corresponding to different energies? It can in fact be shown that *whenever two wavefunctions of a system describe states of different energy, they must be orthogonal.*

1.15 A final dash of physics: angular momentum

How does the electron move in the space around the nucleus? Does it move on a circular orbit? Or does it move along a straight line through the nucleus, accelerating when it is moving toward the nucleus, and slowing down when it is moving away from it? A partial answer to these questions may be provided in terms of the *angular momentum* of the electron.

Consider a particle with mass m moving with velocity v on a circle with radius r. We define a Cartesian coordinate system with the origin at the center and the x- and y-axes in the plane of the circle. See Fig. 1.5A. The magnitude of the angular momentum around the z-axis is given by

$$L_z = rmv = rp$$

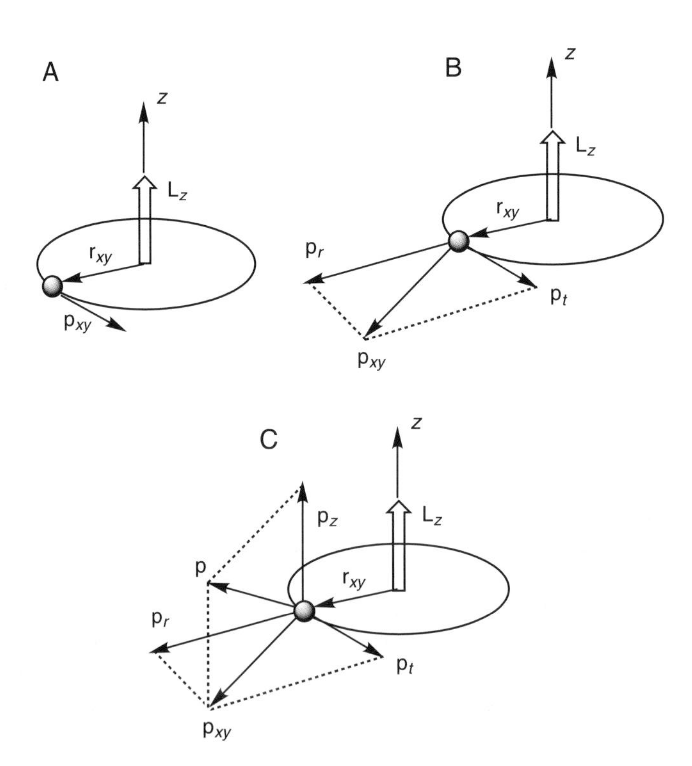

Fig. 1.5. A: The angular momentum of a particle constrained to move on a circle in the xy plane and with the center at the origin of the coordinate system. B: Determination of the angular momentum about the z-axis for a particle free to move in the xy plane. The linear momentum vector $\mathbf{p}$ is decomposed in components in the radial and tangential directions: $\mathbf{L}_z = \mathbf{r}_{xy} \times \mathbf{p}_t$. C: Determination of the angular momentum about the z-axis for a particle free to move in three dimensions: The linear momentum vector $\mathbf{p}$ is decomposed in components along the z-axis and in a horizontal plane parallel to the xy plane. $\mathbf{p}_{xy}$ is then further decomposed in radial and tangential directions: $\mathbf{L}_z = \mathbf{r}_{xy} \times \mathbf{p}_t$.

The angular momentum is given a positive sign if the motion of the particle is counterclockwise when viewed down the positive z-axis, and a negative sign if the motion is clockwise.

The angular momentum may also be regarded as a vector and defined by the vector product

$$\mathbf{L}_z = \mathbf{r}_{xy} \times \mathbf{p}_{xy}$$

where $\mathbf{r}_{xy}$ is a vector from the center of the circle to the particle and $\mathbf{p}_{xy}$ the linear momentum vector of the particle. The direction of $\mathbf{L}_z$ will be along the positive z-axis if the direction of motion is counterclockwise, and along the negative z-axis if the direction of motion is clockwise. See Fig. 1.5A.

We now consider a particle that is free to move in the xy plane. See Fig. 1.5B. The angular momentum about the z-axis may be calculated by decomposing the linear momentum vectors in components in the radial and tangential directions:

$$\mathbf{p} = \mathbf{p}_r + \mathbf{p}_t$$

and then taking the vector product:

$$\mathbf{L}_z = \mathbf{r}_{xy} \times \mathbf{p}_{xy} = \mathbf{r}_{xy} \times \mathbf{p}_t$$

Note that $\mathbf{L}_z$ is equal to zero if the tangential component of $\mathbf{p}$ is equal to zero, i.e. if the particle is moving on a straight line through the origin of the coordinate system.

Finally we consider a particle free to move in three directions. See Fig. 1.5C. While a particle moving in the xy plane will have angular momentum only about the z-axis, the particle which is free to move in three dimensions will generally have angular momenta about all three coordinate axes. In this general case the total angular momentum is defined by

$$\mathbf{L} = \mathbf{L}_x + \mathbf{L}_y + \mathbf{L}_z = \mathbf{r} \times \mathbf{p} \tag{1.27}$$

Note that this definition implies that the total angular momentum is zero if $\mathbf{r} = 0$, i.e. if the particle is at the origin of the coordinate system, or if $\mathbf{r}$ and $\mathbf{p}$ are parallel, i.e. if the particle is moving directly towards or directly away from the origin.

The angular momentum about the z-axis, $\mathbf{L}_z$, may also be calculated by drawing a horizontal plane through the particle and perpendicular to the z-axis. The horizontal distance vector from the z-axis to the particle is denoted by $\mathbf{r}_{xy}$. The linear momentum vector is decomposed in one component along the z-axis and another in the horizontal plane. See Fig. 1.5C. Then

$$\mathbf{L}_z = \mathbf{r}_{xy} \times \mathbf{p}_{xy} = \mathbf{r}_{xy} \times \mathbf{p}_t \tag{1.28}$$

This procedure demonstrates that the angular momentum around the z-axis is zero if $\mathbf{p}_t$ is zero, i.e. if the particle is moving in a plane containing the z-axis.

The wavefunctions listed in Table 1.1 are characterized by the principal quantum number n, the angular quantum number l, and the absolute value of magnetic quantum number m.

When written in this form, the magnitude of the total angular momentum of an electron in the AO is determined by the angular quantum number l:

$$L^2 = |\mathbf{L}|^2 = (l^2 + l)\left(\frac{h}{2\pi}\right)^2 \tag{1.29}$$

where h is the Planck constant. Similarly the magnitude of the angular momentum around the z-axis given by

$$L_z = |\mathbf{L_z}| = |m|\frac{h}{2\pi} \tag{1.30}$$

All s orbitals are characterized by $l = 0$. This means that electrons in s orbitals have zero angular momentum and therefore must be moving back and forth on some straight line through the nucleus (since the probability density has spherical symmetry, we do not know *which* line). If the total angular momentum is zero, so must the component along the z-axis, and the magnetic quantum number m is indeed also zero.

All p orbitals are characterized by $l = 1$. It follows that an electron in a p orbital will never be moving along a straight line through the nucleus and there will always be a component of the linear momentum vector $\mathbf{p}$ which corresponds to motion *around* the nucleus. All p electrons have the same total angular momentum, but they differ in its direction. Electrons in p_z orbitals have magnetic quantum numbers equal to zero; this means that there is no angular momentum around the z-axis: the rotational axis must be lying somewhere in the xy plane. Electrons in p_x or p_y orbitals have $|m| = 1$, which means that they are moving around the z-axis, but since the sign of $\mathbf{L}_z$ is unknown, so is the direction of motion: clockwise or counterclockwise.

All d orbitals are characterized by angular quantum number equal to 2, and will therefore have larger angular momenta than p electrons; a greater part of their kinetic energy is tied up in motion around, as opposed to through, the nucleus.

Atomic orbitals with zero angular momentum around the z-axis are often referred to as sigma orbitals, and p_z and d_z^2 orbitals are denoted by p_σ and d_σ respectively.

Similarly atomic orbitals which are characterized by $|m| = 1$ are referred to as π orbitals, and p and d orbitals characterized by $|m| = 1$ are denoted by p_π and d_π respectively. d orbitals with $|m| = 2$ are denoted by d_δ. Note that π and δ orbitals have zero probability density along the z-axis. When discussing linear molecules, we shall follow convension and define Cartecian coordinate systems in such a manner that all nuclei are situated on the z-axis, and classify the atomic orbitals as σ, π, or δ orbitals as described above.

Problem 1.17 Identify the d orbitals in Table 1.1 as d_σ, d_π, or d_δ orbitals.

When discussing planar molecules we shall classify p orbitals with direction perpendicular to the plane as π, and the p orbitals with direction in the plane as σ orbitals.

1.16 Electron spin and spin orbitals

When the Hamiltonian operator of a one-electron atom equation (1.8) is modified to include relativistic effects, we find on solution of the Schrödinger equation that the electron has an inherent angular momentum of magnitude

$$S = |\mathbf{S}| = \sqrt{\frac{3}{4}}\frac{h}{2\pi} \tag{1.31}$$

This angular momentum comes as an *addition* to any angular momentum due to orbital motion, and is referred to as the electron *spin*.

Experiment and theory agree that there are only two spin states. These states correspond to the same total angular momentum, but different angular momenta about the z-axis:

$$S_z = +\left(\frac{1}{2}\right)\frac{h}{2\pi} \quad \text{or} \quad S_z = -\left(\frac{1}{2}\right)\frac{h}{2\pi} \tag{1.32}$$

The first of these states is often represented by an arrow pointing up, and is said to have positive spin; the second is said to have negative spin and is represented by an arrow pointing downwards.

Whenever we have a state we need a wavefunction. The wavefunctions of the two spin states in (1.32) are denoted by α and β respectively. A wavefunction must be a function of some coordinate, and we say that α and β are functions of a "spin coordinate," s. In the following we shall assume the spin functions to be normalized:

$$\int \alpha^2(s)\,\mathrm{d}s = \int \beta^2(s)\,\mathrm{d}s = 1 \tag{1.33}$$

and orthogonal:

$$\int \alpha(s)\beta(s)\,\mathrm{d}s = 0$$

We shall also assume that the spin coordinate has only two values, $+1/2$ or $-1/2$. $s = +1/2$ corresponds to an electron with positive spin, and $s = -1/2$ corresponds to an electron with negative spin angular momentum.

In order to find the probability that an electron has positive spin, we insert $s = +1/2$ into the spin function and square it. Depending on whether the spin function is α or β, we obtain the probabilities

$$\alpha^2(+1/2) = 1 \quad \text{or} \quad \beta^2(+1/2) = 0$$

Taking the square roots we obtain

$$\alpha(+1/2) = 1 \quad \text{and} \quad \beta(+1/2) = 0 \tag{1.34a}$$

Similarly we write

$$\alpha(-1/2) = 0 \quad \text{and} \quad \beta(-1/2) = 1 \tag{1.34b}$$

For many applications an adequate description of a one-electron atom is obtained by combining a wavefunction obtained from the non-relativistic Schrödinger equation with one of the spin functions α or β to give

$$\Psi(r,\theta,\phi)\,\alpha(s) \quad \text{or} \quad \Psi(r,\theta,\phi)\,\beta(s)$$

Such combined wavefunctions for one electron are referred to as *spin orbitals*. The ground state of the hydrogen atom is thus doubly degenerate.

Such a spin orbital may be used to find the probability that the electron is found in the volume element $\mathrm{d}\tau$ with a spin of say $+1/2$:

$$\text{Probability} = \Psi_{1s}^2(r,\theta,\phi)\,\mathrm{d}\tau\;\alpha^2(+1/2) = \Psi_{1s}^2(r,\theta,\phi)\,\mathrm{d}\tau$$

or

$$\text{Probability} = \Psi_{1s}^2(r,\theta,\phi)\,\mathrm{d}\tau\;\beta^2(+1/2) = 0$$

1.17 Spin–orbit coupling

We are now in a position to understand the origin of the discrepancy between calculated and experimental orbital energies noted for the $2s$ and $2p$ states of the H atom. See Section 1.7.

Just like a coil of wire carrying an electric current, an electron with an orbital angular momentum about the nucleus (such as an electron in a p or a d orbital) will create a magnetic field. So will a spinning electron. The energy of two rod-shaped magnets depends on their relative orientation and a parallel orientation with both north poles pointing up has a higher energy than an antiparallel. Similarly the energy of a one-electron atom with the electron in a $2p$ orbital will be higher if the spin and orbital angular momenta have the same sign than if they have opposite signs. This phenomenon is referred to as spin–orbit (L–S) coupling.

In the absence of L–S coupling the six spin orbitals

$$\Psi_{2p_x}(r,\theta,\phi)\,\alpha(s),\quad \Psi_{2p_y}(r,\theta,\phi)\,\alpha(s),\quad \Psi_{2p_z}(r,\theta,\phi)\,\alpha(s),$$
$$\Psi_{2p_x}(r,\theta,\phi)\,\beta(s),\quad \Psi_{2p_y}(r,\theta,\phi)\,\beta(s),\quad \text{and}\quad \Psi_{2p_z}(r,\theta,\phi)\,\beta(s)$$

are all degenerate. Introduction of L–S coupling leads to four degenerate wavefunctions describing atoms in which spin and orbital angular momenta are parallel and two degenerate wavefunctions describing atoms where they are antiparallel. As expected the states with parallel spin and orbital angular momenta have higher energy than the two states where they are antiparallel. (The wavefunctions of the six states resulting from L–S coupling are formed by combination of the six spin orbitals listed above.)

Chapter 2

Atomic properties: two-electron atoms

Introduction

A two-electron atom consists of a nucleus with charge $+Ze$ and two electrons. This class includes the hydride anion H^- ($Z = 1$), the neutral He atom ($Z = 2$), and the series of cations, Li^+ ($Z = 3$), Be^{+2} ($Z = 4$), etc. If we omit the kinetic energy of the nucleus and neglect electron spin, the wavefunction of the atom is a function of the six coordinates determining the positions of the two electrons:

$$\Psi(x_1, y_1, z_1, x_2, y_2, z_2) \quad \text{or} \quad \Psi(\mathbf{r}_1, \mathbf{r}_2)$$

We begin by writing down an expression for the energy of the atom according to classical mechanics, and then convert this energy expression into a Hamiltonian operator. Due to the term in the Hamiltonian operator that represents the electrostatic repulsion between the two electrons, it is impossible to find exact solutions to the Schrödinger equation in a closed form. In the first part of the chapter we use a simple approximate wavefunction written as a product of two $1s$ atomic orbitals to calculate the approximate energy of the ground state.

In the second half of the chapter we search for acceptable wavefunctions for the ground state of the two-electron atom that include electron spin. This search leads us to the so-called Pauli Principle that states that the only acceptable wavefunctions for a system with two or more electrons are those for which exchange of the positions and spins of any two electrons causes the value of the wavefunction to change its sign. Using wavefunctions satisfying the Pauli Principle we find that the energy of a He atom with the excited $1s^1 2s^1$ electron configuration depends on the electron spins: a triply degenerate state in which the electrons have parallel spins has lower energy than a singly degenerate state with antiparallel spins.

2.1 The energy, the Hamiltonian operator and the wavefunctions of a two-electron atom

We assume the nucleus to remain at rest and place it at the origin of our coordinate system. The electronic energy of the atom is then equal to the sum of the kinetic energies of the two electrons plus the electrostatic potential energy:

$$E = \left(\frac{1}{2}\right) m_e v_1^2 + \left(\frac{1}{2}\right) m_e v_2^2 + \mathrm{PE}(r_1, r_2, r_{1-2})$$

where v_1 and v_2 are the velocities of the two electrons. The potential energy PE is the sum of three terms: two terms due to the attraction of the two electrons to the nucleus and one term due to repulsion between them:

$$E = \frac{1}{2}m_e v_1^2 + \frac{1}{2}m_e v_2^2 - \frac{Ze^2}{4\pi\varepsilon_0 r_1} - \frac{Ze^2}{4\pi\varepsilon_0 r_2} + \frac{e^2}{4\pi\varepsilon_0 r_{1-2}} \tag{2.1}$$

Here r_1 and r_2 are the distances from the nucleus to the first and the second electron respectively, and $r_{1\text{-}2}$ is the distance between the two electrons. The expression (2.1) implies that the energy is defined as zero when the two electrons are at rest and infinitely far from each other and from the nucleus.

The Hamiltonian operator of such an atom becomes

$$\begin{aligned} \mathcal{H} = & -\frac{h^2}{8\pi^2 m_e}\left(\frac{\partial^2}{\partial x_1^2} + \frac{\partial^2}{\partial y_1^2} + \frac{\partial^2}{\partial z_1^2}\right) \\ & -\frac{h^2}{8\pi^2 m_e}\left(\frac{\partial^2}{\partial x_2^2} + \frac{\partial^2}{\partial y_2^2} + \frac{\partial^2}{\partial z_2^2}\right) \\ & -\frac{Ze^2}{4\pi\varepsilon_0 r_1} - \frac{Ze^2}{4\pi\varepsilon_0 r_2} + \frac{e^2}{4\pi\varepsilon_0 r_{1-2}} \end{aligned}$$

The first two terms in $\mathcal{H}$ represent the kinetic energies of the two electrons. The Hamiltonian may be written more compactly as

$$\mathcal{H} = -\frac{h^2}{8\pi^2 m_e}\left(\nabla_1^2 + \nabla_2^2\right) + \frac{e^2}{4\pi\varepsilon_0}\left(-\frac{Z}{r_1} - \frac{Z}{r_2} + \frac{1}{r_{1\text{-}2}}\right) \tag{2.2}$$

According to the principles of quantum mechanics, the allowed values of the energy E and the corresponding wavefunctions are found by solving the time-independent Schrödinger equation:

$$-\frac{h^2}{8\pi^2 m_e}\left(\nabla_1^2 + \nabla_2^2\right)\Psi(\mathbf{r}_1, \mathbf{r}_2) + \frac{e^2}{4\pi\varepsilon_0}\left(-\frac{Z}{r_1} - \frac{Z}{r_2} + \frac{1}{r_{1\text{-}2}}\right)\Psi(\mathbf{r}_1, \mathbf{r}_2) = E\Psi(\mathbf{r}_1, \mathbf{r}_2) \tag{2.3}$$

In the following we assume that the wavefunction is normalized:

$$\iint \Psi^2(\mathbf{r}_1, \mathbf{r}_2)\, \mathrm{d}\tau_1\, \mathrm{d}\tau_2 = 1 \tag{2.4}$$

when the integration is carried out over all values of the x, y, and z coordinates of both electrons. The dimension of the normalized wavefunction of a two-electron atom is (volume)$^{-1}$.

Once we know the normalized wavefunction of a system containing two electrons, we can calculate the probability of finding electron number 1 in a small cube with a corner at the point indicated by the vector $\mathbf{r}_a$ and volume $\mathrm{d}\tau_a = \mathrm{d}x_a\, \mathrm{d}y_a\, \mathrm{d}z_a$ and at the same time finding electron number 2 in another small cube with a corner at the point indicated by the vector $\mathbf{r}_b$ and volume $\mathrm{d}\tau_b = \mathrm{d}x_b\, \mathrm{d}y_b\, \mathrm{d}z_b$. See Fig. 2.1:

The probability of electron 1 being in $\mathrm{d}\tau_a$ and electron 2 in $\mathrm{d}\tau_b$ is $\Psi^2(\mathbf{r}_a, \mathbf{r}_b)\, \mathrm{d}\tau_a\, \mathrm{d}\tau_b$ (2.5)

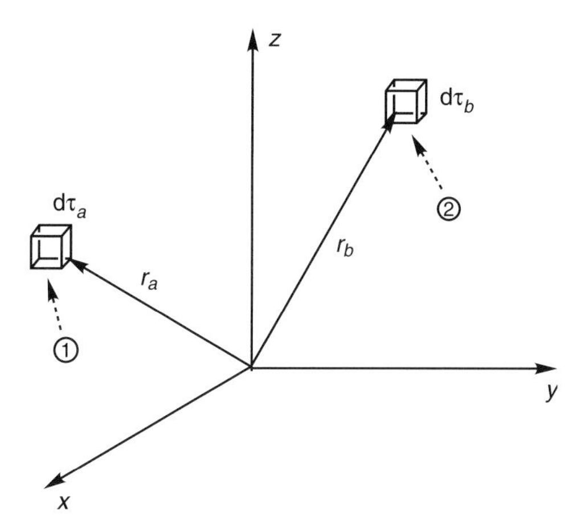

Fig. 2.1. What is the probability of finding electron number 1 in the small cube $d\tau_a$ and electron number 2 in $d\tau_b$?

This probability will be a number between 0 (impossibility) and 1 (certainty).

If we are interested in calculating the probability of finding electron number 1 in $d\tau_a$ regardless of the position of electron number 2, we must integrate the probability expression in (2.5) over all values for the coordinates of electron 2:

$$\text{The probability of electron 1 being in } d\tau_a \text{ is } \int_{r2} \Psi^2(\mathbf{r}_a, \mathbf{r}_2)\, d\tau_a\, d\tau_2 \tag{2.6}$$

Similarly:

$$\text{The probability of electron 2 being in } d\tau_b = \int_{r1} \Psi^2(\mathbf{r}_1, \mathbf{r}_b)\, d\tau_1\, d\tau_b \tag{2.7}$$

where integration is carried out over all values for the coordinates of electron 1.

Finally, we define the *total* electron density in $\mathbf{r}_a$ as the sum of the probability densities of electrons 1 and 2:

$$\text{Total electron density} = \int_{r2} \Psi^2(\mathbf{r}_a, \mathbf{r}_2)\, d\tau_2 + \int_{r1} \Psi^2(\mathbf{r}_1, \mathbf{r}_a)\, d\tau_1 \tag{2.8}$$

2.2 The atomic orbital model

Due to the term in the Hamiltonian operator that represents the electron–electron repulsion energy *it is impossible to find an exact solution to the Schrödinger equation* (2.3) *in closed form.*

A common way to find an *approximate* description of the bound states ($E < 0$) of two-electron atoms is to assume that the wavefunction of the atom may be written as the product of two functions, each of which depends on the coordinates of one electron only:

$$\Psi(\mathbf{r}_1, \mathbf{r}_2) = \Phi_i(\mathbf{r}_1)\Phi_j(\mathbf{r}_2) \tag{2.9}$$

Like the wavefunctions for the bound states of a one-electron atom, Φ_i and Φ_j are functions of the coordinates of *one* electron moving in the vicinity of *one* nucleus, and they are therefore referred to as atomic orbitals (AOs).

For an approximate description of the ground state we assume that both electrons occupy AOs which resemble the 1s orbitals of a one-electron atom with nuclear charge $+Z'e$:

$$\Phi_i(\mathbf{r}) = \Phi_j(\mathbf{r}) = \Phi(\mathbf{r}) = K \exp\left(\frac{Z'r}{a_0}\right) \tag{2.10}$$

where K is a normalization constant, $K = (1/\sqrt{\pi})(Z'/a_0)^{3/2}$, and Z' is an adjustable parameter which may take on non-integral values.

Let us consider the special case of a He atom. Each electron feels the attraction of the two protons in the nucleus and the repulsion from the other electron. If the repulsion from the other electron was negligible, the electron would feel the same force as in the He^+ ion and Z' would be equal to 2. If repulsion from the other electron should cancel the attraction due to one of the nuclear protons, Z' would be equal to 1 as in the H atom. We may guess, therefore, that the best value for Z' will lie somewhere between 1.0 and 2.0.

It is important to remember that equation (2.9) does *not* represent an exact solution to the Schrödinger equation. Why this is so is most easily understood by comparing the probability of finding both electron 1 and electron 2 in the same volume element $\mathrm{d}\tau_a$ with the probability of finding electron 1 in $\mathrm{d}\tau_a$ and electron 2 in $\mathrm{d}\tau_b$ (see Fig. 2.1):

$$\text{The probability of finding 1 in } \mathrm{d}\tau_a \text{ and 2 in } \mathrm{d}\tau_b \text{ is } \Phi_i^2(\mathbf{r}_a)\,\mathrm{d}\tau_a \Phi_j^2(\mathbf{r}_b)\,\mathrm{d}\tau_b \tag{2.11a}$$

$$\text{The probability of finding both 1 and 2 in } \mathrm{d}\tau_a \text{ is } \Phi_i^2(\mathbf{r}_a)\,\mathrm{d}\tau_a \Phi_j^2(\mathbf{r}_a)\,\mathrm{d}\tau_a \tag{2.11b}$$

Now assume that the atom is in the ground state, i.e. that both electrons reside in 1s orbitals: $\Phi_i(\mathbf{r}) = \Phi_j(\mathbf{r}) = \Phi_{1s}(\mathbf{r})$. If the two volume elements are equally large ($\mathrm{d}\tau_a = \mathrm{d}\tau_a = \mathrm{d}\tau$), and if they are equally far from the nucleus ($r_a = r_b = r$) then

$$\text{The probability of finding 1 in } \mathrm{d}\tau_a \text{ and 2 in } \mathrm{d}\tau_a \text{ is } \Phi_{1s}^2(r)\,\mathrm{d}\tau \Phi_{1s}^2(r)\,\mathrm{d}\tau \tag{2.12a}$$

$$\text{The probability of finding both 1 and 2 in } \mathrm{d}\tau_a \text{ is } \Phi_{1s}^2(r)\,\mathrm{d}\tau \Phi_{1s}^2(r)\,\mathrm{d}\tau \tag{2.12b}$$

The two probabilities are equal! This means that the two electrons are just as likely to be found close together as far apart. More accurate solutions of the Schrödinger equation show, however, that the electrostatic repulsion between the two electrons tend to keep them far apart: the probability of finding 1 in $\mathrm{d}\tau_a$ and 2 in $\mathrm{d}\tau_b$ increases with increasing distance between the two volume elements.

The description of an atom containing two or more electrons in terms of a wavefunction that is written as a simple product of atomic orbitals is therefore *inherently wrong*. Or, to put it more gently, it represents a model of the atom. This model may be manipulated in a computer, and in some cases leads to numerical results in reasonable agreement with experimental values. But like all models, it may lead to erroneous conclusions, and the limits of its applicability must be established through comparison with experimental results.

2.3 The ground state of a two-electron atom: calculation of the energy

We now assume both electrons to reside in 1*s* type AOs of the form indicated in equation (2.10) and proceed to determine the best value of the adjustable parameter Z', that means the value that gives the energy in best agreement with experiment.

We begin by rewriting the Hamiltonian operator of the atom, equation (2.2), as

$$\mathcal{H} = h_1^0 + h_2^0 + \frac{e^2}{4\pi\varepsilon_0 r_{1\text{-}2}} \tag{2.13}$$

where h_1^0 and h_2^0 represent the kinetic energy plus the electron–nucleus attraction energy for electrons 1 and 2, respectively. The third term represents the electron–electron repulsion energy.

We have noted that $\Psi(\mathbf{r}_1, \mathbf{r}_2)$ is *not* a solution to the Schrödinger equation. This means that it does *not* describe a stationary state, i.e. a state where the energy is fixed and unchanging. According to the postulates of quantum mechanics, measurement of the energy will then give varying results. However, if several measurements are made, the average energy will be given by the integral

$$E = \iint \Psi(\mathbf{r}_1, \mathbf{r}_2)\, \mathcal{H}\Psi(\mathbf{r}_1, \mathbf{r}_2)\, \mathrm{d}\tau_1\, \mathrm{d}\tau_2 \tag{2.14}$$

or

$$\begin{aligned} E = &\iint \Psi(\mathbf{r}_1, \mathbf{r}_2) h_1^0 \Psi(\mathbf{r}_1, \mathbf{r}_2)\, \mathrm{d}\tau_1\, \mathrm{d}\tau_{22} \\ &+ \iint \Psi(\mathbf{r}_1, \mathbf{r}_2) h_2^0 \Psi(\mathbf{r}_1, \mathbf{r}_2)\, \mathrm{d}\tau_1\, \mathrm{d}\tau_2 \\ &+ \iint \Psi(\mathbf{r}_1, \mathbf{r}_2) \left[\frac{e^2}{(4\pi\varepsilon_0 r_{1\text{-}2}}) \right] \Psi(\mathbf{r}_1, \mathbf{r}_2)\, \mathrm{d}\tau_1\, \mathrm{d}\tau_2 \end{aligned} \tag{2.15}$$

or

$$\begin{aligned} E = &\int \Phi(\mathbf{r}_1) h_1^0 \Phi(\mathbf{r}_1)\, \mathrm{d}\tau_1 + \int \Phi(\mathbf{r}_2) h_2^0 \Phi(\mathbf{r}_2)\, \mathrm{d}\tau_2 \\ &+ \iint \Phi(\mathbf{r}_1)\Phi(\mathbf{r}_2) \left[\frac{e^2}{(4\pi\varepsilon_0 r_{1\text{-}2})}\right] \Phi(\mathbf{r}_1)\Phi(\mathbf{r}_2)\, \mathrm{d}\tau_1\, \mathrm{d}\tau_2 \end{aligned} \tag{2.16}$$

The first and the second integrals in equation (2.16) are equal to the sum of the kinetic energy and the electron–nucleus attraction energy for electrons 1 and 2 respectively. Each of the two integrals is equal to $(Z'^2 - 2ZZ')\mathcal{R}$. The third integral in (2.16) represents the electron–electron repulsion energy and is equal to $1.25\, Z'\mathcal{R}$:

$$E = [2(Z'^2 - 2ZZ') + 1.25Z']\mathcal{R} \tag{2.17}$$

The average energy obtained from several measurements must necessarily be greater than the energy of the lowest possible state, i.e. the ground state. This means that the energy that has been calculated from an arbitrary wavefunction with the true Hamiltonian operator will always be higher than – or equal to – the true energy of the ground state:

$$E = [2(Z'^2 - 2ZZ') + 1.25Z']\mathcal{R} > E_0$$

This, in turn, means that the value of Z' that yields the lowest energy, also yields the energy that is the closest to the real ground state energy. This optimal value of Z' is determined by setting the derivative of E with respect to Z' equal to zero:

$$4Z' - 4Z + 1.25 = 0$$

$$Z' = Z - 0.3125 \tag{2.18a}$$

Since the second derivative is equal to 4, this value of Z' does indeed correspond to a minimum value of E.

The best value of Z' is referred to as the "effective nuclear charge," Z_{eff}:

$$Z_{eff} = Z - 0.3125 \tag{2.18b}$$

The difference between the true and the effective nuclear charge is due to the repulsion between the two electrons, and turns out to be equally large for all two-electron atoms. It is sometimes said that each electron screens the nucleus for the other. Since the effective nuclear charge is smaller than the true nuclear charge Z, the atomic orbitals in (2.10) indicate that the electron density in a two-electron atom is spread out over a larger volume than in the one-electron atom with the same nucleus.

In order to find the best (optimal) value for the energy, we go back to equation (2.17), replace Z' by Z_{eff} and Z by ($Z_{eff} + 0.3125$) and obtain

$$E_{opt} = [2Z_{eff}^2 - 4(Z_{eff} - 0.3125Z_{eff}) + 1.25Z_{eff}]\mathcal{R} = -2Z_{eff}^2\mathcal{R} \tag{2.19}$$

Comparison with equation (1.12) shows that our estimate for the energy of the ground state of a two-electron atom with nuclear charge $+Ze$ is twice as large as the energy of the ground state of a fictitious one-electron atom with nuclear charge equal to $(Z - 0.3125)e$.

According to equation (2.19) the energy of a He atom in the ground state is equal to $-7475\,\mathrm{kJ\,mol^{-1}}$, which is $148\,\mathrm{kJ\,mol^{-1}}$ or about 2% higher than the experimental value, $-7623\,\mathrm{kJ\,mol^{-1}}$.

2.4 Atomic size

We have shown that the best atomic orbitals for a two-electron atom in the ground state are equal to the $1s$ orbital of a fictitious one-electron atom with nuclear charge $+Z_{eff}e$. The classical boundary of such an atom is given by

$$r_{max} = \frac{2a_0}{Z_{eff}} \tag{2.20}$$

In the following we shall use this radius as a measure for the size of the real two-electron atom.

We have seen that the classical radius of a hydrogen atom in the ground state is equal to $2a_0$ while the increased nuclear charge reduces the radius of the He^+ cation to half that value (a_0). When a second electron is added to the cation to form the neutral He atom, repulsion between the electrons increases the atomic radius from a_0 to $2a_0/Z_{eff} = 1.18a_0$.

Problem 2.1 Estimate the classical radii of the hydride anion, H^-, and the Li^+ cation. Compare with the radius of He. Compare the radius of H^- with that of the neutral H atom.

2.5 First ionization energies

The first ionization energy (IE1) of a neutral atom A is defined as the energy required for the process

$$A(g) = A^+(g) + e^-(g) \tag{2.21}$$

when reactants and products are at rest and the products infinitely far apart. The energy required is equal to the difference between the energies of products and reactants:

$$\mathrm{IE1} = E(A^+) + E(e^-) - E(A) = E(A^+) - E(A) \tag{2.22}$$

The second ionization energy of the atom is defined as the energy that is required to remove a second electron:

$$A^+(g) = A^{+2}(g) + e^-(g)$$

$$\mathrm{IE2} = E(A^{+2}) + E(e^-) - E(A^+) = E(A^{+2}) - E(A^+)$$

If the neutral atom contains more than two electrons, third and higher order ionization energies may be defined in a similar manner.

We now attempt to calculate IE1 for the He atom from equation (2.22). The He^+ ion is a one-electron atom and the energy of the ground state is exactly $-Z^2\mathcal{R}$. The helium atom, on the other hand, is a two-electron atom, and the energy is only approximately equal to $-2Z_{\mathrm{eff}}^2\mathcal{R}$. Combining these values we obtain

$$IE1 = (-Z^2 + 2Z_{\mathrm{eff}}^2)\mathcal{R} = 2226\,\mathrm{kJ\,mol^{-1}}.$$

Comparing with the experimental value (2372 kJ mol^{-1}) we find that the calculated ionization energy is 6% too low. The reason for the discrepancy is clear: our estimate for the energy of the reactant (He) is too high.

We shall now show how we can improve the agreement with the experimental ionization energy by making two errors instead of one!

Before ionization, both electrons are described by AOs of the form of equation (2.10) with $Z_{\mathrm{eff}} \approx 1.69$. After ionization, the remaining electron is left in the 1s orbital of a He^+ cation. *When one electron is removed, the wavefunction of the remaining electron is changed*: it no longer feels the repulsion from the other electron and moves closer to the nucleus. This process is referred to as "relaxation."

We now calculate the ionization energy of He from equation (2.22) making the two assumptions:

(a) As before we assume that the energy of the He atom is given by $2Z_{\mathrm{eff}}^2\mathcal{R}$.

(b) We neglect relaxation effects. This means that the energy of the He^+ ion is equal to the first integral in equation (2.16):

$$E(\mathrm{He}^+) = \int \Phi(\mathbf{r}_1) h_1^0 \Phi(\mathbf{r}_1)\, \mathrm{d}\tau_1 = (Z'^2 - 2ZZ')\mathcal{R}$$

Replacement of Z' by Z_{eff} and of Z by $(Z_{\mathrm{eff}} + 0.3125)$ gives

$$E(\mathrm{He}^+) = \left(-Z_{\mathrm{eff}}^2 - 0.0625\, Z_{\mathrm{eff}}\right)\mathcal{R}$$

Using the new value for the energy of the He^+ ion, we obtain

$$\begin{aligned} \mathrm{IE1} &= E(\mathrm{He}^+) - E(\mathrm{He}) \\ &= \left(-Z_{\mathrm{eff}}^2 - 0.0625\, Z_{\mathrm{eff}}\right)\mathcal{R} - \left(-2Z_{\mathrm{eff}}^2 \mathcal{R}\right) \\ &= \left(Z_{\mathrm{eff}}^2 - 0.625\, Z_{\mathrm{eff}}\right)\mathcal{R} \qquad (2.23) \end{aligned}$$

Insertion of numbers yields $\mathrm{IE1} = 2352\,\mathrm{kJ\,mol^{-1}}$. By increasing the number of errors from one to two, we have reduced the disagreement between calculations and experiment from about 6% to about 0.8%! Our overestimate of the energy of He has to a large extent been compensated by a similar overestimate of the energy of He^+.

2.6 Orbital energies and Koopman's theorem

For atoms containing more than one electron, *we define the orbital energy as the kinetic energy of an electron in the orbital plus the energy of attraction to the nucleus and the energies of repulsion to all other electrons*. Thus the $1s$ orbital energy of a two-electron atom in the ground state is given by the sum of the first and third integrals in (2.16):

$$\varepsilon(1\mathrm{s}) = (Z'^2 - 2ZZ')\mathcal{R} + 1.25Z'\mathcal{R} \qquad (2.24)$$

Replacement of Z' by Z_{eff} and of Z by $(Z_{\mathrm{eff}} + 0.3125)$ yields

$$\varepsilon(1\mathrm{s}) = (-Z_{\mathrm{eff}}^2 + 0.625Z_{\mathrm{eff}})\mathcal{R} \qquad (2.25)$$

Comparison with equation (2.23) shows that the first ionization energy of He is equal to minus the orbital energy.

The general statement that *the energy required to remove an electron from an atom or a molecule, is equal to minus the orbital energy of the electron that is removed*, is known as "Koopman's theorem." The general theorem can be "proven" if we use an atomic or molecular orbital model and assume that ionization is *not* followed by relaxation. Comparison with experiments shows that predictions based on Koopman's theorem usually are accurate to about 10%. The accuracy would be even poorer if there was not a fortunate cancellation of errors!

Problem 2.2 Use equation (2.25) and Koopman's theorem to predict the second ionization energy of Li and the electron affinity of the H atom. (The electron affinity of H is equal to the energy required to remove an electron from the hydride anion H^-). Compare with the experimental values IE2(Li) $= 7300\,\text{kJ mol}^{-1}$ and EA(H) $= 74\,\text{kJ mol}^{-1}$.

Since the orbital energy includes the energy of repulsion to the other electrons in the molecule, the orbital energy will change if we move one of the other electrons to another orbital or if we remove it completely (ionization). *The orbital energies are not constant for a given atom but must be calculated separately for each electron configuration.*

What is the relationship between the total energy of an atom and the sum of the orbital energies of the constituent electrons? According to equation (2.19) the energy of the atom is given by

$$E = -2Z_{\text{eff}}^2\mathcal{R}$$

Adding the orbital energies of the two electrons in He we obtain

$$2\varepsilon(1\text{s}) = (-2Z_{\text{eff}}^2 + 1.25Z_{\text{eff}})\mathcal{R}$$

This means that

$$E = 2\varepsilon(1\text{s}) - 1.25Z'_{\text{eff}}\mathcal{R}$$

The second term on the right-hand side is equal to the electron–electron repulsion energy: When we calculate the energy the according to equation (2.19) electron–electron repulsion is included *once*; when we add the two orbital energies it is counted *twice*.

This is an awkward result, and the reader may wonder why the definition of orbital energies is not modified in such a manner that their sum is equal to the total energy of the atom? The reason is that such a modification would destroy the simple relationship between orbital and ionization energies.

The result for two-electron atoms may be generalized: *The total energy of an atom or molecule with more than one electron is not equal to the sum of the occupied orbital energies*, but *is equal to the sum of occupied orbital energies minus all electron–electron repulsion energies*:

$$E = \Sigma\varepsilon_i - (\text{all electron–electron repulsion energies})$$

2.7 The ground state of a two-electron atom: beginning the search for an acceptable wavefunction including the electron spin

The ground state of a one-electron atom is described by one of the two spin orbitals

$$\Phi_1^{\text{s}}(\mathbf{r}, s) = \Phi_{1s}(\mathbf{r})\alpha(s) \quad \text{or} \quad \Phi_2^{\text{s}}(\mathbf{r}, s) = \Phi_{1s}(\mathbf{r})\beta(s) \tag{2.26}$$

According to the Exclusion Principle two electrons can only occupy the same AO if they have different spins. A reasonable wavefunction for the ground state of a two-electron atom might therefore be the simple product of the two spin orbitals

$$\Psi^0(\mathbf{r}_1, s_1, \mathbf{r}_2, s_2) = \Phi_{1s}(\mathbf{r}_1)\alpha(s_1)\Phi_{1s}(\mathbf{r}_2)\beta(s_2)$$

or, written more simply

$$\Psi^0(1,2) = \Phi_{1s}(1)\alpha(1)\Phi_{1s}(2)\beta(2) \tag{2.27}$$

where the numbers 1 and 2 denote the appropriate coordinates, space coordinates (**r**), spin coordinates (s), or both, of atoms number 1 or 2, respectively. As it turns out, *a simple product of spin orbitals does not provide an adequate description of atoms containing two or more electrons.* Let us therefore examine $\Psi^0(1,2)$ more closely.

See Fig. 2.1. Assume that we wish to compute the probability of finding one electron in $d\tau_a$ with spin $= +1/2$ and the other in $d\tau_b$ with spin $= -1/2$. The first step is to calculate the value of the wavefunction for the appropriate values of the space and spin coordinates. You first decide to refer to the electron in $d\tau_a$ as number 1 and the electron in $d\tau_b$ as number 2:

$$\Psi^0(1,2) = \Phi_{1s}(\mathbf{r}_a)\alpha(1/2)\Phi_{1s}(\mathbf{r}_b)\beta(-1/2) = \Phi_{1s}(\mathbf{r}_a)\Phi_{1s}(\mathbf{r}_b) \tag{2.28a}$$

Since $1s$ AOs are positive everywhere, both $\Psi_0(1,2)$ and the probability density $\Psi_0^2(1,2)$ will be greater than zero.

Then you decide to refer to the electron in $d\tau_a$ as number 2 and the electron in $d\tau_b$ as number 1:

$$\Psi^0(1,2) = \Phi_{1s}(\mathbf{r}_b)\alpha(-1/2)\Phi_{1s}(\mathbf{r}_a)\beta(1/2) = 0 \tag{2.28b}$$

This is a disturbing result: The probability of finding an electron in $d\tau_a$ with spin $= +1/2$ and the other in $d\tau_b$ with spin $= -1/2$ depends on the way you choose to number the electrons! The wavefunction in (2.27) is clearly unacceptable.

According to the principles of quantum mechanics the probability density must be independent of the indexes assigned to the two electrons:

$$\Psi^2(\mathbf{r}_a, s_a, \mathbf{r}_b, s_b) = \Psi^2(\mathbf{r}_b, s_b, \mathbf{r}_a, s_a) \tag{2.29}$$

Since we are dealing with real wavefunctions, this means that

$$\Psi(\mathbf{r}_a, s_a, \mathbf{r}_b, s_b) = \pm\Psi(\mathbf{r}_b, s_b, \mathbf{r}_a, s_a) \tag{2.30}$$

2.8 The search for acceptable wavefunctions including the electron spin: the Pauli Principle

Wavefunctions for the ground state of a two-electron atom that satisfy the condition (2.30) may be obtained by combination of the two simple products

$$\Phi_{1s}(1)\alpha(1)\Phi_{1s}(2)\beta(2) \text{ and } \Phi_{1s}(2)\alpha(2)\Phi_{1s}(1)\beta(1)$$

Note that the two products differ only in the number assigned to the electrons occupying the $\Phi_{1s}\alpha$ and $\Phi_{1s}\beta$ spin orbitals respectively. Linear combinations of the two simple products yield the wavefuntions:

$$\begin{aligned}\Psi_+(1,2) &= (1/\sqrt{2})[\Phi_{1s}(1)\alpha(1)\Phi_{1s}(2)\beta(2) + \Phi_{1s}(2)\alpha(2)\Phi_{1s}(1)\beta(1)] \\ &= (1/\sqrt{2})\Phi_{1s}(1)\Phi_{1s}(2)[\alpha(1)\beta(2) + \alpha(2)\beta(1)]\end{aligned} \tag{2.31}$$

and

$$\begin{aligned}\Psi_-(1,2) &= (1/\sqrt{2})[\Phi_{1s}(1)\alpha(1)\Phi_{1s}(2)\beta(2) - \Phi_{1s}(2)\alpha(2)\Phi_{1s}(1)\beta(1)] \\ &= (1/\sqrt{2})\Phi_{1s}(1)\Phi_{1s}(2)[\alpha(1)\beta(2) - \alpha(2)\beta(1)] \end{aligned} \tag{2.32}$$

where the factors $(1/\sqrt{2})$ are normalization constants.

We now explore the mathematical properties of the two functions: evaluation of the function $\Psi_+(1,2)$ when the electron in $d\tau_a$ is referred to as number 1 (and the electron in $d\tau_b$ as number 2), that is with $(\mathbf{r}_1, s_1, \mathbf{r}_2, s_2) = (\mathbf{r}_a, +1/2, \mathbf{r}_b, -1/2)$ yields

$$\begin{aligned}\Psi_+(1,2) &= (1/\sqrt{2})\Phi_{1s}(\mathbf{r}_a)\Phi_{1s}(\mathbf{r}_b)\,[\alpha(1/2)\beta(-1/2) + \alpha(-1/2)\beta(+1/2)] \\ &= (1/\sqrt{2})\Phi_{1s}(\mathbf{r}_a)\Phi_{1s}(\mathbf{r}_b)[1 \times 1 + 0 \times 0] \\ &= (1/\sqrt{2})\Phi_{1s}(\mathbf{r}_a)\Phi_{1s}(\mathbf{r}_b) \end{aligned} \tag{2.33a}$$

Evaluation of the function $\Psi_+(1,2)$ when the electron in $d\tau_a$ is referred to as number 2, that is with $(\mathbf{r}_1, s_1, \mathbf{r}_2, s_2) = (\mathbf{r}_b, -1/2, \mathbf{r}_a, +1/2)$ yields

$$\begin{aligned}\Psi_+(1,2) &= (1/\sqrt{2})\Phi_{1s}(\mathbf{r}_b)\Phi_{1s}(\mathbf{r}_a)\,[\alpha(-1/2)\beta(+1/2) + \alpha(+1/2)\beta(-1/2)] \\ &= (1/\sqrt{2})\Phi_{1s}(\mathbf{r}_b)\Phi_{1s}(\mathbf{r}_a)[0 \times 0 + 1 \times 1] \\ &= (1/\sqrt{2})\Phi_{1s}(\mathbf{r}_a)\Phi_{1s}(\mathbf{r}_b) \end{aligned} \tag{2.33b}$$

Replacing the coordinates $(\mathbf{r}_a, s_a, \mathbf{r}_b, s_b)$ with the coordinates $(\mathbf{r}_b, s_b, \mathbf{r}_a, s_a)$ corresponds to *exchanging the positions and spins of the two electrons*. Comparing the expressions (2.33a) and (2.33b) we see that exchange of the two electrons leaves the value of both the wavefunction $\Psi_+(\mathbf{r}_1, s_1, \mathbf{r}_2, s_2)$ and the probability density $\Psi_+^2(\mathbf{r}_1, s_1, \mathbf{r}_2, s_2)$ unchanged.

Evaluation of the function $\Psi_-(1,2)$ when the electron in $\mathbf{r}_a$ is referred to as number 1, i.e. with $(\mathbf{r}_1, s_1, \mathbf{r}_2, s_2) = (\mathbf{r}_1, s_1, \mathbf{r}_2, s_2) = (\mathbf{r}_a, +1/2, \mathbf{r}_b, -1/2)$ yields

$$\begin{aligned}\Psi_-(1,2) &= (1/\sqrt{2})\Phi_{1s}(\mathbf{r}_a)\Phi_{1s}(\mathbf{r}_b)[\alpha(+1/2)\beta(-1/2) - \alpha(-1/2)\beta(+1/2)] \\ &= (1/\sqrt{2})\Phi_{1s}(\mathbf{r}_a)\Phi_{1s}(\mathbf{r}_b)[1 \times 1 - 0 \times 0] \\ &= (1/\sqrt{2})\Phi_{1s}(\mathbf{r}_a)\Phi_{1s}(\mathbf{r}_b) \end{aligned} \tag{2.34a}$$

Evaluation of the function (2.31) when the electron in $\mathbf{r}_b$ is referred to as "electron 1", i.e. with $(\mathbf{r}_1, s_1, \mathbf{r}_2, s_2) = (\mathbf{r}_b, -1/2, \mathbf{r}_a, +1/2)$ yields

$$\begin{aligned}\Psi_-(1,2) &= (1/\sqrt{2})\Phi_{1s}(\mathbf{r}_b)\Phi_{1s}(\mathbf{r}_a)\,[\alpha(-1/2)\beta(+1/2) - \alpha(+1/2)\beta(-1/2)] \\ &= (1/\sqrt{2})\Phi_{1s}(\mathbf{r}_b)\Phi_{1s}(\mathbf{r}_a)[0 \times 0 - 1 \times 1] \\ &= (-1/\sqrt{2})\Phi_{1s}(\mathbf{r}_a)\Phi_{1s}(\mathbf{r}_b) \end{aligned} \tag{2.34b}$$

Comparing the expressions (2.34a) and (2.34b) we see that exchange of the positions and spins of the two electrons changes the *sign* of the wavefunction $\Psi_-(\mathbf{r}_1, s_1, \mathbf{r}_2, s_2)$ but leaves the probability density $\Psi_-^2(\mathbf{r}_1, s_1, \mathbf{r}_2, s_2)$ unchanged.

Thus the functions Ψ_+ and Ψ_- both satisfy the condition that

$$\Psi^2(\mathbf{r}_a, s_a, \mathbf{r}_b, s_b) = \Psi^2(\mathbf{r}_b, s_b, \mathbf{r}_a, s_a) \quad \text{or} \quad \Psi^2(1,2) = \Psi^2(2,1)$$

We have succeeded in constructing two wavefunctions for which probability densities are independent of the way we number the electrons. Are they both acceptable? The answer is NO:

The vector representing the magnitude and direction of the total spin angular momentum of a two-electron atom is defined as the sum of the two vectors representing the spin angular momenta of the two electrons. Examination of the wavefunction Ψ_- using the tools of quantum mechanics reveals that this wavefunction describes an atom in which the total spin angular momentum vector is equal to zero. Since the effects of the two electron spins cancel completely, we say that they have opposite, or antiparallel, spins. All two-electron atoms that have been examined experimentally have been found to have antiparallel spins in the ground state.

A similar examination of the wavefunction Ψ_+ shows that this wavefunction describes an atom in which the total spin angular momentum is nearly twice as large as that of a single spinning electron. Such an atom is described as having two electrons of (approximately) parallel spins. As we have pointed out above no two-electron atom has been found to have parallel electron spins in the ground state.

We conclude that the *only* acceptable wavefunction for the ground state of the two-electron atom is given by

$$\begin{aligned}\Psi_-(1,2) &= (1/\sqrt{2})[\Phi_{1s}(1)\alpha(1)\Phi_{1s}(2)\beta(2) - \Phi_{1s}(2)\alpha(2)\Phi_{1s}(1)\beta(1)] \\ &= (1/\sqrt{2})\Phi_{1s}(1)\Phi_{1s}(2)[\alpha(1)\beta(2) - \alpha(2)\beta(1)] \qquad (2.32)\end{aligned}$$

This result for the ground state of two-electron atoms may be generalized:

The only acceptable wavefunctions for an atom or molecule with two or more electrons are those for which the exchange of the positions and spins of any two electrons causes the value of the wavefunction to change its sign.

The general statement is known as the *Pauli Principle*.

2.9 The Pauli Principle and Slater determinants

Using the spin orbitals defined in (2.26) we may rewrite equation (2.32) as

$$\Psi_-(\mathbf{r}_1, s_1, \mathbf{r}_2, s_2) = 1/\sqrt{2}[\Phi_1^s(\mathbf{r}_1, s_1)\Phi_2^s(\mathbf{r}_2, s_2) - \Phi_1^s(\mathbf{r}_2, s_2)\Phi_2^s(\mathbf{r}_1, s_1)] \qquad (2.35a)$$

or as a determinant

$$\Psi_-(\mathbf{r}_1, s_1, \mathbf{r}_2, s_2) = (1/\sqrt{2})\begin{vmatrix}\Phi_1^s(\mathbf{r}_1, s_1)\Phi_2^s(\mathbf{r}_1, s_1) \\ \Phi_1^s(\mathbf{r}_2, s_2)\Phi_2^s(\mathbf{r}_2, s_2)\end{vmatrix} \qquad (2.35b)$$

Note that the first line in the determinant contains the two spin orbitals evaluated for the coordinates of the first electron. The second line consists of the spin orbitals listed in the same order, but evaluated for the coordinates of the second electron. Such a determinant is referred to as a Slater determinant.

What happens if we assume the two electrons to occupy the same atomic orbital and to have the same spin? In this case $\Phi_1^s(\mathbf{r}, s) = \Phi_2^s(\mathbf{r}, s)$, the two columns are identical, and the determinant is equal to zero for all values of both space and spin coordinates. *No acceptable wavefunction can be constructed if we assume to electrons to occupy the same AO with the same spin*. The Exclusion Principle is therefore included in the more general Pauli Principle.

Since $\Psi_-(\mathbf{r}_1, s_1, \mathbf{r}_2, s_2)$ is the only satisfactory wavefunction for the $1s^2$ electron configuration, the ground state of a two-electron atom is non-degenerate.

Earlier in this chapter we used the simple product $\Psi(\mathbf{r}_1, \mathbf{r}_1) = \Phi_{1s}(\mathbf{r}_1)\Phi_{1s}(\mathbf{r}_2)$ to calculate the energy of the ground state of a two-electron atom. There are two reasons why this was possible:

(a) Neither the classical energy expression nor the Hamiltonian operator contained terms depending on the spin coordinates.

(b) The complete wavefunction that includes electron spin $\Psi_-(\mathbf{r}_1, s_1, \mathbf{r}_2, s_2)$ (2.32) is neatly factored into two parts: the first factor containing only the space functions, the second containing only the spin functions.

When calculating the energy according to

$$E = \int \Psi_-(\mathbf{r}_1, s_1, \mathbf{r}_2, s_2)\mathcal{H}\Psi_-(\mathbf{r}_1, s_1, \mathbf{r}_2, s_2)\,\mathrm{d}\tau_1\,\mathrm{d}\tau_2\,\mathrm{d}s_1\,\mathrm{d}s_2$$

we begin by integration over the spin coordinates and obtain the simple expression (2.14):

$$E = \int \Phi_{1s}(1)\,\Phi_{1s}(2)\,\mathcal{H}\,\Phi_{1s}(1)\,\Phi_{1s}(2)\,\mathrm{d}\tau_1\,\mathrm{d}\tau_2$$

The atoms in Groups 2, 12 or 18, where all atomic orbitals are completely filled (or completely empty) in the ground state, are referred to as closed shell atoms. If such an atom contains N electrons, combination of $N/2$ atomic orbitals with α or β spin functions will produce N spin-orbitals. The ground state is non-degenerate, and a wavefunction satisfying the Pauli Principle is easily obtained by writing down the $N \times N$ Slater determinant:

$$\Psi(\mathbf{r}_1, s_1, \ldots \mathbf{r}_N, s_N) = \sqrt{(N!)}\begin{vmatrix} \Phi_1^s(\mathbf{r}_1, s_1) & \Phi_2^s(\mathbf{r}_1, s_1) & \Phi_1^s(\mathbf{r}_1, s_1) & - & - & \Phi_N^s(\mathbf{r}_1, s_1) \\ \Phi_1^s(\mathbf{r}_2, s_2) & \Phi_2^s(\mathbf{r}_2, s_2) & \Phi_1^s(\mathbf{r}_2, s_2) & - & - & \Phi_N^s(\mathbf{r}_2, s_2) \\ - & - & - & - & - & - \\ - & - & - & - & - & - \\ \Phi_1^s(\mathbf{r}_N, s_N) & \Phi_2^s(\mathbf{r}_N, s_N) & \Phi_1^s(\mathbf{r}_N, s_N) & - & - & \Phi_N^s(\mathbf{r}_N, s_N) \end{vmatrix} \tag{2.36}$$

The first line of the determinant consists of the N spin orbitals evaluated for the coordinates of the first electron. The second line consists of the spin orbitals listed in the same order evaluated for the coordinates of the second electron, and so on until the last line where the spin orbitals are evaluated for the coordinates of the Nth electron. $\sqrt{(N!)}$ is a normalization constant.

We have seen that the Slater determinant wavefunction for the ground state of the He atom may be written as a product of two parts, one factor containing only space functions, the second only spin functions. For this reason a wavefunction consisting of simple product of the atomic orbitals, $\Phi_{1s}(1)\Phi_{1s}(2)$, was sufficient for the calculation of the energy of the atom. Such a factorization of a Slater determinant is only possible if all electrons occupy the same atomic orbital or have the same spin function. This means that reasonably accurate calculations on the ground states of atoms containing more than two electrons are only possible with the use of Slater determinant wavefunctions. Calculations based on such Slater determinant wavefunctions are sometimes referred to as Hartee–Fock (HF) calculations, sometimes as Self-Consistent Field (SCF) calculations.

The metals in Group 1 have one unpaired electron in the valence shell s orbital. This AO may be combined with α or β spin functions to form two spin orbitals. It is therefore possible to write down two Slater determinant wavefunctions for these atoms: the ground states are doubly degenerate. Atoms that contain two or more unpaired electrons are more difficult to describe. For these species one is forced to form wavefunctions by linear combination of two or more Slater determinants. In the next section we shall deal with the simplest atom of this kind, viz. a helium atom excited to the $1s^1 2s^1$ electron configuration.

2.10 The lowest excited states of the helium atom: finding acceptable wavefunctions

The electron configuration corresponding to the first excited state of the He atom is $1s^1 2s^1$. We begin our search for acceptable wavefunctions by combining the $1s$ and $2s$ atomic orbitals with α and β spin functions to form four spin orbitals:

$$\begin{aligned} \Phi^{s}_{1s\alpha}(\mathbf{r}, s) &= \Phi_{1s}(\mathbf{r})\,\alpha(s) \qquad & \Phi^{s}_{1s\beta}(\mathbf{r}, s) &= \Phi_{1s}(\mathbf{r})\,\beta(s) \\ \Phi^{s}_{2s\alpha}(\mathbf{r}, s) &= \Phi_{2s}(\mathbf{r})\,\alpha(s) \qquad & \Phi^{s}_{2s\beta}(\mathbf{r}, s) &= \Phi_{2s}(\mathbf{r})\,\beta(s) \end{aligned} \tag{2.37}$$

We first disregard the Pauli Principle and form the four "zero'th-order" wavefunctions

$$\begin{aligned} \Psi^0_{\alpha\alpha}(1,2) &= \Phi^{s}_{1s\alpha}(1)\,\Phi^{s}_{2s\alpha}(2) \qquad & \Psi^0_{\beta\beta}(1,2) &= \Phi^{s}_{1s\beta}(1)\,\Phi^{s}_{2s\beta}(2) \\ \Psi^0_{\alpha\beta}(1,2) &= \Phi^{s}_{1s\alpha}(1)\,\Phi^{s}_{2s\beta}(2) \qquad & \Psi^0_{\beta\alpha}(1,2) &= \Phi^{s}_{1s\beta}(1)\,\Phi^{s}_{2s\alpha}(2) \end{aligned} \tag{2.38}$$

Each zero-order wavefunction indicates that one electron occupies a $1s$ and the other a $2s$ AO. Note that the first zero-order wavefunction $\Psi^0_{\alpha\alpha}(1,2)$ indicates that both electrons have α and the second that both have β spin. The third zero-order wavefunction indicates that the $1s$ electron has α spin and the $2s$ electron β spin while the fourth indicates that the $1s$ electron has β while the $2s$ electron has α spin.

The energy corresponding to the wavefunction $\Psi^0_{\alpha\alpha}$ may be calculated from

$$\begin{aligned} E^0_{\alpha\alpha} &= \int \Psi^0_{\alpha\alpha}(1,2)\,\mathcal{H}\,\Psi^0_{\alpha\alpha}(1,2)\,\mathrm{d}\tau_1\,\mathrm{d}\tau_2\,\mathrm{d}s_1\,\mathrm{d}s_2 \\ &= \int \Phi_{1s}(1)\,\alpha(1)\,\Phi_{2s}(2)\,\alpha(2)\,\mathcal{H}\,\Phi_{1s}(1)\,\alpha(1)\,\Phi_{2s}(2)\,\alpha(2)\,\mathrm{d}\tau_1\,\mathrm{d}\tau_2\,\mathrm{d}s_1\,\mathrm{d}s_2 \end{aligned}$$

Integration over the spin coordinates yields

$$E^0 = \int \Phi_{1s}(1)\ \Phi_{2s}(2)\ \mathcal{H}\, \Phi_{1s}(1)\ \Phi_{2s}(2)\ \mathrm{d}\tau_1\ \mathrm{d}\tau_2 \tag{2.39}$$

The same energy expression is obtained for $\Psi^0_{\beta\beta}(1,2)$, $\Psi^0_{\alpha\beta}(1,2)$, and $\Psi^0_{\beta\alpha}(1,2)$: the simple product wavefunctions in (2.38) thus indicate that the $1s^1 2s^1$ electron configuration gives rise to four degenerate states. This conclusion is *not* in agreement with experiments which show that *though there are four states, they are not all degenerate.*

We now construct Slater determinants to obtain "first-order" wavefunctions satisfying the Pauli Principle. Combining the $\Phi^{\mathrm{s}}_{1s\alpha}$ and $\Phi^{\mathrm{s}}_{2s\alpha}$ spin oribtals which both describe electrons with α-spin, we obtain:

$$\Psi_{\alpha\alpha}(1,2) = (1/\sqrt{2}) \begin{vmatrix} \Phi^{\mathrm{s}}_{1s\alpha}(1) & \Phi^{\mathrm{s}}_{2s\alpha}(1) \\ \Phi^{\mathrm{s}}_{1s\alpha}(2) & \Phi^{\mathrm{s}}_{2s\alpha}(2) \end{vmatrix}$$

This determinant describes an atom in the $1s^1 2s^1$ electron configuration where the two electrons have parallel spin ($s = +1/2$). If we expand the determinant and collect terms we find that

$$\Psi_{\alpha\alpha}(1,2) = (1/\sqrt{2})\ [\Phi_{1s}(1)\ \Phi_{2s}(2) - \Phi_{1s}(2)\ \Phi_{2s}(1)]\ \alpha(1)\ \alpha(2)$$

The next determinant describes a state where both electrons have spin $-1/2$:

$$\begin{aligned} \Psi_{\beta\beta}(1,2) &= (1/\sqrt{2}) \begin{vmatrix} \Phi^{\mathrm{s}}_{1s\beta}(1) & \Phi^{\mathrm{s}}_{2s\beta}(1) \\ \Phi^{\mathrm{s}}_{1s\beta}(2) & \Phi^{\mathrm{s}}_{2s\beta}(2) \end{vmatrix} \\ &= (1/\sqrt{2})\ [\Phi_{1s}(1)\ \Phi_{2s}(2) - \Phi_{1s}(2)\ \Phi_{2s}]\ \beta(1)\ \beta(2) \end{aligned}$$

Combining spin orbitals with different spin functions one obtains the determinants

$$\Psi_{\alpha\beta}(1,2) = (1/\sqrt{2}) \begin{vmatrix} \Phi^{\mathrm{s}}_{1s\alpha}(1) & \Phi^{\mathrm{s}}_{2s\beta}(1) \\ \Phi^{\mathrm{s}}_{1s\alpha}(2) & \Phi^{\mathrm{s}}_{2s\beta}(2) \end{vmatrix}$$

and

$$\Psi_{\beta\alpha}(1,2) = (1/\sqrt{2}) \begin{vmatrix} \Phi^{\mathrm{s}}_{1s\beta}(1) & \Phi^{\mathrm{s}}_{2s\alpha}(1) \\ \Phi^{\mathrm{s}}_{1s\beta}(2) & \Phi^{\mathrm{s}}_{2s\alpha}(2) \end{vmatrix}$$

At first glance it would appear that $\Psi_{\alpha\beta}(1,2)$ and $\Psi_{\beta\alpha}(1,2)$ describe states where the electrons have opposite spins, but closer examination (which we shall not carry out) reveals that these functions do not correspond to proper spin states at all. Proper spin wavefunctions are obtained by forming linear combination of the Slater determinants:

$$\begin{aligned} \Psi_{\mathrm{t}}(1,2) &= (1/\sqrt{2})\ [\Psi_{\alpha\beta}(1,2) + \Psi_{\beta\alpha}(1,2)] \\ &= (1/\sqrt{2})\ [\Phi_{1s}(1)\ \Phi_{2s}(2) - \Phi_{1s}(2)\ \Phi_{2s}(1)]\ (1/\sqrt{2})[\alpha(1)\ \beta(2) + \alpha(2)\ \beta(1)] \end{aligned}$$

and

$$\Psi_\text{s}(1,2) = (1/\sqrt{2})\ [\Psi_{\alpha\beta}(1,2) - \Psi_{\beta\alpha}(1,2)]$$
$$= (1/\sqrt{2})\ [\Phi_{1s}(1)\ \Phi_{2s}(2) + \Phi_{1s}(2)\ \Phi_{2s}(1)]\ (1/\sqrt{2})[\alpha(1)\ \beta(2) - \alpha(2)\ \beta(1)]$$

(The subscripts to the wavefunctions Ψ_t and Ψ_s stand for *triplet* and *singlet* respectively. See below.) Note that the space part of Ψ_t is identical to the space parts of $\Psi_{\alpha\alpha}$ and $\Psi_{\beta\beta}$. Indeed, closer investigation shows that Ψ_t, like $\Psi_{\alpha\alpha}$ and $\Psi_{\beta\beta}$, describes a state where the two electrons have parallel spins. Only Ψ_s describes a state where the two electrons have opposite spins! Note also that the spin part of Ψ_s is identical to the spin part of the ground state wavefunction, equation (2.32).

Problem 2.3 Use the space functions $(1/\sqrt{2})\ [\Phi_{1s}(1)\ \Phi_{2s}(2) \pm \Phi_{1s}(2)\ \Phi_{2s}(1)]$ and show that for both functions the probability of finding electron 1 in $d\tau_a$ is given by $1/2\ [\Phi_{1s}^2(\mathbf{r}_a) + \Phi_{2s}^2(\mathbf{r}_a)]\ d\tau_a$. Since the probability of finding electron number 2 in $d\tau_a$ can be shown to be equally large, this means that the total probability densities for Ψ_s and Ψ_t both are given by $[\Phi_{1s}^2(\mathbf{r}_a) + \Phi_{2s}^2(\mathbf{r}_a)]$.

The result obtained in Problem 2.3 may be generalized: When we use a Slater determinant as wavefunction for an atom or molecule, the total electron density is equal to the sum of the probability densities of all occupied orbitals:

$$\text{Total electron density} = \Sigma \nu \Phi^2(\mathbf{r})$$

where the sum extends over all occupied orbitals Φ and ν denotes the number of electrons (one or two) in the orbital.

2.11 The lowest excited states of the He atom: calculation of the energies

The energy corresponding to each of the four states $\Psi_{\alpha\alpha}$, $\Psi_{\beta\beta}$, Ψ_t, and Ψ_s may be calculated from the integrals

$$E = \int \Psi(1,2)\ \mathcal{H}\Psi(1,2)\ d\tau_1\ d\tau_2\ ds_1\ ds_2 \tag{2.40}$$

where the Hamiltonian operator is given by equation (2.2). Just as in the case of the ground state, we may begin by integrating over the spin coordinates. The energies of the states $\Psi_{\alpha\alpha}$, $\Psi_{\beta\beta}$, and Ψ_t are then all given by the integral

$$E_\text{t} = (1/2) \int [\Phi_{1s}(1)\ \Phi_{2s}(2) - \Phi_{1s}(2)\ \Phi_{2s}(1)]$$
$$\mathcal{H}[\Phi_{1s}(1)\ \Phi_{2s}(2) - \Phi_{1s}(2)\ \Phi_{2s}(1)]\ d\tau_1\ d\tau_2$$

The three states where the two electrons have parallel spins are thus degenerate. Taken together the three states are referred to as a triplet. Writing out the integrand we obtain

$$
\begin{aligned}
E_t = (1/2)\int & \Phi_{1s}(1)\,\Phi_{2s}(2)\,\mathcal{H}\,\Phi_{1s}(1)\,\Phi_{2s}(2)\,d\tau_1\,d\tau_2 \\
&+ (1/2)\int \Phi_{1s}(2)\,\Phi_{2s}(1)\,\mathcal{H}\,\Phi_{1s}(2)\,\Phi_{2s}(1)\,d\tau_1\,d\tau_2 \\
&- (1/2)\int \Phi_{1s}(1)\,\Phi_{2s}(2)\,\mathcal{H}\,\Phi_{1s}(2)\,\Phi_{2s}(1)\,d\tau_1\,d\tau_2 \\
&- (1/2)\int \Phi_{1s}(2)\,\Phi_{2s}(1)\,\mathcal{H}\,\Phi_{1s}(1)\,\Phi_{2s}(2)\,d\tau_1\,d\tau_2 \qquad (2.41)
\end{aligned}
$$

Comparison with equation (2.39) shows that each of the two first integrals is equal E^0, the energy calculated for the "zero-order" wavefunction. Since the third integral is equal to the fourth

$$
\begin{aligned}
E_t = \int & \Phi_{1s}(1)\,\Phi_{2s}(2)\,\mathcal{H}\,\Phi_{1s}(1)\,\Phi_{2s}(2)\,d\tau_1\,d\tau_2 \\
&- \int \Phi_{1s}(1)\,\Phi_{2s}(2)\,\mathcal{H}\,\Phi_{1s}(2)\,\Phi_{2s}(1)\,d\tau_1\,d\tau_2
\end{aligned}
$$

or

$$E_t = E^0 - K \qquad (2.42)$$

Starting with the singlet wavefunction Ψ_s, equation (2.34), we obtain

$$E_s = E^0 + K \qquad (2.43)$$

The integral E^0, like the corresponding terms in equation (2.16), contains two terms representing the kinetic energy of the two electrons, two terms representing their potential energy due to attraction to the nucleus, and a fifth term due to electron–electron repulsion.

Note that in the product of two atomic orbitals in *front* of the Hamiltonian in integral K, electron 1 is in the $1s$ and electron 2 in the $2s$ orbital, but that the electrons have changed orbitals in the product *behind* the Hamiltonian. K is therefore commonly referred to as the "exchange integral." The presence of the integral K in the energy expression is due to the formulation of the wavefunction as a Slater determinant, in other words to the requirement that a proper wavefunction must satisfy the Pauli Principle.

Equations (2.42) and (2.43) show that the energy of the states in which the electron spins are antiparallel is different from energy of a state in which they are parallel.

Since we are mainly interested in the difference between the triplet and singlet states, we concentrate on the evaluation of the exchange integral:

$$K = \int \Phi_{1s}(1)\, \Phi_{2s}(2)\, \mathcal{H}\, \Phi_{1s}(2)\, \Phi_{2s}(1)\, \mathrm{d}\tau_1\, \mathrm{d}\tau_2$$

$$= \int \Phi_{1s}(1)\, \Phi_{2s}(2)\, (\hat{h}_1^0 + \hat{h}_2^0) \Phi_{1s}(2)\, \Phi_{2s}(1)\, \mathrm{d}\tau_1\, \mathrm{d}\tau_2$$

$$+ \int \Phi_{1s}(1)\, \Phi_{2s}(2) \left(\frac{e^2}{4\pi\varepsilon_0 r_{1-2}} \right) \Phi_{1s}(2)\, \Phi_{2s}(1)\, \mathrm{d}\tau_1\, \mathrm{d}\tau_2$$

where $\hat{h}_1^0$ and $\hat{h}_2^0$ have the same meaning as in equation (2.13). Since the $1s$ and $2s$ orbitals are orthogonal, the first of these integrals is equal to zero, and we are left with

$$K = \int \Phi_{1s}(1)\, \Phi_{2s}(2) \left(\frac{e^2}{4\pi\varepsilon_0 r_{1-2}} \right) \Phi_{1s}(2)\, \Phi_{2s}(1)\, \mathrm{d}\tau_1\, \mathrm{d}\tau_2$$

The exchange integral obviously has something to do with the electron–electron repulsion energy. You might think of it as a correction to the electron–electron repulsion energy included in E^0. The exchange integral turns out to be positive. *This means that the energy of the singlet state is increased by K while the energy of the triplet is reduced.* This conclusion is confirmed by experiment: The energy of the triplet state is found to be 1912 kJ mol^{-1} above the energy of the ground state, and the energy of the singlet state is found to be 77 kJ mol^{-1} above that of the triplet. The singlet–triplet energy difference can only be reproduced by calculations if the wavefunctions satisfy the Pauli Principle. The Pauli Principle may at first glance appear to be no more than abstract hairsplitting, but the implications are real and important!

2.12 The rule of maximum spin

The result which we have just obtained may be generalized to all atoms or molecules: *Whenever the electron configuration of an atom or molecule is such that two or more unpaired electrons occupy different orbitals, the energy is not determined by the electron configuration alone, but depends on their spin functions. The lowest energy state is that in which all unpaired electrons have parallel spins.* This statement is known as the rule of maximum spin.

Chapter 3

Atoms with more than two electrons: atomic properties and periodic trends

Introduction

We begin this chapter with a brief description of the ground states of lithium, beryllium and boron atoms: the Li atom has the electron configuration $1s^22s^1$ and the ground state is doubly degenerate; the Be atom has the electron configuration $1s^22s^2$ and the ground state is non-degenerate; the B atom has the electron configuration $1s^22s^22p^1$ and the ground state is sixfold degenerate. For these atoms the energy is completely determined by the electron configuration. This is, however, not the case for carbon, nitrogen, or oxygen atoms (or for the heavier elements in Groups 14, 15 or 16). For C the ground state $1s^22s^22p^2$ electron configuration gives rise to 15 states: the ninefold degenerate ground state plus a fivefold and a singly degenerate excited state. The ground state electron configuration of the N atom gives rise to 20 states and the ground state electron configuration of O to 15. For these and similar atoms we define the "average configuration energy" of an atom with a given electron configuration as the average of all states corresponding to that configuration.

A plot of the first ionization energies (IE1) of the atoms in the second period shows a general increase of IE1 with increasing atomic number. There is however, a marked reduction of IE1 between Be and B which is due to the higher energy of p electrons, and a second reduction between N and O due to the particular stability accorded to N by the presence of three unpaired electron with parallel spins. In this chapter we define the ionization energies of s and p electrons not as the difference between the ground state energies of atoms and cations but as the difference between their average configuration energies. Finally we define the mean valence shell ionization energie (MIE) of an atom as the weighted average of the ionization energies of the valence shell s and p electrons.

The energy difference between valence shell s and p electrons in an atom is of secondary importance for our description of a molecule if the atom uses both s and p electrons for bond formation, or if we describe the molecule using molecular orbitals formed from sp hybridized atomic orbitals. The mean valence electron ionization energies may therefore be used to define a set of electronegativity coefficients. The magnitudes of these electronegativity coefficients are found to increase across a period, and – with a few exceptions – to decrease down a group.

Finally we provide the basis for later discussions of the charge distributions in molecules by defining the electric dipole moment of an atom and showing how it may be calculated from the charge density.

3.1 The ground state of the lithium atom

The lithium atom has three electrons. If we neglect the kinetic energy of the nucleus, the energy and the Hamiltonian operator both consist of nine terms:

- three terms representing the kinetic energy of the electrons,
- three terms representing the potential energy of the electrons relative to the nucleus, and
- three terms representing repulsion between three different electron pairs, 1-2, 1-3 and 2-3.

A reasonably accurate calculation of the energy of the atom is possible only if we use a 3×3 Slater determinant as wavefunction rather than a simple product of atomic orbitals. Before we can construct such a wavefunction, we need, however, to decide on the electron configuration: two electrons obviously occupy a $1s$ orbital, but what about the third electron?

One way to answer this question is to determine the angular momentum of the unpaired electron in the ground state of the atom. Experiments show that the angular momentum is zero, corresponding to the $1s^2 2s^1$ electron configuration. The energy of the $1s^2 2p^1$ configuration is in fact 178 kJ mol^{-1} above the ground state. Thus, while the $2s$ and $2p$ orbitals are degenerate in the one-electron ion Li^{+2}, the degeneracy has been removed in the neutral atom.

The $1s$ AO of Li may be chosen to be of the same form as in the ground state of the He atom:

$$\Phi_{1s}(\mathbf{r}) = K_1 \exp\left(\frac{-Z'_{1s} r}{a_0}\right)$$

The $2s$ AO is often chosen to have the form

$$\Phi_{2s}(\mathbf{r}) = K_2 \left(\frac{Z'_{2s} r}{a_0}\right) \exp\left(\frac{-Z'_{2s} r}{2a_0}\right)$$

K_1 and K_2 are normalization constants and Z'_{1s} and Z'_{2s} adjustable constants. Note that the radial part of the $2s$ AO differs from that of the $2s$ orbital of a one-electron atom: the factor $(2 - Z_{2s} r/a_0)$ has been replaced by $(Z'_{2s} r/a_0)$.

Combination of the $1s$ and $2s$ atomic orbitals with α or β spin functions yields four spin orbitals; for construction of a Slater determinant we need only three. The energy calculated from the wavefunction is the same whether we use the spin orbital $\Phi_{2s}(\mathbf{r})\alpha(s)$ or $\Phi_{2s}(\mathbf{r})\beta(s)$. This means that the ground state of the Li atom (or any other Group 1 metal), like the ground state of the H atom, is doubly degenerate.

Problem 3.1 Write down an appropriate Slater determinant for the ground state of the Li atom in terms of the space functions Φ_{1s} and Φ_{2s} and the spin functions α and β.

The best values for Z'_{1s} and Z'_{2s} are obtained by variation calculations, i.e. by calculating the energy and adjusting the constants to find its minimum value. This approach yields the best values $Z'_{1s} = 2.70$ and $Z'_{2s} = 1.30$. The $1s$ electrons are close to and feel nearly the full attraction of the nucleus. The $2s$ electron is at a much larger average distance from the nucleus, and the attraction is to a large extent cancelled by repulsion from the two $1s$ electrons. In the following we shall refer to the best value of the Z' parameter of the highest occupied atomic orbital as the effective nuclear charge, Z_{eff}, of the atom

3.2 The beryllium atom

The Be atom has four electrons. If we neglect the kinetic energy of the nucleus, the classical energy and the Hamiltonian operator both consist of fourteen terms:

- four terms representing the kinetic energy of the electrons,
- four terms representing the potential energy of the electrons relative to the nucleus, and
- six terms representing repulsion between six different electron pairs, 1-2, 1-3, 1-4, 2-3, 2-4 and 3-4.

A reasonably accurate calculation of the energy of the ground state of the atom is possible only if we use a 4×4 Slater determinant as wavefunction. The $1s$ and $2s$ atomic orbitals are often assumed to have the same form as for the Li atom:

$$\Phi_{1s}(\mathbf{r}) = K_1 \exp\left(\frac{-Z'_{1s} r}{a_0}\right) \quad \text{and} \quad \Phi_{2s}(\mathbf{r}) = K_2 \left(\frac{Z'_{2s} r}{a_0}\right) \exp\left(\frac{-Z'_{2s} r}{2a_0}\right),$$

but the best values for the constants Z'_{1s} and Z'_{2s} will, of course, be different.

Combination of the two AOs with the spin functions α and β produces four spin orbitals each containing one electron. This means that only one wavefunction (Slater determinant) can be written down: the ground state of Be (or any other Group 2 atom) is non-degenerate.

The best values for Z'_{1s} and Z'_{2s} obtained by variation calculations are $Z'_{1s} = 3.70$ and $Z'_{2s} = 1.95$. The $1s$ electrons are close to and feel nearly the full pull of the nucleus. The $2s$ electrons are at a much larger average distance from the nucleus which is effectively "screened" by the two $1s$ electrons.

3.3 Lithium and beryllium: ionization energies and atomic sizes

In Chapter 1 we found that the ionization energy and classical radius of a one-electron atom are given by

$$\text{IE} = \left(\frac{Z^2}{n^2}\right)\mathcal{R} \quad \text{and} \quad r = 2\left(\frac{n^2}{Z}\right)a_0 \tag{3.1}$$

For two-electron atoms in the ground state ($n = 1$) we derived the approximate values

$$\text{IE1} \approx (Z_{\text{eff}}^2 - 0.625\, Z_{\text{eff}})\mathcal{R} \quad \text{and} \quad r \approx 2\left(\frac{1}{Z_{\text{eff}}}\right)a_0 \tag{3.2}$$

In the following we shall associate the size of an atom with the size of the highest occupied atomic orbital, and as suggested by equations (3.1) and (3.2) we shall assume that the *radius of the atom increases with the increasing principal quantum number of the highest occupied orbital and decreases with increasing effective nuclear charge*. Similarly we shall assume that *the ionization energy increases with increasing values of Z_{eff} and decreases with increasing values of n.*

When we go from He to Li the principal quantum number of the valence electrons increases from 1 to 2, while the effective nuclear charge decreases from 1.69 to 1.30. Both changes should tend to decrease the ionization energy, and this is confirmed by experiment:

$$\text{IE1(He)} = 2372\,\text{kJ mol}^{-1} \quad \text{and} \quad \text{IE1(Li)} = 520\,\text{kJ mol}^{-1}.$$

We also expect the Li atom to be much larger than the He atom. It is not easy to find direct experimental confirmation of this expectation, but since we have already shown that the H atom is larger than He, it should be sufficient to show that the Li atom is larger than H. And indeed, the bond distance in gaseous LiH is 160 pm as compared to a bond distance of 74 pm in gaseous H_2.

On going from Li to Be, the principal quantum number remains unchanged, but the effective nuclear charge increases from 1.30 to 1.95. This indicates that the Be atom should have a larger ionization energy and a smaller radius than Li. Both expectations are fulfilled: IE1(Be) = 900 kJ mol^{-1} and the bond distance in gaseous BeH = 134 pm.

3.4 The *p*-block elements from boron to neon: average configuration energies

The ground state electron configuration of the boron atom is $1s^2 2s^2 2p^1$. Depending on the choice of $2p$ orbital and spin function, it is possible to write down six Slater determinants. Calculation of the energy using a Hamiltonian operator containing five kinetic energy terms, five electron–nucleus attraction energy terms and ten electron–electron repulsion terms indicates that the six Slater determinants describe a sixfold degenerate ground state. Comparison with experiment shows that this is not strictly true; the ground state is doubly degenerate while the energy of a fourfold degenerate state is 0.2 kJ mol^{-1} higher. This splitting of the ground state is due to spin–orbit (*L*–*S*) coupling. See Section 1.17. In the case of the B atom, the spin–orbit interaction energy is negligibly small, for all practical purposes we may regard the ground state as sixfold degenerate. However, the *L*–*S* coupling energy increases rapidly as Group 13 is descended; in the heaviest element in the group, thallium, the energy difference between the two levels has grown to 93 kJ mol^{-1}.

Before going on we pause to note that the treatment of the Li, Be and B atoms outlined in the preceding sections showed that the ground state energies of the atoms are completely determined by their electron configurations. This is not the case for the C, N or O atoms.

The ground state electron configuration of C, $1s^2 2s^2 2p^2$, allows the construction of 15 different Slater determinants! These must then be combined to form 15 proper wavefunctions that will *not* all be degenerate. Indeed the energy of the highest of the 15 states corresponding to the $1s^2 2s^2 2p^2$ electron configuration, is about 260 kJ mol^{-1} above that of the lowest!

The number of states possible for the ground state electron configuration of N reaches 20, for O there are 15 states (as for C) and for F six (as for B). Only when we reach the noble gas Ne do we find that only one Slater determinant can be constructed from the electron configuration $1s^2 2s^2 2p^6$. Like He and Be, Ne and the other noble gases have non-degenerate ground states.

Consider an atom A with electron configuration $[\text{Ng}]ns^k np^l$ where Ng is one of the noble gases. We shall refer to the average energy of all states corresponding to the stated configuration as the average configuration energy: $E_{\text{av}}(\text{A};s^k p^l)$. The average energy of N atoms with the ground state electron configuration, $E_{\text{av}}(\text{N};s^2 p^3)$ is obtained by averaging the energy of 20 states. $E_{av}(C; s^2p^2)$ and $E_{av}(O; s^2p^4)$ both represent the average of 15 states. If we neglect the minute spin–orbit coupling energy of boron, $E_{av}(\mathrm{B}; s^2p^1)$ is equal to the energy of the (sixfold degenerate) ground state. The same applies to the F atoms: $E_{av}(F; s^2p^5)$ is equal to the energy of the ground state. So are the average configuration energies of Li, Be and Ne.

Problem 3.2 Calculate the total electron density set up by one electron in a $2p_x$ orbital and one electron in a $2p_y$ orbital. Show that this density is independent of ϕ, and thus has cylinder symmetry around the z-axis. Make a sketch of the density in the xz plane and in the xy plane.

Problem 3.3 Calculate the total electron density set up by one electron in each of the three orbitals $2p_x$, $2p_y$ and $2p_z$. Show that the density is independent of both θ and ϕ and thus has spherical symmetry.

3.5 The *aufbau* model and the ground state electron configuration of the elements

Most textbooks on elementary chemistry describe a simple, but not infallible, model for prediction of the ground state electron configuration of the elements. The model has two components. The first is a schematic orbital energy level diagram for atoms which indicates that orbital energies increase in the order

$$1s < 2s < 2p < 3s < 3p < 4s < 3d < 4p < 5s < 4d < 5p < 6s, \text{ etc.} \tag{3.3}$$

The second component is the assertion that the ground state electron configuration of an atom is obtained by filling each atomic orbital with two electrons, from the bottom up, beginning with the $1s$ orbital (the *aufbau* or building-up principle).

The general validity of the model depends on three assumptions which are not always made explicit:

(i) The energy of a polyelectron atom is equal to the sum of the occupied orbital energies.
(ii) The sequence of atomic orbital energies (3.3) is valid for all atoms as far as the orbitals are filled in the ground state.
(iii) The orbital energies of a given atom are not changed when electrons are transferred from one orbital to another.

Considering that none of these assumptions are valid, it is hardly surprising that the model is fallible.

Examination of the first 20 elements, from H to Ca, nevertheless shows that for each element the occupied orbital energies do increase in the order indicated and that the ground state electron configuration of each element is that predicted by the *aufbau* principle. The 20th element Ca has the electron configuration $1s^22s^22p^63s^23p^64s^2$ or $[\mathrm{Ar}]4s^2$, and the energy required to promote one electron from the $4s$ to a $3d$ orbital is considerable: 243 kJ mol^{-1}.

The problems begin with the 21st element, scandium, which is found to have the ground state electron configuration $[\mathrm{Ar}]4s^23d^1$. At first glance this would seem to be in perfect agreement with the model, but accurate calculations show that the $4s$ orbital energy is in fact *higher* than the $3d$! This means that the electrons in the valence shell are *not* filled into the orbitals with the lower energy as assumed by the model. Why are not the $4s$ electrons

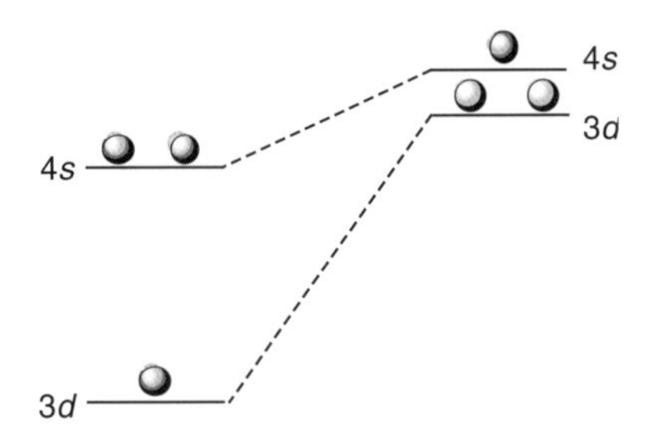

Fig. 3.1. The atomic orbital energies associated with the ground state [Ar]$4s^2 3d^1$ and excited state [Ar]$4s^1 3d^2$ electron configurations of scandium. [1]

transferred into the lower energy $3d$ orbitals? The reason for the unexpected behavior is found in the electron–electron repulsion energies: the $4s$ electrons are spread out over a larger volume than the $3d$ electrons. This means that the two electrons in the $4s$ orbital are – on the average – much further apart than two electrons in $3d$ orbitals (even if they occupy different d orbitals). $4s$-$4s$ electron–electron repulsion energies are therefore much smaller than $3d$-$3d$ repulsion energies, while $4s$-$3d$ repulsions are of intermediate magnitude. Thus, if an electron is transferred from the $4s$ to a $3d$ orbital the increased electron–electron repulsion leads to higher $3d$ and $4s$ orbital energies, and a corresponding increase of the total energy of the atom [1]. The result of an analysis of the orbital energies are shown in Fig. 3.1. When one electron is transferred from the $4s$ to the lower $3d$ orbital, the increase of the $3d$ orbital energy is in fact greater than the original $4s$-$3d$ energy gap.

As we move further to the right from Sc to Ti and the following elements in the fourth period, the $3d$ orbital energy continues to fall slowly relative to the $4s$. All the elements from Sc to Ni ($Z = 28$) violate the *aufbau* principle insofar as they have one or two electrons in the higher energy $4s$ orbitals. Only when we reach element number 29, copper, do we find a ground state electron configuration, [Ar]$3d^{10}4s^1$ in accordance with the *aufbau* principle. In the following element zinc ($Z = 30$) with the electron configuration [Ar]$3d^{10}4s^2$ the energy of the $3d$ electrons is so low that they are no longer counted among the valence electrons. The electron configurations of the following six elements, from Ga to Kr, are in agreement with the simple *aufbau* model.

We shall not discuss the electron configurations of all the elements in the fifth and sixth periods, but be content to point out that the ground state electron configurations of all main group elements in Groups 1, 2 and from 12 to 18 are in accord with the *aufbau* model.

3.6 The ionization energies of the first 20 elements

The first ionization energies of the elements from hydrogen to calcium are listed in Table 3.1 and displayed in Fig. 3.2.

Problem 3.4 Discuss the first ionization energies, IE1, of the elements from Li to K in terms of effective nuclear charges, principal quantum numbers, the energy difference between valence shell s and p orbitals, and the particular stabilty of electron configurations characterized by two or more unpaired electrons with parallel spins.

Table 3.1. First ionization energies IE1, valence shell *s* electron ionization energies IE(*ns*), and mean valence shell ionization energies MIE (all in kJ mol^{-1}) of the first 20 elements.

	Group 1	Group 2	Group 13	Group 14	Group 15	Group 16	Group 17	Group 18
	H							He
IE1	1313							2372
IE(*ns*)	1313							2372
MIE	1313							2372
	Li	Be	B	C	N	O	F	Ne
IE1	520	900	801	1086	1404	1314	1681	2081
IE(*ns*)	520	900	1247	1601	1968	2748	3653	4677
MIE	520	900	1170	1452	1749	2061	2393	2732
	Na	Mg	Al	Si	P	S	Cl	Ar
IE1	496	738	578	787	1059	1000	1256	1521
IE(*ns*)	496	738	1025	1299	1687	1949	2373	2821
MIE	496	738	921	1093	1286	1477	1637	1850
	K	Ca						
IE1	419	590						
IE(*ns*)	419	590						
MIE	419	590						

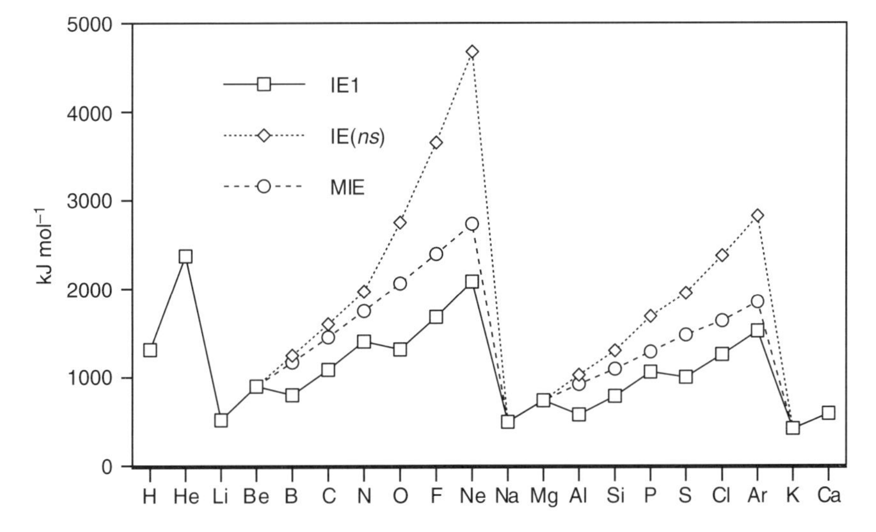

Fig. 3.2. First ionization energies, IE1, valence shell *s* electron ionization energies, IE(*ns*), and mean valence shell ionization energies, MIE, of the first 20 elements, all in kJ mol^{-1}.

3.7 Valence shell *s*-electron ionization energies

While the first ionization energy of Li or Be is equal to the minimum energy required to remove a valence shell *s* electron, the ionization energy of B or the following elements in Groups 14 to 18 represents the energy required to remove a valence shell *p* electron. It is seen that the first ionization energy of B is smaller than that of Be, presumably because the $2p$ orbital energy is higher than the $2s$ orbital energy.

The minimum energy required to remove a $2s$ electron from B, IE(ns), is equal to the first ionization energy plus the energy ΔE required to promote the B^+ cation from the [He]$2s^2$ ground state to the lowest of several states with the electron configuration [He]$2s^1 2p^1$:

$$\mathrm{IE}(ns) = \mathrm{IE1} + \Delta E$$

This promotion energy may be calculated from experimental atomic energy level data [2]. The valence shell *s*-electron ionization energies of the first 20 elements obtained in this manner are included in Table 3.1 and displayed in Fig. 3.2. Note that the valence shell *s*-electron ionization energies of the elements in groups 13 through 18 are much higher than the first ionization energies. Note also that the difference *increases* with increasing group number. Clearly the difference between *ns* and *np* orbital energies increases as we go from left to right along a period.

3.8 Configuration-averaged ionization energies and mean valence shell ionization energies

Removal of an *s* electron from an atom A with valence shell electron configuration $s^k p^l$ yields a cation A^+ with $s^{k-1}p^l$ configuration. We shall define the configuration-averaged ionization energy of the valence shell *s* electrons as the difference between the average configuration energies of the cation A^+ and the neutral atom A:

$$i\varepsilon_\mathrm{s} = E_\mathrm{av}(\mathrm{A}^+; s^{k-1}p^l) - E_\mathrm{av}(\mathrm{A}; s^k p^l) \tag{3.4}$$

For elements in groups 1, 2 or 12, $i\varepsilon_s = IE1$.

Similarly the configuration-averaged ionization energy of the valence shell *p* electrons is defined as the difference between the average configuration energies of the cation and the neutral atom:

$$i\varepsilon_\mathrm{p} = E_\mathrm{av}(\mathrm{A}^+; s^k p^{l-1}) - E_\mathrm{av}(\mathrm{A}; s^k p^l) \tag{3.5}$$

Allen [3] has defined the mean valence shell ionization energy or MIE of an atom as the weighted mean of the configuration-averaged ionization energies of valence shell *s* and *p* electrons:

$$\mathrm{MIE} = \frac{k\, i\varepsilon_\mathrm{s} + l\, i\varepsilon_\mathrm{p}}{k + l}. \tag{3.6}$$

For elements in Groups 1, 2 or 12, $l = 0$ and MIE = IE1. The configuration energies of the first 20 elements are listed in Table 3.1 and displayed in Fig. 3.2. For the elements in Groups

13 to 18 the MIEs lie between the valence shell *s*- and *p*-electron ionization energies and show a nearly linear increase along each period.

3.9 Periodic variation of the mean valence electron shell ionization energies of main group elements

In these notes we shall use the term "main group element" to include the elements in Groups 1, 2 and 12 through 18. In Table 3.2 we list MIEs for all main group elements up to Bi. The mean valence shell ionization energies for main group elements in the second, third, fourth, fifth and sixth periods (except the noble gases in Group 18) are also displayed in Fig. 3.3.

The MIEs increase monotonically across each period, presumably because of the increasing effective nuclear charge. The ionization energy then drops sharply as each new period begins with an alkali metal, presumably because of the increase of the principal quantum number *n* (the last electron has entered a new shell at a larger distance from the nucleus) and a decreased effective nuclear charge.

Inspection of Table 3.2 also shows that there is a general tendency for MIEs to decrease as a group is descended, presumably because the principal quantum number of the valence electrons increases. The variation down a group is, however, much smaller than across a period, and there are four exceptions to the general trend: the MIEs of Ne, Ga, Ge and Hg are all greater than that of the element above.

When Mendeljev worked out the first version of the periodic table at the end of the 1860s, he divided about 25 known elements into eight groups according to their chemical

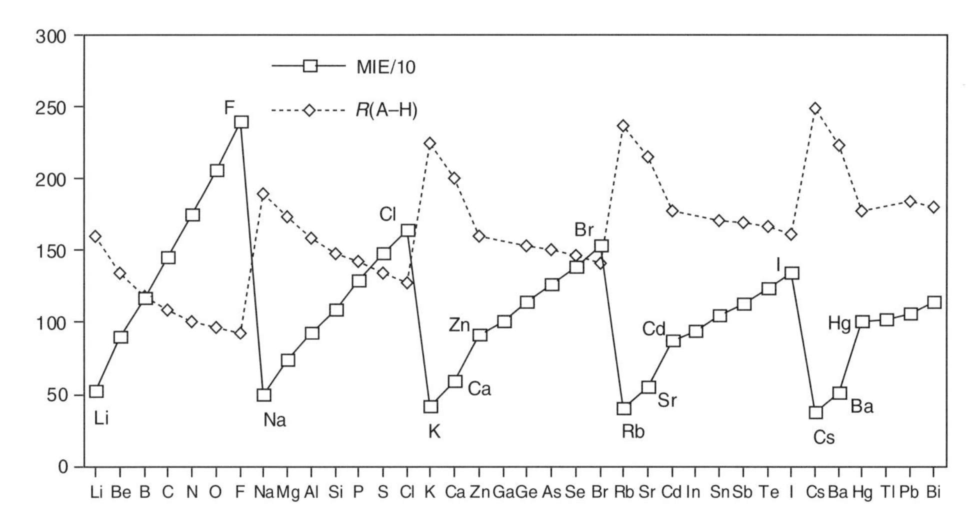

Fig. 3.3. The mean valence shell ionization energies, MIE, and A–H bond distances in hydrogen compounds of main group elements (Group 18 elements excluded). The numbers on the perpendicular axis refer to MIEs in kJ mol^{-1} multiplied by 10^{-1}, and to bond distances *R*(A–H) in pm. The experimental bond distances have been taken from the gaseous, monomeric compounds AH, $\cdot$AH, $\cdot$AH, $\cdot AH_2$, AH_4, AH_3, AH_2 and HA of the elements in Groups 1, 2, 12, 13, 14, 15, 16 and 17 respectively. Note that bond distances for $\cdot GaH_2$, $\cdot InH_2$ and $\cdot TlH_2$ are missing.

Table 3.2. Mean valence shell ionization energies MIE (in kJ mol^{-1}) and Allen electronegativity coefficients χ of elements in Groups 1, 2, and 12 through 18.

	Group 1	Group 2	Group 12	Group 13	Group 14	Group 15	Group 16	Group 17	Group 18
	H								He
MIE	1313								2372
χ	2.30								4.16
	Li	Be		B	C	N	O	F	Ne
MIE	520	900		1170	1452	1749	2061	2393	2732
χ	0.91	1.58		2.05	2.54	3.07	3.61	4.19	4.79
	Na	Mg		Al	Si	P	S	Cl	Ar
MIE	496	738		921	1093	1286	1477	1637	1850
χ	0.87	1.29		1.61	1.92	2.25	2.59	2.87	3.24
	K	Ca	Zn	Ga	Ge	As	Se	Br	Kr
MIE	419	590	906	1003	1139	1262	1384	1532	1692
χ	0.73	1.03	1.59	1.76	1.99	2.21	2.42	2.69	2.97
	Rb	Sr	Cd	In	Sn	Sb	Te	I	Xe
MIE	403	550	867	945	1041	1133	1231	1346	1473
χ	0.71	0.96	1.52	1.66	1.82	1.98	2.16	2.36	2.58
	Cs	Ba	Hg	Tl	Pb	Bi			
MIE	375	503	1007	1020	1058	1148			
χ	0.66	0.88	1.76	1.79	1.85	2.01			

properties. After the electron had been discovered at the very end of the nineteenth century and as the structure of the atom became better understood during the first decade of the twentieth, it became clear that in most cases the members of each group had the same number of valence electrons. This, however, is not true for the He atom which has two valence electrons, while Ne and the other elements in the group have eight. If we compare He with the second-period element with the same number of valence electrons, viz. Be, we find that the increase of the principal quantum number from 1 to 2 leads to the expected decrease of the MIE (from 2372 to 900 kJ mol^{-1}). When going from Be to Ne along the second period the principal quantum number remains unaltered, but the nuclear charge is increased by +6 and six electrons are added to the $2p$ orbitals. The increased nuclear charge is not compensated by the screening effect of the six additional electrons, the effective nuclear charge increases from 1.95 to 5.20 and the MIE from 900 to 2732 kJ mol^{-1}.

Inspection of Fig. 3.3 shows that there is an abrupt increase of the mean ionization energy on going from Ca to Zn, from Sr to Cd and from Ba to Hg. Between Ca and Zn and between Sr and Cd the nuclear charge is increased by +10 while ten electrons are filled into the $(n-1)\,d$ orbitals. The screening by the ten additional electrons does not compensate for the increased nuclear charge, and the mean ionization energy increases accordingly. On going from Zn to Ga, and from Ga to Ge there is a normal increase of the ionization energy, but the effect of the increased nuclear charge lingers on; these elements have larger ionization energies than the elements above. The ionization energies of As, Se and Br, however, are normal in the sense that they are smaller than that of the element above.

On going from Ba to Hg the nuclear charge increases by +24 while 24 electrons are filled into the 5*d*- and 4*f*-orbitals. The screening by the additional electrons does not compensate for the additional nuclear charge; as a result the MIE increases even more than between Ca and Zn or between Sr and Cd.

3.10 The atomic sizes and bonding radii of main group elements

Experimentally determined bond distances are available for gaseous hydrides AH of the Group 1 elements, as well for compounds of composition AH_4 for the Group 14 elements, AH_3 for the Group 15 elements, AH_2 for the Group 16 elements and HA for the Group 17 elements. If we regard the bond distance as the sum of the bonding radii of the elements, and the bonding radius of H as a constant, the variation of the A–H bond distances reflects the changing size of the atoms A. The observed bond distances are plotted in Fig. 3.3. Since experimental values are lacking for most gaseous monomeric dihydrides of the Group 2 and 12 elements and for the gaseous monomeric trihydrides of the Group 13 elements, the plot includes experimentally determined distances in the monohydrides ·AH for the Group 2 and Group 12 metals and in the dihydrides ·AH_2 for A = B and Al. (We hope that the effect of a single, unpaired electron on A is about equivalent to the effect of two shared electrons.) Values for Ga–H, In–H and Tl–H bond distances are, unfortunately, not available.

Figure 3.3 shows that the bonding radii and mean valence shell ionization energies are strongly correlated: mean valence shell ionization energies increase as a period is traversed, bonding radii decrease; Mean ionization energies decrease as a group is descended; bonding radii increase: "Small atoms have high ionization energies."

Like the increasing ionization energies, the decreasing bonding radii of the atoms across a period reflect the increasing effective nuclear charges. The abrupt increase of the bonding radii before each Group 1 element, like the decrease of the ionization energy, reflects the fact that the valence electron has entered a new shell at a greater distance from the nucleus.

While the bond distances decrease smoothly across the second period from Li to F and across the third from Na to Cl, there is an abrupt decrease of the A–H bond distance between Ca and Zn in the third period and between Sr and Cd in the fifth. This phenomenon is referred to as the "*d*-block contraction." Like the corresponding increase of the ionization energy between the same elements, it is due to an increased effective nuclear charge.

Inspection of Fig. 3.3 shows that the A–H bond distances decrease much more slowly between Zn and Br or between Cd and I than they did between the Group 2 and Group 17 elements in the preceding periods. The explanation is sought in the behavior of the *d* orbitals: As we move further to the right across the period from Zn or Cd the $(n-1)$ *d* orbitals contract and the electrons in them move closer to the nucleus. As they do so, their screening of the 10 additional nuclear charges becomes more effective. By the time we have moved across the fourth period from Zn to Br and into the fifth from Rb to Sr, the effect of the *d*-block contraction has been eliminated: the 3*d* electrons are now so close to the nucleus that they completely compensate for 10 positive charges on the nucleus.

We shall return to a discussion of the abrupt decrease of the A–H bond distance between the sixth-period elements Ba and Hg in Chapter 10.

3.11 Electron affinities

Many gaseous atoms are able to capture an additional electron and form a stable monatomic anion. The electron affinity of an atom A is defined as the energy of the process

$$A^-(g) \rightarrow A(g) + e^-(g) \tag{3.7}$$

when the reactant and products are at rest.

Electron affinities of some main group elements are listed in Table 3.3. Electron affinities are seen to be particularly large for the elements in Groups 16 and 17. The noble gases in Group 18, however, do not form gaseous anions.

The table contains some surprises: Since the reduction potentials of the halogens in water decreases monotonically from F_2 to I_2, most chemists would probably have expected the electron affinities decrease in order

$$F > Cl > Br > I$$

Inspection of Table 3.3, however, shows that the electron affinities gaseous atoms vary in the order

$$Cl > F \approx Br > I.$$

Secondly, since compounds containing alkali metal cations are very common, and compounds containing alkali metal anions are very rare, most chemists would probably assume that their electron affinities are negligible. Consultation with the table shows that this is not the case.

Finally we note that no atoms is known to be able to bind two electrons in the gas phase, not even oxygen, though the O^{-2} ion is believed to exist in solid metal oxides.

Problem 3.5 Use the electron affinities listed in Table 3.3 and the bond energies listed in Table 4.1 to calculate the energies of the reactions $F_2 + 2e^- = 2\,F^-$ and $Cl_2 + 2\,e^- = 2\,Cl^-$ in the gas phase.

Table 3.3. The electron affinities (in kJ mol^{-1}) of some main group elements [4].

Group 1	Group 13	Group 14	Group 15	Group 16	Group 17	Group 18
H 73						He 0
Li 60	B 27	C 122	N 0	O 141	F 328	Ne 0
Na 53	Al 43	Si 134	P 72	S 200	Cl 349	Ar 0
K 48	Ga 28	Ge 115	As 78	Se 195	Br 325	Kr 0
Rb 47		Sn 120	Sb 103	Te 190	I 295	Xe 0

3.12 Electronegativity and electronegativity coefficients

The electronegativity of an element is defined as the ability of the atom to hold on to its own electrons and to attract more electrons when it forms part of a molecule. (Elements with low electronegativity are described as electropositive.) It is not obvious that the electronegativity of an atom will be the same in all molecules, so electronegativity is a somewhat nebulous, but useful concept. Much thought and work has gone into the establishment of numerical quantities, "electronegativity coefficients", which would permit us to assess the relative electronegativity of two elements quantitatively or semiquantitatively[1].

It would seem logical to correlate the ability of an atom to hold on to its own electrons with the first ionization energy. This correlation indicates that electronegativity coefficients, χ, should increase from left to right along a period, and that in each period the element with the highest χ would be the noble gas. The chemical inertness of the noble gases may – at least in part – be rationalized as a consequence of their unwillingness to form cations or to share their electrons with other atoms.

It is not obvious that the ability of an atom to hold on to its own electrons correlates with its ability to attract additional electrons. Indeed, as we have seen, the electron affinities of the noble gases are zero. It would seem logical to correlate the ability of an atom to attract excess electrons in a molecule with the electron affinity. The difficulty is that the EAs in Table 3.2 lead to trends that are in poor agreement with the expectations based on a large amount of indirect chemical evidence. Such evidence indicates that the electronegativities of O, S and the halogen atoms decrease in the order

$$\mathrm{F} > \mathrm{O} > \mathrm{Cl} > \mathrm{Br} \approx \mathrm{S} > \mathrm{I}$$

while the electron affinities of the atoms indicates the order

$$\mathrm{Cl} > \mathrm{F} \approx \mathrm{Br} > \mathrm{I} > \mathrm{S} > \mathrm{O}$$

Allen has suggested that the electronegativity coefficient χ should be defined in terms of the mean valence shell ionization energy [3]:

$$\chi = \mathrm{MIE}(1.752 \times 10^{-3}\ \mathrm{mol\,kJ^{-1}}) \tag{3.8}$$

The conversion factor has been chosen to bring the coefficients on a scale comparable to those suggested earlier by other workers.

Compared to other definitions, Allen's has the advantage of being both clear and simple *and* yielding values that correspond to accumulated chemical experience. Mean valence

[1] The concept of electronegativity was the object of lively discussion in the Soviet Union in the 1950s and 60s. The proponents of the concept stated that they found it useful and no more nebulous than many other concepts that are used in chemistry. The opponents claimed that the concept "electronegativity" is unscientific since the definition does not allow it to be observed or calculated. The discussion led to a two-day meeting in April 1962 that was organized by the Institute of General and Inorganic Chemistry of the Academy of Sciences of the USSR and Moscow State University and attracted more than 300 participants. A report from the conference (in English) was published in *Journal of Structural Chemistry USSR*, 3 (1962) 478. The discussion continued, however, for several years after the conference. Though this is not apparent in the conference report, the author of this book has been told that the discussion acquired ideological overtones: Was the concept in agreement with Marxist materialism?

shell ionization energies for elements in Groups 1 and 2 and from 12 to 18 and the electronegativity coefficients derived from them, are listed in Table 3.3.

3.13 Net atomic charges and atomic dipoles

The net electric charge on an isolated atom or monatomic ion may be calculated from

$$\text{Net charge} = Q_{\text{net}} = \int \rho(x, y, z)\, d\tau \tag{3.9}$$

where $\rho(x, y, z)$ is the charge density (in coulomb per m^3) at the point x, y, z and the integration is carried out over all space.

The integral may be divided into two parts in such a way that positive and negative charges are counted separately:

$$Q_{\text{net}} = \int \rho_+(x, y, z) d\tau + \int \rho_-(x, y, z) d\tau \tag{3.10}$$

where ρ_+ and ρ_- represent positive and negative charge densities respectively. According to the electron cloud model the total negative charge density is given by

$$\rho_-(x, y, z) = -e \sum \nu \Phi^2(x, y, z) \tag{3.11}$$

where the sum extends over all occupied AOs Φ, and ν denotes the number of electrons (one or two) in the orbital.

The first integral in (3.10) is equal to the total positive charge, i.e. to the charge on the nucleus, $Q_+ = +Ze$, the second is equal to the total negative charge, Q_-, which in turn is equal to $-Ne$ where N is the total number of electrons. The net charge on the atom or ion is

$$Q_{\text{net}} = Q_+ + Q_- = +Ze - Ne$$

The *electric dipole moment* of an atom or atomic ion is a vector. If we use a Cartesian coordinate system, the dipole moment in the z-direction is defined by

$$\mu_{\text{el},z} = \int \rho(x, y, z) z\, d\tau \tag{3.12}$$

The dipole moment components in the x and y directions are defined by similar integrals.

Again the integral may be separated into two in such a way that positive and negative charges are counted separately:

$$\mu_{\text{el},z} = \int \rho_+(x, y, z) z\, d\tau + \int \rho_-(x, y, z) z\, d\tau \tag{3.13}$$

The first integral is equal to $(+Ze)z_+ = Q_+ z_+$ where z_+ is the z-coordinate of the nucleus. If we place the origin of our coordinate system at the nucleus as usual, $z_+ = 0$ and the first integral is equal to zero. The second integral in (3.13) may be written as $Q_- z_-$ where z_- is the z-coordinate of *the centre of gravity* of negative charge, or the centre of gravity of the "electron cloud". The atomic dipole is then

$$\mu_{\mathrm{el},z} = Q_{-}z_{-} = (-Ne)z_{-} \tag{3.14}$$

Similar equations may be derived for the x and y component of the dipole. *If the centre of gravity of the electron cloud of a neutral atom or monatomic ion coincides with the position of the nucleus, for instance if the electron cloud is spherical, the three components of the atomic dipole are zero by symmetry.* Such atoms or ions are described as *non-polar*. If one or more components of μ_{el} are different from zero, the atom or ion is described as *polar*.

3.14 Atomic polarizabilities

In the absence of an external electric field all atoms and monatomic ions will have spherical charge distributions in their ground state. When such an atom is placed in an electric field, the nucleus and the electrons are pulled in opposite directions, the atom is deformed and acquires an induced dipole moment which has the same direction as the field. As a first approximation we assume that the magnitude of the induced dipole is proportional to the strength of the electric field $\mathcal{E}$:

$$\mu^* = \alpha\mathcal{E} \tag{3.15}$$

The constant α is referred to as the *polarizability* of the atom. Polarizabilities thus carry information about how easily the atom is deformed by an electric field: the larger the polarizability, the larger the induced dipole moment.

Polarization represents a deformation of the atom and requires energy

$$E_{\mathrm{pol}} = \frac{\mu^{*2}}{2\alpha} = \frac{1}{2}\alpha\mathcal{E}^2 \tag{3.16}$$

The fraction $\alpha' = \alpha/(4\pi\varepsilon_0)$ has the dimension (length)^3 and is referred to as the polarizability volume. This is the parameter that is normally listed in reference works. The polarizabilities atoms or monatomic ions in the gas phase are not accurately known, but recommended values for polarizability volumes of neutral alkali metal and halogen atoms and ions are listed in Table 3.4.

Table 3.4. Recommended values for the polarizability volumes α' (in 10^{-30} m^3) of some atoms and monatomic ions in the gas phase [5].

M	α'(M)	α'(M$^+$)	X	α'(X)	α'(X$^-$)
Li	24.3	0.03	F	0.56	2.81
Na	23.6	0.15	Cl	2.18	4.37
K	43.4	0.79	Br	3.05	6.36
Rb	47.3	1.35	I	4 ± 2	8.28
Cs	59.6	2.34			

These values illustrate some common trends: polarizabilities of neutral atoms decrease strongly across a period, and increase down a given group: large atoms in which the electrons are loosely held are easy to deform.

Table 3.4 also contains the polarizability volumes of some common monatomic ions. It is seen that removal of the most loosely held electron in an alkali metal atom leads to a dramatic reduction of the polarizability. Similarly, addition of an electron to a halogen atom to form an anion leads to an increase of the polarizability. Comparing cations and anion, we see that halogen anions have considerably higher polarizabilities than alkali metal cations.

References

[1] L. G. Vanquickenborn, K. Pierloot, and D. Devoghel, *Inorg. Chem.*, 1989, 28, 1805 and *J. Chem. Ed.*, 1994, 71, 469.

[2] First ionization energies and data required for calculation of valence shell *s*-electron energies were taken from the standard collection of experimental values of atomic energy levels: C. E. Moore, *Atomic Energy Levels*, National Bureau of Standards, Circular 467, Washington, 1949.

[3] L. C. Allen, *J. Am. Chem. Soc.*, 1989, 111, 9003; J. B. Mann, T. L. Meek and L. C. Allen, *J. Am. Chem. Soc.*, 2000, 122, 2780. Allen and coworkers refer to the configuration-averaged ionization energy of the *s* electrons defined by equation (3.4) as simply the "*s*-electron energy" or ε_s, and to the configuration-averaged ionization energy of the *p* electrons defined by equation (3.5) as the "*p*-electron energy" or ε_p. Similarly they refer to the mean valence shell ionization energy defined by equation (3.6) simply as the "configuration energy" or CE. We have changed the nomenclature in order to make it clear to the reader that we are referring to the energy changes associated with ionization processes rather than ground states energies relative to separated electrons and nuclei at rest.

[4] Electron affinities have been taken from H. Hotop and W. C. Lineberger, *J. Phys. Chem. Ref. Data*, 1985, 14, 731 and (Ge and Sn) from W. E. Dasent, *Inorganic Energetics*, Cambridge University Press, 1982.

[5] Recommended values for the polarizability volumes of gaseous atoms have been taken from T. M. Miller and B. Bederson, *Adv. At. Mol. Phys*, 1977, 13, 1; polarizability volumes for alkali metal cations and halide anions from S. Hati, B. Datta and D. Datta, *J. Phys. Chem.*, 1996, 100, 19808.

Chapter 4

Homonuclear diatomic molecules

Introduction

In this chapter we define the terms "electronic energy" and "potential energy" for a molecule, and introduce the "adiabatic approximation" which allows us to represent the potential energy of a diatomic molecule as a function of only one parameter, namely the internuclear distance R. We then use this curve to define the equilibrium bond distance R_e and the dissociation energy of the non-vibrating molecule D_e. We show how the vibrational frequency of a diatomic molecule depends on the force constant and the nuclear masses. Finally we define the dissociation energy at zero K, D_0, and show that the dissociation energies at 0 and 298 K differ by less than 1 kJ mol^{-1}.

4.1 Bond strength

The defining property of a chemical bond is the bond strength: two atoms will only be described as being joined by a chemical bond, if it requires a certain amount of energy to separate them. Usually one will only use the term if the required energy is above a certain threshold, say 20 or 40 kJ mol^{-1}. The threshold is chosen according to the circumstances, so as to simplify discussion of the phenomena under consideration.

For a diatomic molecule A_2 the strength of the bond may be defined as the standard dissociation energy, that is the standard energy of the chemical reaction

$$A_2(g) = 2A(g); \quad D = \Delta U^0$$

Standard dissociation energies may be calculated from tabulated values of standard energies or enthalpies of formation:

$$D = 2\Delta U_f^0(A(g)) - \Delta U_f^0(A_2(g)) = 2\Delta H_f^0(A(g)) - \Delta H_f^0(A_2(g)) - RT \tag{4.1}$$

The standard dissociation energies of H_2, the dihalogen molecules X_2 and the dialkalimetal molecules M_2 at 298 K are listed in Table 4.1 [1]. As you know, chemists agree that in each of these molecules, the atoms are held together by a single bond. The strongest bond is found in the H_2 molecule, and the X–X bonds in the dihalogens are stronger than the M–M bonds in the dialkalimetals. The bond strength in the M_2 molecules decreases monotonically from Li_2 to Cs_2, the bond strength in the X_2 molecules decreases from Cl_2 to I_2. This is in agreement with a general observation that *the strength of a bond between the atom pair AB tends to decrease when A or B – or both – are exchanged with a heavier atom in the same group.* In chemistry there are, however, few rules without exceptions: the Cl–Cl and Br–Br bonds are both stronger than the F–F bond.

Table 4.1. Gaseous homonuclear diatomic molecules, A_2(g); standard dissociation energies at 298 K, D_{298}; equilibrium bond distances, R_e; vibrational wavenumbers, ω; dissociation energies at zero K, D_0; reduced masses, μ_M; and force constants, f_R.

A_2(g)	D_{298} (kJ mol^{-1})	Isotopomer	R_e (pm)	ω (cm^{-1})	D_0 (kJ mol^{-1})	μ_M (a.u.)	f_R (N m^{-1})
H–H	433.5	$^1H_2$	74.1	4401	432.1	0.504	575
		$^1H^2H$	74.1	3813	435.5	0.672	575
Li–Li	105.2	$^7Li_2$	267.3	351.4	99.8	3.508	25.5
Na–Na	74.6	$^{23}Na_2$	307.9	159.1	71.1	11.50	17.2
K–K	48.5	$^{39}K_2$	392.4	92.0	49.6	19.48	9.7
Rb–Rb	46.0	$^{85}Rb_2$	417.0	57.3	46.8	42.46	8.2
Cs–Cs	44.2	$^{133}Cs_2$	465.8	42.0	43.3	66.45	6.9
F–F	155.3	$^{19}F_2$	141.2	916.6	154.6	9.499	470
Cl–Cl	240.1	$^{35}Cl_2$	198.8	559.7	239.2	17.48	322
Br–Br	190.3	$^{79}Br_2$	228.1	325.3	190.2	39.46	246
I–I	148.6	$^{127}I_2$	266.6	214.5	148.8	63.45	172

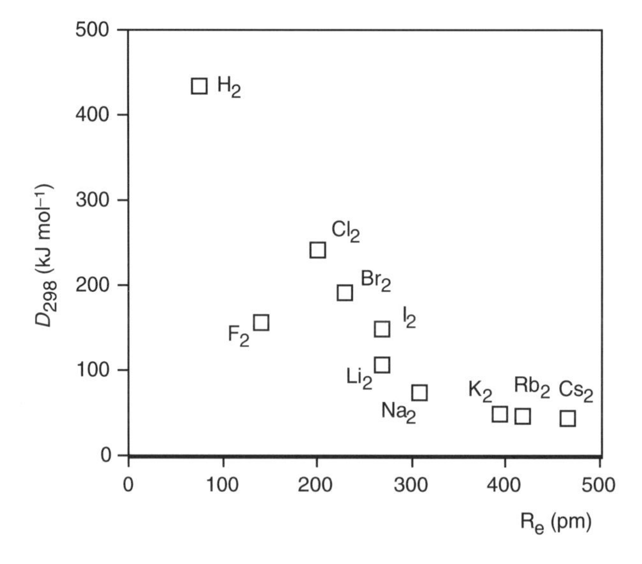

Fig. 4.1. Standard dissociation energies of homonuclear diatomic molecules at 298 K (in kJ mol^{-1}) as a function of the equilibrium bond distance (in pm).

In Fig. 4.1 we present a plot of the dissociation energies of the homonuclear diatomic molecules H_2, X_2 and M_2 as a function of the bond distances. The bond distances reflect the sizes of the atoms involved. Indeed, one might define the bonding radius of the atom A as half the single bond distance A–A. It is seen that – except for the rule-breaker F_2 – the bond energy decreases with increasing bond distance: *large atoms form weak bonds.*

Our sample is, however, restricted to homonuclear molecules. In the following chapters we shall consider heteronuclear diatomic molecules, and it will become clear that there is a second factor that strongly influences the bond strength, namely the bond polarity.

The standard dissociation energies listed in the second column of Table 4.1 pertain to substances with the natural isotopic composition. The remainder of the entries in the table have been derived from spectroscopic measurements performed on molecules consisting of specific isotopes as indicated in the third column [2].

4.2 The potential energy curve

In earlier chapters we defined the electronic energy of an atom as the sum of the kinetic energies of the electrons plus all electron–nucleus attraction and all electron–electron repulsion energies. Similarly we define the electronic energy of a molecule as the sum of the kinetic energies of all electrons, all electron–nucleus attraction energies, and all electron–electron and nucleus–nucleus repulsion energies. Note that the electronic energy excludes the kinetic energy of the nuclei, but includes the energy due to repulsion between them. It is clear that the electronic energy depends on the distances between the nuclei, but in order to simplify our thinking, we assume that it is independent of the nuclear masses and velocities. This assumption is referred to as the "adiabatic approximation" or as the Born–Oppenheimer approximation. The validity of the approximation depends on the velocity of the electrons being much greater than the velocity of the nuclei, so that the wavefunction of the electrons, "the shape of the electron cloud," always is perfectly adjusted to the instantaneous positions of the nuclei. There is no lag: the shape does not depend on whether the atoms are moving apart or towards each other.

The electronic energy of a molecule is defined as zero when all electrons and nuclei are infinitely far apart and at rest. When discussing the properties of the molecule it is, however, often more convenient to define the energy as zero when all bonds are broken, that is when the constituent atoms are infinitely far apart and in their ground states. We therefore define the "potential energy" of the molecule as the difference between its electronic energy and the sum of the electronic energies of the atoms:

$$V = E(\text{molecule}) - \Sigma E(\text{atoms})$$

If the adiabatic approximation is valid, we may draw a "potential energy curve" for a diatomic molecule that shows how the potential energy depends on a single variable, the internuclear distance R. See Fig. 4.2.

4.3 The equilibrium bond distance and the dissociation energy of a diatomic molecule

The value of R corresponding to the minimum of the potential energy curve is referred to as the equilibrium bond distance, R_e. The perpendicular distance from the minimum of the curve to the asymptotic value corresponding to two separated atoms, i.e. the energy required to increase the internuclear distance from R_e to infinity, is referred to as the dissociation energy of the non-vibrating molecule, D_e.

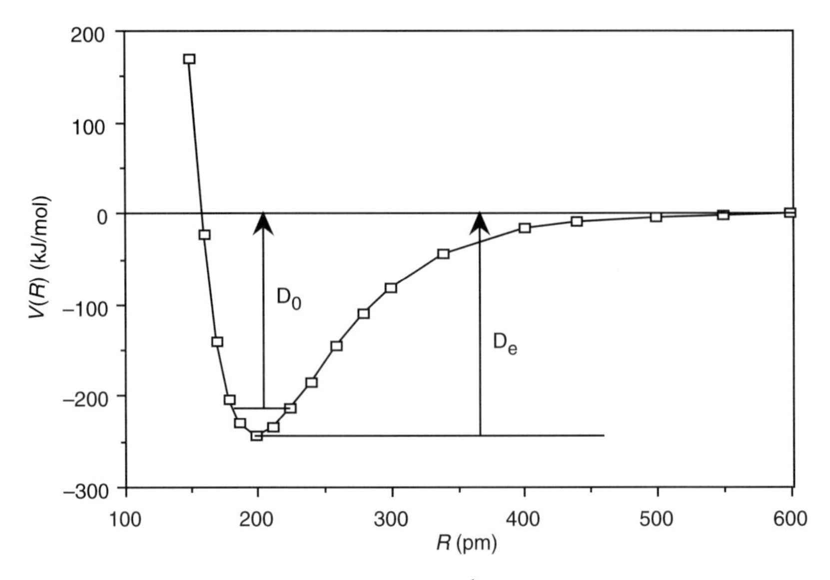

Fig. 4.2. The potential energy curve $V(R)$ (in kJ mol^{-1}) for a diatomic molecule. The equilibrium bond distance R_e and the dissociation energy D_e correspond to those of the Cl_2 molecule, but for clarity the zero point vibrational energy has been arbitrarily increased from 3.3 kJ mol^{-1} to about 25 kJ mol^{-1}.

The potential energy curve of a diatomic molecule is often assumed to have the form

$$V(R) = D_e\{\exp[-2a(R - R_e)] - 2\exp[-a(R - R_e)]\} \tag{4.2}$$

where the value of the constant a, like D_e and R_e, differs from one molecule to another. This curve, which is called the Morse potential, has the correct general shape, and is relatively easy to manipulate mathematically, but does not reproduce real potential energy curves with great accuracy.

Problem 4.1 What are the values of the Morse potential at $R = 0$, $R = R_e$ and $R \to \infty$?

4.4 Vibrational energy levels

According to quantum mechanics the two nuclei in a diatomic molecule cannot remain at a constant distance from each other: the molecule must vibrate. The vibrational energy levels are determined by the masses of the vibrating nuclei and by the shape of the potential energy curve.

The lower vibrational energy levels are determined by the shape of the potential energy curve in the region around R_e. The problem is greatly simplified if $V(R)$ in this region is assumed to be given by the simple expression

$$V(R) = -D_e + \frac{1}{2}f_R(R - R_e)^2 \tag{4.3}$$

where the force constant f_R is given by

$$f_R = \left(\frac{d^2V}{dR^2}\right)_{Re}$$

The dimension of the force constant is (force/length), in the SI system N m^{-1}. If the adiabatic approximation is valid, the force constant should be independent of the masses of the nuclei, i.e. have the same value for isotopomers like $^1H_2$ and $^1H^2H$.

Problem 4.2 Calculate the force constant from the Morse potential.

According to classical mechanics the vibrational frequency of a molecule with force constant f_R is given by

$$\nu = (1/2\pi)\sqrt{f_R/\mu_M} \tag{4.4}$$

where the parameter μ_M is the reduced mass of the molecule which is defined in terms of the masses of the two atoms:

$$\mu_M = \frac{M_A \times M_B}{M_A + M_B} \tag{4.5}$$

Somewhat surprisingly the vibrational frequency turns out to be independent of the amplitude. Such a molecule is described as a harmonic oscillator.

According to quantum mechanics the allowed vibrational energies of such a molecule is given by

$$E(u) = \left(u + \frac{1}{2}\right)h\nu \tag{4.6}$$

where u is a quantum number equal to zero or a positive integer, h is the Planck constant and ν the classical frequency defined above.

The wavenumber of the oscillator is defined as

$$\omega = \nu/c$$

where c is the speed of light. The values of the vibrational wavenumbers in the molecules under consideration are listed in the fifth column of Table 4.1. The difference between the vibrational wavenumbers of the isotopomers $^1H_2$ and $^1H^2H$ is due to the difference between their reduced masses.

Problem 4.3 Estimate the vibrational wavenumber of the molecule $^2H_2$.

The lowest possible vibrational energy, corresponding to a vibrational quantum number $u = 0$, is referred to as the zero point energy, E_0. In the harmonic approximation

$$E_0 = \frac{1}{2}h\nu = \frac{1}{2}h\omega c \tag{4.7}$$

for one molecule. Multiplication with the Avogadro number gives the zero point energy for one mole:

$$E_0 = \frac{1}{2}N_A h\nu = \frac{1}{2}N_A h\omega c$$

According to equation (4.6) the higher energy levels should be equally spaced, the distance from any level and the one above it being equal to $\Delta E = h\nu$. In reality it is found that the distance between energy levels decreases with increasing energy. The reason for the breakdown of equation (4.6) is that the discrepancy between the harmonic potential (4.3) and the real $V(R)$ curve increases as the vibrational energy increases and the molecule vibrates further away from the equilibrium bond distance.

The energy levels of a real, anharmonic molecule are given much more accurately by the expression

$$E(u) = \left(u + \frac{1}{2}\right) h\nu - \left(u + \frac{1}{2}\right)^2 h\nu x \qquad (4.8)$$

where x is a small positive number. For the $^1H_2$ molecule $x = 0.027$, while for the $^{35}Cl_2$ molecule $x = 0.0047$.

The number of vibrational energy levels is not infinite: if the vibrational energy exceeds D_e, the molecule dissociates. The $^1H_2$ molecule has 14 vibrational levels below the dissociation threshold. Since the spacing between the energy levels is smaller in $^1H^2H$, this molecule has 17 levels below the dissociation threshold [3].

Problem 4.4 Use the data in Table 4.1 and equation (4.6) to obtain a rough estimate of the number of vibrational energy levels below the dissociation threshold in F_2, Cl_2, I_2, Li_2, and Cs_2.

Problem 4.5 The ratio between the number of molecules in the first excited vibrational state, N_1, and in the ground state, N_0, is given by

$$\frac{N_1}{N_0} = \exp\left(-\frac{h\nu}{kT}\right)$$

where k is the Boltzmann constant

$$k = R/N_A = 1.381 \times 10^{-23} \text{J K}^{-1}.$$

Calculate the temperature at which the fraction is 0.01 for H_2, F_2, Cl_2, Br_2 and I_2.

4.5 The standard dissociation energy at zero kelvin

While the lowest possible vibrational energy of a diatomic molecule is given by (4.7), the lowest possible rotational and translational energies of a gaseous molecule are both zero. The energy of a molecule at zero kelvin may therefore be represented by a horizontal line on the potential energy curve, E_0 above the minimum. The dissociation energy of the molecule at zero kelvin, D_0, is equal to the vertical distance from this line to the asymptotic value of the potential energy curve:

$$D_0 = 2\text{E}(\text{A(g)}) - \text{E}(\text{A}_2\text{(g)}) = 2 \times 0 - (-D_e + E_0) = D_e - E_0 \qquad (4.9)$$

See Fig. 4.2.

Values of D_0 calculated from spectroscopic measurements are listed in the sixth column of Table 4.1. The difference between the D_0 values of the isotopomers $^1H_2$ and $^1H^2H$ is due to the difference between the zero point energies, which in turn is due to the difference between the reduced masses of the two molecules.

Naturally occurring hydrogen gas contains about 99.96% $^1H_2$ and about 0.04% $^1H^2H$ (and less than 0.000 02% $^2H_2$). The dissociation energy of one mole of the mixture at zero kelvin

$$D_0(H_2) = 0.9996\, D_0(^1H_2) + 0.0004\, D_0(^1H^2H)$$

differs from $D_0(^1H_2)$ by less than 0.01 kJ mol^{-1}.

The difference between the zero point energies of isotopomers of the other molecules in the table is much smaller than between the hydrogen isotopomers. Thus the difference between the zero point energies of $^6Li^7Li$ and $^7Li_2$ is less than 0.1 kJ mol^{-1}: unless accuracies greater than 0.1 kJ mol^{-1} are required, we may disregard isotope effects on dissociation energies for molecules that do not contain hydrogen.

4.6 Comparison of the standard dissociation energies at 0 K and 298 K

Comparison of the dissociation energies at zero, D_0, with the standard dissociation energies at 298 K, D_{298}, shows that, with the exceptions of Li_2 and Na_2, the numbers agree to within one or two kJ mol^{-1}. See Table 4.1. At first glance the agreement is surprising since the two dissociation energies refer to very different temperatures. When the temperature of one mole of A_2 gas is increased from zero to $T = 298$ K, the translational energy increases by $\frac{3}{2}RT$, and the rotational energy by RT. Some of the molecules are promoted to higher vibrational energy levels and the vibrational energy increases by a certain amount ΔE_v. The kinetic energy of two moles of A atoms at increases by $3RT$. The dissociation energy at 298 K becomes:

$$\begin{aligned} D_{298} &= 2U^0_{298}(A(g)) - U^0_{298}(A_2(g)) \\ &= (3RT) - (-D_0 + 3RT/2 + RT + \Delta E_v) \\ &= D_0 + \left(\frac{1}{2}RT - \Delta E_v\right) \end{aligned} \tag{4.10}$$

Closer examination by the tools of statistical thermodynamics shows that the absolute value of the difference $\left(\frac{1}{2}RT - \Delta E_v\right)$ always will be less than $\frac{1}{2}RT$ or about 1.2 kJ mol^{-1} at 298 K: in practice, the difference between the dissociation energies at zero and 298 K is usually smaller than their combined experimental uncertainties.

4.7 Force constants

The observed vibrational wavenumbers and the reduced masses listed in Table 4.1 may be used to calculate the force constants under the harmonic approximation, equation (4.4). A large force constant indicates that the bond is very stiff, a small force constant indicates that the molecule is soft or easily deformed. It is seen that the H–H bond which is the strongest bond in the table, is also the stiffest. Furthermore the force constants decrease from Li_2 to Cs_2 and from Cl_2 to I_2 just like the bond strength. There is therefore some truth in the general – and rather imprecise – statement that a weak bond also tends to be a soft bond. Note, however, that even though the Cl–Cl bond is stronger than the F–F bond, the force constant is smaller.

References

[1] Thermochemical data for calculation of standard dissociation energies at 298 K have been taken from *NIST-JANAF Thermochemical Tables*, fourth edition, M. W. Chase, Jr., Ed., *J. Phys. Chem. Ref. Data*, Monograph No. 9, 1998.

[2] Most of the data derived from spectroscopic measurements have been taken from K. P. Huber and G. Herzberg, *Molecular Spectra and Molecular Structure*, Vol. IV, *Constants of Diatomic Molecules*, Van Nostrand, New York, 1979. Some values for equilibrium bond distances and dissociation energies of dialkalimetal molecules have been taken from G. Dotelli, E. Lombardi and L. Jansen, *J. Mol. Struct. (Theochem)*, 276 (1992) 159.

[3] The information about the number of vibrational levels of ${}^1H_2$ and ${}^1H^2H$ is found in W. Kolos, K. Szalewicz and H. J. Monkhorst, *J. Chem. Phys.*, 84 (1968) 3278.

Chapter 5

Gaseous alkali metal halides: ionic bonds

Introduction

The alkali metal halides, MX, form solids at room temperature. The best known member of the family, NaCl, melts at 801°, the vapor pressure reaches 1 torr at 865°, and the melt boils at 1413°. The dominant species in gaseous alkali metal halides is the monomeric formula unit, MX(g), but smaller amounts of dimers (M_2X_2), trimers (M_3X_3) and tetramers (M_4X_4) have also been detected by mass spectroscopy. Information about the monomeric molecules obtained by spectroscopic studies at high temperatures is collected in Table 5.1 [1].

Solid and molten alkali metal halides are described as consisting of M^+ and X^- ions. How should we describe the molecular units in the gas phase? Do they consist of a pair of ions, M^+X^-, which are held together by Coulomb attraction?

We first assume that the monomeric gas molecules, the dimers and the tetramers consist of spherical ions with integer charges, M^+ and X^-, that touch, but do not penetrate each other. The Coulomb interaction between the ions may be calculated from classical physics. In addition we assume the ions to repel one another at close range. This simple "spherical ion model" leads to calculated electric dipole moments which are on the average more than 30% higher than the observed. The calculated dissociation energies of the monomers are, however, in reasonable agreement with experiment. The spherical ion model also permits the calculation of M–X bond distances in the dimers, tetramers and in the crystalline phase from the bond distances in the monomers. Both calculated and experimental bond distances are found to increase with increasing aggregation.

When the model is refined to include polarization effects, i.e. to allow for that fact the electric field set up by each ion in the monomer deforms the electron cloud in the other, calculated molecular dipole moments are in significantly better agreement with the observed values. Polarization effects do not, however, lead to significant improvement of the fit between calculated and experimental dissociation energies.

5.1 The electric dipole moments of molecules

The electric dipole moment of a molecule, like the dipole moment of an atom, is a vector. The component in the z-direction is defined by the integral

$$\mu_{\text{el},z} = \int \rho(x, y, z) z \, d\tau \tag{5.1}$$

where the integration is carried out over all space. The components in the x and y directions are defined by similar integrals.

The integral (5.1) may be divided into two parts in such a way that positive and negative charge is counted separately:

$$\mu_{\text{el},z} = \int \rho_+(x, y, z) z \, d\tau + \int \rho_-(x, y, z) z \, d\tau \tag{5.2}$$

Table 5.1. Gaseous monomeric alkalimetal halides, MX(g): experimental electric dipole moments, μ_{el}; electric dipole moments predicted by the spherical ion model, μ_{el}(calc); equilibrium bond distances, R_e; vibrational wavenumbers, ω; dissociation energies at zero K, D_0; reduced masses, μ_M; force constants, f_R; dissociation energies calculated from the spherical ion model according to equation (5.16a), D_0(calc).

MX(g)	μ_{el} (D)	μ_{el}(calc) (D)	R_e (pm)	ω (cm^{-1})	D_0 (kJ mol^{-1})	μ_M (a.u.)	f_r (N m^{-1})	D_0(calc) (kJ mol^{-1})
Li–F	6.28	7.51	156.4	910.3	570	5.124	250	597
Li–Cl	7.09	9.70	202.1	643.3	467	5.844	142	436
Li–Br	7.23	10.42	217.0	563.2	418	6.443	120	375
Li–I	7.43	11.49	239.2	498.2	342	6.648	97.2	292
Na–F	8.12	9.25	192.6	536.0	476	10.40	176	475
Na–Cl	8.97	11.34	236.1	366.0	408	13.87	109	374
Na–Br	9.09	12.01	250.2	302.4	361	17.80	95.9	325
Na–I	9.21	13.02	271.1	258.0	290	19.46	76.3	257
K–F	8.56	10.43	217.1	428.0	489	12.77	138	480
K–Cl	10.24	12.81	266.7	281.0	419	18.43	85.7	392
K–Br	10.60	13.55	282.1	213.0	377	26.08	69.7	347
K–I	10.80	14.64	304.8	186.5	319	29.81	61.1	284
Rb–F	8.51	10.90	227.0	376.0	483	15.52	129	472
Rb–Cl	10.48	13.38	278.7	228.0	419	24.77	75.9	388
Rb–Br	10.86	14.14	294.5	169.5	376	40.90	69.2	345
Rb–I	11.48	15.26	317.7	138.5	318	50.87	57.5	284
Cs–F	7.85	11.26	234.5	352.6	497	16.62	122	481
Cs–Cl	10.36	13.95	290.6	214.2	442	27.68	74.8	397
Cs–Br	10.82	14.75	307.2	149.7	402	49.52	65.4	354
Cs–I	11.69	15.92	331.5	119.2	344	64.92	54.3	295

The first integral is equal to the product of the magnitude of the total positive charge in the molecule and the z-coordinate of its center of gravity, z_+:

$$\int \rho_r(x,y,z)z\mathrm{d}\tau = (\text{total positive charge}) \times z_+ = (\Sigma Z_i e)z_+$$

where Z_i is the atomic number of the i'th atom and the sum extends over all atoms in the molecule. Similarly the second integral is equal to the magnitude of the total negative charge in the molecule multiplied by the z-coordinate of its centre of gravity, z:

$$\int \rho_-(x,y,z)z\mathrm{d}\tau = (\text{total negative charge}) \times z_- = (-\mathrm{N}e)z_-$$

where N represents the total number of electrons in the molecule. For a neutral molecule $(-\mathrm{N}e) = -(\Sigma Z_i e)$ and

$$\mu_{el,z} = (\Sigma Z_i e)(z_+ - z_-) \tag{5.3}$$

Similar expressions may be written down for the x and y components of μ_{el}. *We see that if the centers of gravity of positive and negative charge in a molecular unit coincide, then the three components of the dipole moment are equal to zero.*

In a homonuclear diatomic molecule, A_2, the centers of gravity of positive and negative charge must both fall at the midpoint of the A–A bond; such molecules are *non-polar*. The bond would sometimes be described as perfectly covalent.

For heteronuclear diatomic molecules we define the direction of the z-axis in such a way that it runs through both nuclei. The centers of gravity of positive and negative charge must both lie on the z-axes, but will not coincide. The x- and y-components of the dipole moment will be zero by symmetry, but not the z-component. These molecules are *polar*.

5.2 The spherical ion model

Let us assume that a molecular unit consists of two or more spherical, monatomic ions for which the electron density becomes zero at a certain distance from the nucleus (according to quantum mechanics there are no such atoms), and that the ions touch, but *do not penetrate each other*. The integration over space in equation (5.1) may then be divided in such a way that it extends over one atom at a time:

$$\begin{aligned} \mu_{el,z} = & \int \rho_+(x,y,z)z\,\mathrm{d}\tau + \int \rho_-(x,y,z)z\,\mathrm{d}\tau \qquad \text{(integrated over atom A)} \\ & + \int \rho_+(x,y,z)z\,\mathrm{d}\tau + \int \rho_-(x,y,z)z\,\mathrm{d}\tau \qquad \text{(integrated over atom B)} \end{aligned}$$

The first integral in each line is equal to $(+Z_ie)z_i$ where z_i is the z-coordinate of nucleus i. Since we have assumed the electron density of the ion to be spherical, the center of gravity of negative charge falls at the nucleus and the second integral becomes equal to $(-N_ie)z_i$ where N_i is the number of electrons on the atomic ion. The sum of the two integrals in each line is

$$(+Z_ie)z_i - (-N_ie)z_i = Q_{net,i}z_i$$

where $Q_{net,i}$ is the net charge on the i'th ion.

Thus, under the assumption that the molecule consists of spherical and non-penetrating ions we may write

$$\mu_{el,z} = \Sigma Q_{net,i}z_i \tag{5.4}$$

where the sum extends over all ions in the molecular unit.

In a neutral diatomic molecule AB, the net positive charge on one atom must be balanced by the net negative charge of the other:

$$Q_{net,B} = -Q_{net,A}$$

Denoting the positively charged atom by A and using equation (5.4) we obtain:

$$\mu_{el,z} == Q_{net,A}z_A - Q_{net,A}z_B = Q_{net,A}(z_A - z_B) \tag{5.5}$$

Note that the sign of the dipole moment depends on the direction of the z-axis: if the z-axis runs from the atom carrying a net negative to the atom carrying a net positive charge, i.e. if $z_A > z_B$, the dipole moment is positive. If the z-axis runs from the atom carrying a net positive to the atom carrying a net negative charge, i.e. if $z_B > z_A$, the dipole moment is negative. The dipole moment is therefore sometimes represented by an arrow pointing from the negative to the positive charge.

The magnitude of the electric dipole moment is given by

$$|\mu_{\mathrm{el}}| = Q_{\mathrm{net,A}}|z_A - z_B| = Q_{\mathrm{net,A}}R \tag{5.6}$$

where R is the bond distance.

The experimental determination of the magnitude of the electric dipole moment of a diatomic molecule is relatively straightforward, but it is unfortunately more difficult to determine the sign.

If charges are measured in coulombs and distances in meters, the dimension of a dipole moment is equal to C m. Traditionally, however, molecular dipole moments have been given in Debye units which are defined by

$$1 \text{ Debye} = 1 \text{ D} = 3.336 \times 10^{-30} \text{ C m}. \tag{5.7}$$

Problem 5.1 Calculate the electric dipole moment (in Debyes) of a molecule consisting of two spherical ions with charges equal to $\pm e$ at a distance of 100 pm.

5.3 The electric dipole moments of the gaseous monomeric alkali metal halides

The electric dipole moments obtained by spectroscopic measurements on the diatomic molecules MX; M = alkali metal, X = halogen, are listed in the second column of Table 5.1. The dipole moments range from 6.28 Debye in LiF(g) to 11.7 Debye in CsI(g).

How do these dipole moments compare with those calculated for the spherical ion model? In the third column of Table 5.1 we list the electric dipole moments calculated from equation (5.6) with net ionic charges equal to $\pm e$ and the experimentally determined bond distances:

$$\mu_{\mathrm{el}}(\text{calc}) = eR_e$$

It is seen that all the calculated dipole moments are *higher* than the true, experimental values. The average deviation between estimate and experiment amounts to 32% of the experimental value, so it is clear that the charge distribution in alkali metal halide molecules deviates significantly from that predicted from the spherical ion model.

The assumption that the ions will remain spherical when they are close to each other is in fact unreasonable: The net charge on each ion will create an electric field which will induce a dipole moment on the other, and the direction of the induced dipole moments is such as to reduce the total dipole moment of the molecule. We shall return to this point in Section 5.13.

5.4 Bond strength

The bond in a gaseous alkali metal halide molecule may be ruptured in such a way that one obtains two neutral atoms:

$$MX(g) \rightarrow M(g) + X(g)$$

or in such a way that one obtains two separated ions:

$$MX(g) \rightarrow M^+(g) + X^-(g)$$

The first process is referred to as homolytic, the second as heterolytic, rupture.

Since the ionization energies of all elements are higher than the highest electron affinity, the process

$$M(g) + X(g) \rightarrow M^+(g) + X^-(g)$$

will always require energy. This means that even if an alkali metal halide molecule MX consists of M^+ and X^- ions, minimum energy dissociation will yield neutral M and X atoms.

According to Table 4.1 the dialkali metal molecule with the strongest bond (Li_2) has a dissociation energy of 100 kJ mol^{-1}, while the dihalogen molecule with the strongest bond (Cl_2) has a dissociation energy of 239 kJ mol^{-1}. The homolytic bond dissociation energies of alkali metal halide molecules at zero kelvin are listed in Table 5.1. They range from 290 kJ mol^{-1} in NaI(g) to 570 kJ mol^{-1} in LiF(g). It is clear that *the ionic bonds in monomeric alkali metal halides are very strong compared to the covalent bonds in the dialkali metal molecules* M_2 *or in the dihalogen molecules* X_2.

The dissociation energy of each MX(g) molecules is in fact larger than the *sum* of the dissociation energies of the corresponding $M_2(g)$ and $X_2(g)$ molecules:

$$D_0(MX) > D_0(M_2) + D_0(X_2)$$

for all combinations of M and X.

This means that the chemical reactions

$$M_2(g) + X_2(g) = 2MX(g)$$

are all very exothermal. The gaseous alkali metal halides are thermodynamically much more stable than the elements, the reason why we do not find them under normal laboratory conditions is that they have exceptionally high sublimation enthalpies: the crystalline solids are even more stable than the gaseous molecules.

Problem 5.2 Are the bond energies of the alkali metal halides in agreement with the expectation that the strength of a bond between an atom pair AB tends to decrease when A or B – or both – are replaced by a heavier atom in the same group?

5.5 Bond distances in alkali metal halide molecules

The equilibrium bond distances in gaseous, monomeric alkali metal halides are listed in Table 5.1. In Fig. 5.1 we plot the bond distances in the fluorides, chlorides, bromides and

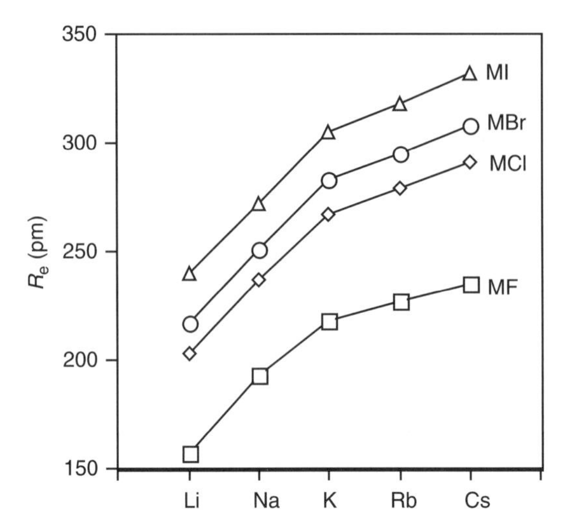

Fig. 5.1. The equilibrium bond distances in gaseous, monomeric alkali metal fluorides, chlorides, bromides and iodides as a function of the metal cation as Group 1 is descended.

iodides as a function of the metal cation as Group 1 is descended. The shapes of the four curves are remarkably similar. If we assume that each M–X bond distance is equal to the sum of the radii of M^+ and X^- ions, then the shape of each of curve reflects the variation of the radius of the metal cation. It is seen that the radii of the metal cations increase linearly from Li to K and from K to Cs. Between K and Rb we have passed through the d-block, and the expected increase of the atomic or ionic radius is presumably reduced by the lingering effect of the d-block contraction. See Section 3.10. The vertical distance between any two curves is presumably equal to the difference between the radii of the two anions. This means that the radius of the halide anion increases more between F and Cl than between Cl and Br: between Cl and Br we have again passed through the d-block.

Comparison of bond distances in Tables 4.1 and 5.1 shows that the bond distances in the gaseous alkali metal halides are shorter than the average of the bond distances in the corresponding M_2(g) and X_2(g) molecules:

$$R(\mathrm{M–X}) < 1/2[R(\mathrm{M–M}) + R(X–X)]$$

for any combination of M and X. The difference between the two sides of the inequality ranges from 16 pm in NaI(g) to 69 pm in CsF(g).

If we define the bonding radius of an atom A as half the bond distance in the A_2(g) molecule, we obtain

$$R(\mathrm{M–X}) < r(\mathrm{M}) + r(\mathrm{X}).$$

Ionic bonds are much shorter than the sum of the bonding radii of the two neutral atoms.

5.6 Force constants

Comparison of the force constants in Tables 4.1 and 5.1 shows that even though the ionic bonds in MX(g) are stronger and relatively shorter than the covalent bonds in both M_2(g) and X_2(g), the force constants are *smaller* than in the dihalogens: ionic bonds are short and strong, but soft!

5.7 The potential energy curve of a spherical ion pair

We shall now return to the spherical ion model and use it to construct the potential energy curve of an ion pair.

When the separated ions are formed from the separated atoms

$$\mathrm{M(g) + X(g) \rightarrow M^+(g) + X^-(g)},$$

the energy increases with the ionization energy of M minus the electron affinity of X, IE(M)–EA(X). If the ions are allowed to move towards each other, the energy will decrease due to the electrostatic attraction:

$$V(R) = \mathrm{IE(M)} - \mathrm{EA}(X) - \frac{e^2}{4\pi\varepsilon_0 R}$$

At some point $V(R)$ becomes equal to zero, or

$$\frac{e^2}{4\pi\varepsilon_0 R} = \mathrm{IE(M)} - \mathrm{EA(X)} \tag{5.8}$$

At this point the energy of the ion pair is exactly equal to the energy of the separated atoms. For NaCl this occurs at a distance of about 1000 pm. See Fig. 5.2.

Problem 5.3 Calculate the distance at which equation (5.8) is satisfied for NaCl. For which alkali metal halide is the distance the shortest, and for which is the distance the longest? Calculate these distances.

This means that the character of the chemical system "NaCl" changes radically at around 1000 pm: when the distance between the nuclei is greater than 1000 pm, the system consists of two neutral atoms; when the distance is shorter, it consists of two ions.

When the internuclear distance is reduced below 1000 pm, the energy continues to fall. The approach of the ions is, however, stopped at some point, presumably because the spherical ions repel each other at close range, much like two rubber balls when they are pressed together. The close range repulsion between ions is a topic of which we do not know very much today, even less was known when the spherical ion model was developed in the early 1920s.

The energy of repulsion was therefore assumed to be given by the simple expression

$$\text{“Born repulsion energy”} = B_{\mathrm{MX}}/R^n \tag{5.9}$$

where B_{MX} is a constant (the "Born constant") which depends on the size and hardness of the M^+ and X^- ions, and the exponent n is chosen in the range between 8 and 14.

Equation (5.9) implies that the repulsion energy rises steeply when R becomes smaller than the equilibrium bond distance. See Fig. 5.2. The total potential energy of the molecule is then given by

$$V(R) = \text{IE(M)} - \text{EA(X)} - \frac{e^2}{4\pi\varepsilon_0 R} + \frac{B_{\text{MX}}}{R^n} \tag{5.10}$$

While the Born constant is not accurately known, the bond distance has been determined experimentally with high accuracy. We therefore eliminate B_{MX} by using the fact that the derivative of $V(R)$ must be zero when $R = R_e$:

$$\left(\frac{\mathrm{d}V(R)}{\mathrm{d}R}\right)_{\text{Re}} = \frac{e^2}{4\pi\varepsilon_0 R_e^2} - \frac{nB_{\text{MX}}}{R_e^{n+1}} = 0 \tag{5.11}$$

Rearrangement of (5.11) yields

$$B_{\text{MX}} = \left[\frac{e^2}{4\pi\varepsilon_0 R_e}\right]\frac{R_e^n}{n} \tag{5.12}$$

Inserting (5.12) in (5.10) we obtain

$$V(R) = \text{IE(M)} - \text{EA(X)} - \frac{e^2}{4\pi\varepsilon_0 R}\left[1 - \frac{1}{n}\left(\frac{R_e}{R}\right)^n\right] \tag{5.13}$$

The potential energy for the NaCl molecule calculated from (5.13) is shown in Fig. 5.2.

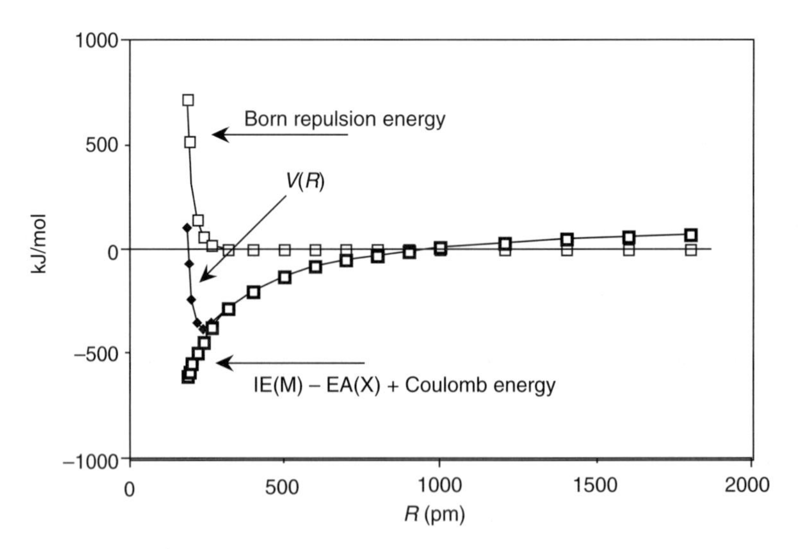

Fig. 5.2. The potential energy curve of a gaseous alkali metal halide according to the spherical ion pair model, equation (5.13). The magnitudes of IE(M), EA(X) and the equilibrium bond distance correspond to NaCl(g) and the exponent has been chosen $n = 10$.

5.8 The dissociation energy of a spherical ion pair at zero kelvin

Insertion of $R = R_e$ in (5.13) gives

$$V(R_e) = \mathrm{IE(M)} - \mathrm{EA(X)} - \frac{e^2}{4\pi\varepsilon_0 R_e}\left(1 - \frac{1}{n}\right) \tag{5.14}$$

Since we have chosen to define the energy as zero for the separated neutral atoms,

$$D_e = 0 - V(R_e)$$

or

$$D_e = -\mathrm{IE(M)} + \mathrm{EA(X)} + \frac{e^2}{4\pi\varepsilon_0 R_e}\left(1 - \frac{1}{n}\right) \tag{5.15}$$

Introduction of the zero point vibrational energy E_0 yields the dissociation energy at 0 K:

$$D_0 = D_e + E_0 = -\mathrm{IE(M)} + \mathrm{EA(X)} + \frac{e^2}{4\pi\varepsilon_0 R_e}\left(1 - \frac{1}{n}\right) - E_0 \tag{5.16}$$

Insertion of the ionization energy of *one* metal atom, the electron affinity of *one* halogen atom and the zero point energy of *one* molecule yields the dissociation energy of *one* MX molecule. Alternatively insertion of the ionization energy and electron affinity per mole atoms and multiplication of the last two terms by the Avogadro number yield the dissociation energy per mole MX(g):

$$D_0 = -\mathrm{IE(M)} + \mathrm{EA(X)} + N_A \frac{e^2}{4\pi\varepsilon_0 R_e}\left(1 - \frac{1}{n}\right) - N_A E_0 \tag{5.16a}$$

The dissociation energies calculated from equations (5.16) or (5.16a) depend on the value chosen for the exponent n, but the reader may convince herself that a change of n from 10 to 12 has little effect.

The dissociation energy D_0 calculated from (5.16a) with the exponent $n = 10$, and using the experimental vibrational frequencies to obtain the zero point vibrational energies, are listed in the last column of Table 5.1. For the LiF(g) molecule the spherical ion pair model leads to an overestimate of the dissociation energy by 5%. All other dissociation energies are underestimated by an average of about 9%. The largest deviations between experimental and estimated dissociation energies, 15%, are found for two iodides, LiI and CsI. The spherical ion model thus leads to dissociation energies in reasonable agreement with experiment. It would take another 50 years before the development of large computers allowed us to calculate the dissociation energies of the covalently bonded M_2 or X_2 molecules with a similar accuracy.

According to equation (5.16a) formation of a strong ionic bond is favored by a large Coulomb stabilization energy, $N_A[e^2/(4\pi\varepsilon_0 R_e)]$, and a high electron affinity of the halogen, and opposed by a high ionization energy of the metal atom and a high Born repulsion energy $N_A[e^2/(4\pi\varepsilon_0 R_e)](1/n)$. (The zero point vibrational energy is much smaller than the other terms.) For the NaCl molecule the Coulomb stabilization energy is 583 kJ mol^{-1}, EA(X) = 347 kJ mol^{-1}, IE(M) = 496 kJ mol^{-1} and the Born repulsion energy is 58 kJ mol^{-1}. The dominant term is the Coulomb stabilization energy which increases with decreasing bond distance R_e: the strongest ionic bonds are formed by combination of the smallest ions.

Problem 5.4 LiH is a colorless solid at room temperature. The melting point is about 700° and the melt conducts electricity. Both the solid and the liquid are therefore believed to consist of Li^+ and H^- (hydride) ions. The electric dipole moment of the gaseous LiH is 5.88 D, the dissociation energy at zero K is 234 kJ mol^{-1}, the bond distance is 160 pm, and the vibrational wavenumber 1406 cm^{-1}. Calculate the electric dipole moment from the spherical ion model. Show that the dissociation energy calculated from the spherical ion model is in good agreement with the experimental if the exponent in the Born repulsion term B_{LiH}/R^n, is reduced from $n = 10$ to 5. Is there any reason to believe that the H^- ion is softer than F^- or Cl^- ions? Would you describe LiH(g) as an ionic molecule?

Problem 5.5 The gaseous HF molecule has a dipole moment of 1.83 Debye. The dissociation energy $D_0 = 566$ kJ mol^{-1}, the bond distance is 92 pm and the vibrational wavenumber 4138 cm^{-1}. Use the spherical ion model to estimate the dipole moment and the dissociation energy (in kJ mol^{-1}) at 0 K. Would you describe HF as an ionic molecule?

Problem 5.6 The molecular structure of KO_2 (potassium superoxide) in solid argon has been determined by vibrational spectroscopy. The molecule is T-shaped [2]. Can this result be understood if we assume the molecule to consist of K^+ and O_2^- ions? Hint: The total electron density of the O_2^- ion is probably very similar to that of the F_2 molecule. See Fig. 9.3. Where is the center of gravity of the negative charge on the anion? Where would you place the cation to maximize the Coulomb stabilization energy?

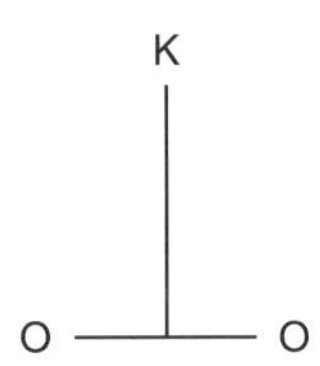

5.9 The potential energy of alkali metal halide dimers

The mass spectra of gaseous alkali metal halides show that in addition to the monomeric species discussed above, the gas may contain significant quantities of dimers (M_2X_2) and minor amounts of trimers (M_3X_3) or tetramers (M_4X_4). Gas electron diffraction studies have shown that the dimers have planar rhombic structures, but failed to yield accurate values for the M–X bond distances or the ∠XMX valence angles. Quantum chemical calculations on Na_2Cl_2 indicate that the Na–Cl bond distance is 16 pm or about 7% longer than in the monomer, and that the angles ∠ClNaCl and ∠NaClNa are about 103° and 77°, respectively. Since the Cl anion is known to be much larger than the Na cation, the wider angle at Na is presumably due to Born repulsion between the anions. Indeed, similar calculations on the nine dimers M_2X_2 with M=Li, Na and K and X=F, Cl or Br indicate that the largest ∠XMX angle is found when the largest anion, Br^- is combined with the smallest cation,

Li^+; $\angle$BrLiBr = 110°. When the smallest anion, F^-, is combined with the largest cation, K^+, the $\angle$XMX angle is calculated to be smaller than the $\angle$MXM angle [3].

We define the dissociation energy of a dimer as the standard energy of the reaction

$$M_2X_2(g) = 2MX(g);\ \Delta U^\circ_{\text{dis}} = 2\Delta U^\circ_f(MX(g)) - \Delta U^\circ_f(M_2X_2(g))$$

The dissociation energy of Na_2Cl_2 at 298 K is equal to 201 kJ mol^{-1}.

We shall now show how the spherical ion model may be used to estimate both the MX bond distance and the dissociation energy of the dimer, and begin by writing down an expression for the potential energy, i.e. the energy of the dimer relative to the energy of the four constituent atoms. The energy required to generate the four ions from the separated, neutral atoms is equal to

$$2\text{IE(M)} - 2\text{EA(X)},$$

while the Coulomb energy of the dimer is given by

$$\frac{-e^2}{4\pi\varepsilon_0}\left(\frac{4}{R} - \frac{1}{R_{\text{MM}}} - \frac{1}{R_{\text{XX}}}\right)$$

where R is the M–X bond distance, R_{MM} is the distance between the cations and R_{XX} the distance between the anions.

If we neglect Born repulsion between the two anions and between the two cations, i.e. if we assume that only the nearest neighbors touch one another, the Born repulsion energy is given by

$$\frac{4B_{\text{MX}}}{R^n}$$

where the magnitude of the Born constant is given by equation (5.12).

The total energy of the dimer relative to the energy of four separated atoms is

$$V_2(R) = 2\text{IE(M)} - 2\text{EA(X)} - \frac{e^2}{4\pi\varepsilon_0}\left(\frac{4}{R} - \frac{1}{R_{\text{MM}}} - \frac{1}{R_{\text{XX}}}\right) + \frac{4B_{\text{MX}}}{R^n}$$

This energy will be at a minimum when the molecule is square, i.e. when $R_{\text{MM}} = R_{\text{XX}} = \sqrt{2}R$:

$$\begin{aligned} V_2(R) &= 2\text{IE(M)} - 2\text{EA(X)} - \frac{e^2}{4\pi\varepsilon_0}\left(\frac{4}{R} - \frac{\sqrt{2}}{R}\right) + \frac{4B_{\text{MX}}}{R^n} \\ &= 2\left\{\text{IE(M)} - \text{EA(X)} - \frac{e^2}{4\pi\varepsilon_0 R}\left(2 - \frac{1}{\sqrt{2}}\right) + \frac{2B_{\text{MX}}}{R^n}\right\} \\ &= 2\left\{\text{IE(M)} - \text{EA(X)} - \frac{e^2}{4\pi\varepsilon_0 R}M_2 + C_2\frac{B_{\text{MX}}}{R^n}\right\} \end{aligned} \quad (5.17)$$

The four terms in the wavy parentheses represent the energy per MX unit. The Born repulsion energy per MX unit is given by C_2B_{MX}/R^n where $C_2 = 2$ is the coordination number of both cation and anion. The Coulomb energy per MX unit is given by $[e^2/(4\pi\varepsilon_0 R)]M_2$ where M_2 is the "Madelung constant." M_2 is the sum of two terms, the first is equal to the coordination number, i.e. to the number of stabilizing M–X Coulomb interactions per metal ion; the second term is equal to the number of destabilizing Coulomb interaction between the two halide ions bonded to the metal ion divided by a factor $\sqrt{2}$ to account for the fact that the X$\cdots$X is larger than the M–X distance.

The potential energy of the monomer, equation (5.10), may be rewritten as:

$$V_1(R) = \text{IE(M)} - \text{EA(X)} - M_1\frac{e^2}{4\pi\varepsilon_0 R} + C_1\frac{B_{MX}}{R^n} \tag{5.10a}$$

where $M_1 = C_1 = 1$. Comparison of equation (5.17) and (5.10a) shows that the two expressions have the same form:

$$V_k(R) = k\left\{\text{IE(M)} - \text{EA(X)} - M_k\frac{e^2}{4\pi\varepsilon_0 R} + C_k\frac{B_{MX}}{R^n}\right\} \tag{5.18}$$

where k is the degree of association: 1 for the monomer and 2 for the dimer.

5.10 Alkali metal halide dimers: calculation of equilibrium bond distances and dissociation energies

Since the value of the Born constant is known, see (5.12), we may use equation (5.17) to calculate both the equilibrium M–X bond distance and the dissociation energy of the dimer.

We first use the fact that the derivative of $V_2(R)$ with respect to R must be zero when R is equal to the equilibrium bond distance which we shall denote by R_2, and derive the equations

$$\frac{C_2B_{MX}}{R_2^n} = \frac{M_2e^2}{4\pi\varepsilon_0 R_2}\left(\frac{1}{n}\right) \tag{5.19}$$

and

$$R_2^{(n-1)} = \left(\frac{C_2}{M_2}\right)nB_{MX}\frac{4\pi\varepsilon_0}{e^2} \tag{5.20}$$

Replacement of R_e in equation (5.12) by R_1 and rearrangement gives the corresponding equation for the monomer:

$$R_1^{(n-1)} = nB_{MX}\frac{4\pi\varepsilon_0}{e^2} \tag{5.21}$$

Division of (5.20) by (5.21) yields

$$\left(\frac{R_2}{R_1}\right)^{n-1} = \frac{C_2}{M_2} \tag{5.22}$$

As we have seen above, the presence of the term representing the Coulomb repulsion between the halide ions reduces the Madelung constant of the dimer from $C_2 = 2$ to 1.293. This means that the bond distance in the dimer will be *larger* than in the monomer. If $n = 10$ we obtain

$$\frac{R_2}{R_1} = 1.05 \tag{5.23}$$

Our calculation based on the spherical ion model thus indicates that M–X bond distances in all dimers should be 5% longer than in the monomers. Quantum chemical calculations, on the other hand, indicate an elongation by about 7%.

We now turn our attention to the calculation of the dissociation energy. The energy of the dimer at the equilibrium bond distance is given by

$$V_2(R_2) = 2\left\{\mathrm{IE(M)} - \mathrm{EA(X)} - M_2\frac{e^2}{4\pi\varepsilon_0 R_2} + \frac{C_2 B_{\mathrm{MX}}}{R_2^n}\right\}$$

or using equation (5.19):

$$V_2(R_2) = 2\left\{\mathrm{IE(M)} - \mathrm{EA(X)} - M_2\frac{e^2}{4\pi\varepsilon_0 R_2}\left(1 - \frac{1}{n}\right)\right\} \tag{5.24}$$

Replacement of R_e in equation (5.14) by R_1 gives the corresponding expression for the energy of the monomer at its equilibrium bond distance:

$$V_1(R_1) = \mathrm{IE(M)} - \mathrm{EA(X)} - \frac{e^2}{4\pi\varepsilon_0 R_1}\left(1 - \frac{1}{n}\right) \tag{5.25}$$

If we disregard the vibrational energies of both monomer and dimer, the dissociation energy of the latter is given by

$$\Delta E_{\mathrm{dis}} = 2V_1(R_1) - V_2(R_2)$$

Using (5.23) and (5.24) we obtain:

$$\begin{aligned}\Delta E_{\mathrm{dis}} &= 2\left\{\frac{-e^2}{4\pi\varepsilon_0 R_1}\left(1 - \frac{1}{n}\right)\right\} - 2\left\{-\left(\frac{e^2}{4\pi\varepsilon_0 R_2}\right)M_2\left(1 - \frac{1}{n}\right)\right\} \\ &= 2\left\{\frac{e^2}{4\pi\varepsilon_0}\left(1 - \frac{1}{n}\right)\right\}\left(\frac{M_2}{R_2} - \frac{1}{R_1}\right)\end{aligned} \tag{5.26}$$

Insertion of $M_2 = 1.293$ and $R_2 = 1.05\ R_1$ gives

$$\Delta E_{\mathrm{dis}} = 2\left\{\frac{e^2}{4\pi\varepsilon_0 R_1}\left(1 - \frac{1}{n}\right)\right\}0.231 \tag{5.26a}$$

Equations (5.26) and (5.27) give the dissociation energy of one dimer molecule. Multiplication with the Avogadro number yields the dissociation energy per mole dimer:

$$\Delta E_{\text{dis}} = 2N_{\text{A}} \left\{ \frac{e^2}{4\pi\varepsilon_0 R_1} \left(1 - \frac{1}{n}\right) \right\} 0.231 \tag{5.26b}$$

Insertion of the experimental bond distance of the monomer yields $\Delta E_{\text{dis}} = 240$ kJ mol^{-1}. The agreement with the experimental value 202 kJ mol^{-1} is only moderate. One reason for the discrepancy may be that we have neglected the Born repulsion between the two halide ions and thus overestimated the stability of the dimer.

Problem 5.7 Assume that the structure of the tetramer M_4X_4 is cubic with M^+ and X^- ions occupying alternating corners. Show that the energy is given by equation (5.18) with $k = 4$, $C_4 = 3$ and $M_4 = 1.456$. Note that the Madelung constant contains a third term that represents the stabilizing interactions between a metal atom cation and the halide anion at diagonally opposed corners of the cube. Use this result to estimate the M–X bond distance ratio R_4/R_1.

5.11 Calculation of the M–X bond distance in crystalline alkali metal halides

All the alkali metal halides except CsCl, CsBr and CsI form cubic crystals with "rock-salt structure," where each metal cation is surrounded by six halide anions, and each anion surrounded by six cations at the corners of an octahedron. See Fig. 5.3. Consultation of a textbook on elementary inorganic or physical chemistry should make it clear that the energy of one mole of the crystalline material relative to the separated, neutral atoms is given by

$$V_{\text{c}}(R) = \text{IE(M)} - \text{EA(X)} - N_{\text{A}} M_{\text{c}} \frac{e^2}{4\pi\varepsilon_0 R} + N_{\text{A}} C_{\text{c}} \frac{B_{\text{MX}}}{R^n} \tag{5.27}$$

where $C_{\text{c}} = 6$ and $M_{\text{c}} = 1.748$.

In order to determine the bond distances in the crystalline phase we proceed in the same fashion as for the dimers: Using the fact that the derivative of $V_{\text{c}}(R)$ with respect to R must be zero when R is equal to the equilibrium bond distance which we shall denote by R_{c}, we

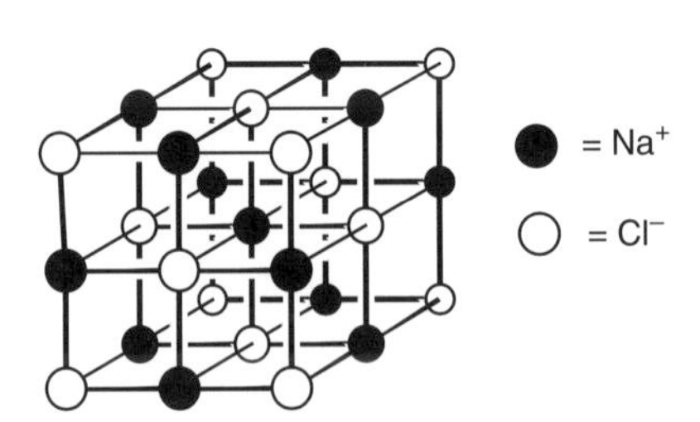

Fig. 5.3. The rock salt structure.

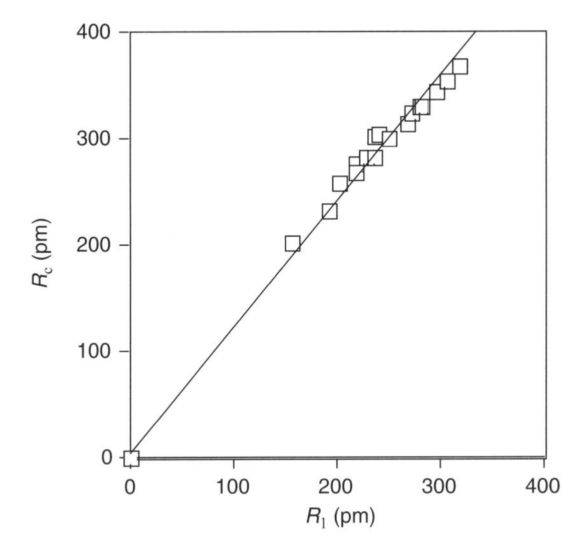

Fig. 5.4. M–X bond distances (R_c) in 17 crystalline alkali metal halides with rock-salt structure as a function of the corresponding bond distance in the gaseous, monomeric MX molecules.

derive the equations

$$\frac{C_c B_{MX}}{R_c^n} = \frac{M_c e^2}{4\pi \varepsilon_0 R_c} \left(\frac{1}{n}\right) \tag{5.28}$$

and

$$R_c^{(n-1)} = \left(\frac{C_c}{M_c}\right) n B_{MX} \frac{4\pi \varepsilon_0}{e^2}. \tag{5.29}$$

Division of (5.29) by equation (5.21) yields

$$\left(\frac{R_c}{R_1}\right)^{n-1} = \frac{C_c}{M_c} \tag{5.30}$$

or

$$\frac{R_c}{R_1} = 1.15$$

Figure 5.4 plots the MX bond distance in the crystal as a function of the bond distance in the monomer. It is seen that the bond distances in the crystal increase linearly with the bond distance in the monomers. The best fit to the points in the figure is, however, obtained by assuming the slope R_c/R_1 to be 1.19 rather than 1.15.

In Table 5.2 we summarize the result of our calculations of bond distances. We see that M–X bond distances increase with increasing aggregation, i.e. with increasing coordination numbers. This elongation is due to Coulomb repulsion between the increasing number of anions around each cation (and to repulsion between the increasing number of cations around each anion).

Table 5.2. Coordination numbers, C, Madelung constants, M, and calculated bond distance ratios, R_k/R_1, for gaseous, monomeric alkali metal halides MX, gaseous square dimers M_2X_2, for cubic tetramers M_4X_4 and for MX crystals with rock-salt structures.

k	C_k	M_k	R_k/R_1
1	1	1	[1]
2	2	1.293	1.05
4	3	1.456	1.08
∞	6	1.748	1.15

Crystalline CsCl, CsBr and CsI all adopt so-called "CsCl structures" in which each cation is surrounded by eight anion and each anion is surrounded by eight cations. The Madelung constants are $M = 1.763$. Calculations similar to those that we have outlined for the rock-salt structure indicate that the bond distances in the crystals should be 18% longer than in the gas phase. The observed elongations decrease from 22% in CsCl to 19% in CsI.

5.12 Ionic radii

When the first crystal structures were determined by X-ray crystallography during the first decades of the twentieth century, several scientists attempted to assign radii to the ions in such a manner that distance between oppositely charged nearest neighbours in the crystal might be estimated from the sum of the ionic radii.

The problem of assigning crystal radii is similar to that of establishing single electrode potentials. Once the standard potential of *one* electrode has been fixed, the magnitude of the others follow. Early estimates were based on the assumption that the radii of six-coordinate O^{-2} and F^- ions are 140 and 132 pm respectively. The latest and most authoritative collection ([6]) is based on the values 126 and 119 pm for the radii of six-coordinate oxide and fluoride ions. These radii differ from traditional ionic radii in being 14 pm smaller for the anions and 14 pm greater for the cations. See Table 5.3.

The bond distances in the *homonuclear* diatomic molecules listed in Table 4.1 indicate that the covalent-bond radii increase in the order

$$\mathrm{F < Cl < Br < I \approx Li < Na < K < Rb < Cs}$$

i.e. that the radius of the largest halogen atom is equal to the radius of the smallest alkali metal atom. Comparison with the ionic radii (for coordination number six) shows that removal of an electron from an alkali metal atom to form a M^+ cation *reduces* the radius by 40 to 50 pm, while addition of an electron to a neutral halogen atom to form an X^- anion *increases* the radius by 50–70 pm. As a result the ionic radii increase in the order

$$\mathrm{Li^+ < Na^+ < K^+ < Rb^+ < Cs^+ \approx F^- < Cl^- < Br^- < I^-}$$

The radius of the largest alkali metal cation is equal to that of the smallest halogen anion.

Problem 5.8 In the crystal structures of the alkali halides the coordination numbers of cations and anions are necessarily equal. Assume that you have a crystal of composition A_aB_b. Assume that each

Table 5.3. Crystal radii for some ions (in pm). The first radius listed for each ion pertains to crystals where the coordination number of the ion is equal to four, the second to crystals where it is equal to six [4]. *Note that these crystal radii differ from of traditional ionic radii by being 14 pm greater for the cations and 14 pm smaller for the anions.*

Li^{+}	Be^{+2}		B^{+3}	O^{-2}	F^{-}
73 / 90	41 / 59		25 / 41	124 / 126	117 / 119
Na^{+}	Mg^{+2}		Al^{+3}	S^{-2}	Cl^{-}
113 / 116	71 / 86		53 / 68	– / 170	– / 167
K^{+}	Ca^{+2}	Zn^{+2}	Ga^{+3}	Se^{-2}	Br^{-}
151 / 152	– / 114	74 / 88	61 / 76	– / 184	– / 182
Rb^{+}	Sr^{+2}	Cd^{+2}	In^{+3}	Te^{-2}	I^{-}
– / 166	– / 132	92 / 109	76 / 94	– / 207	– / 206
Cs^{+}	Ba^{+2}	Hg^{+2}	Tl^{+3}		
– / 181	– / 149	110 / 116	89 / 103		

A atom is surrounded by C_A B atoms, and vice versa: that each B atom is surrounded by C_B A atoms. Find a relation between the coordination numbers C_A and C_B and the stoichiometric coefficients a and b. Hint: One mole of the solid contains a moles of A atoms. How many close contact to B atoms do they have? Write down a similar expression for the number of contacts that the B atoms have to A atoms. What is the relationship between the two numbers?

5.13 The polarizable ion model for the monomeric alkali metal halides

We have seen that the electric dipole moments of the gaseous alkali metal halides estimated from the spherical ion model are significantly larger than the experimental values. The assumption that the ions remain spherical as they are brought close to each other is, in fact, an unreasonable one particularly for the anions which are easily polarized.

In the following we assume that the z-axis runs from the cation to the anion as indicated in Fig. 5.5. This means that the electric dipole moment is negative:

$$\mu_0 = e(z_M - z_X) = -eR$$

The cation then generates an electric field at the anion that is given by

$$\mathcal{E}_a = +\frac{e}{4\pi\varepsilon_0 R^2}$$

(The positive sign indicates that the direction of the field is such as to repel the positive nucleus of the anion and to attract the negative electrons.) Let us assume that this field is constant over the whole anion. The field will then induce an electric dipole moment of

$$\mu_a* = \alpha_a \mathcal{E}_a = \frac{\alpha_a e}{4\pi\varepsilon_0 R^2} \tag{5.31a}$$

where α_a is the polarizability of the anion. A similar expression may be written down for the dipole moment that the anion induces in the cation:

$$\mu_c* = \alpha_c \mathcal{E}_c = \frac{\alpha_c e}{4\pi\varepsilon_0 R^2} \tag{5.31b}$$

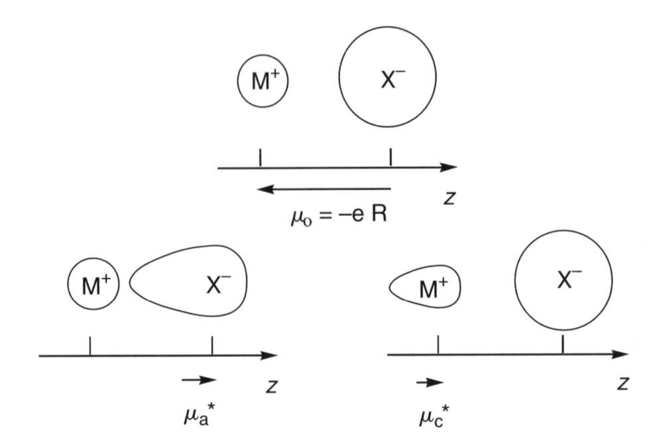

Fig. 5.5. The polarizable ion model. Above: spherical, unpolarized, ions. Below left: polarization of the anion. Below right: polarization of the cation. The direction of the z-axis has been chosen in such a way that the dipole moment of the molecule is negative. This choice of axis, in turn, implies that the induced dipole moments are positive.

Since the polarizabilities of monatomic anions are much larger than the polarizabilities of monatomic cations, the induced dipole moment on the anion is much larger than that induced on the cation.

The total dipole moment of the molecule becomes

$$\mu_{\mathrm{el}} = -eR + \mu_{\mathrm{a}} * + \mu_{\mathrm{c}} * = -eR + \frac{(\alpha_{\mathrm{a}} + \alpha_{\mathrm{c}})e}{4\pi\varepsilon_0 R^2} \tag{5.32}$$

where the signs of μ_{a}^* *and* μ_{c}^* *are such as to reduce magnitude of the total dipole moment of the molecule.* See Fig. 5.5.

Dipole moments calculated from the polarizable ion model, equation (5.32), using the polarizabilities listed in Table 3.4 and experimentally determined bond distances, reproduce the experimental dipole moments with an average deviation of only 5% as compared to 32% for the spherical ion model [5].

Note that since polarization of the anion increases the negative charge density between the ions, polarization may be seen as a first step towards covalent bonding.

5.14 The energy of a pair of polarized ions at infinite distance

We shall now construct a potential energy curve for a polarizable ion pair. The potential energy will be the sum of three terms:

(i) the energy required to produce the polarized ions from the neutral atoms at infinite distance from each other;
(ii) the Coulomb interaction energy of the polarized ions at finite distance;
(iii) the Born repulsion energy when the two ions are close to each other.

We begin by calculating the first term.

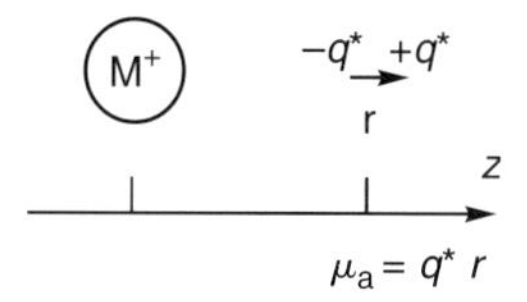

Fig. 5.6. Calculation of the Coulomb interaction energy between the net charge on the cation and the induced dipole moment on the anion. See text.

The energy required to form the unpolarized ions from the neutral atoms is given by IE(M)−EA(X) as before. The energy used to polarize an atom or ion is given by equation (3.16):

$$E_{\text{pol}} = \frac{\mu^{*2}}{2\alpha}$$

The energy required to polarize the anion is thus

$$E_{\text{pol,a}} = \frac{\mu_{\text{a}}^{*2}}{2\alpha_{\text{a}}} = \frac{1}{2}\frac{\alpha_{\text{a}} e^2}{(4\pi\varepsilon_0 R^2)^2} \tag{5.33a}$$

while the energy required to polarize the cation is

$$E_{\text{pol,c}} = \frac{\mu_{\text{c}}^{*2}}{2\alpha_{\text{c}}} = \frac{1}{2}\frac{\alpha_{\text{c}} e^2}{(4\pi\varepsilon_0 R^2)^2} \tag{5.33b}$$

The energy of the polarized ions at infinite distance is thus

$$V(R=\infty) = \text{IE(M)} - \text{EA(X)} + \frac{1}{2}(\alpha_{\text{a}} + \alpha_{\text{c}})\frac{e^2}{(4\pi\varepsilon_0 R^2)^2} \tag{5.34}$$

5.15 The Coulomb interaction energy of two polarized ions at distance *R*

The total Coulomb energy of the ion pair may be written as the sum of four terms:

(i) the Coulomb interaction energy between the net charges of the two atoms; this energy is equal to $-e^2/(4\pi\varepsilon_0 R)$ as before;
(ii) the Coulomb interaction energy between the net charge of the cation and the induced dipole moment on the anion;
(iii) the Coulomb interaction energy between the net charge of the anion and the induced dipole moment on the cation;
(iv) the Coulomb interaction energy between the two induced dipole moments.

In Fig. 5.6 we represent the induced dipole moment on the anion as two charges, $+q^*$ and $-q^*$ separated by a small distance r. Note that the cation is situated on the line through these charges and closer to the negative than to the positive pole.

The energy due to attraction between the cation and the induced dipole moment on the anion may be calculated as follows:

$$E(\mathrm{M}^+, \mu_\mathrm{a}^*) = \frac{e(-q^*)}{4\pi\varepsilon_0(R - r/2)} + \frac{e(+q^*)}{4\pi\varepsilon_0(R + r/2)}$$
$$= \frac{-eq^*[(R + r/2) - (R - r/2)]}{4\pi\varepsilon_0(R + r/2)(R - r/2)}$$
$$\approx \frac{-eq^*r}{4\pi\varepsilon_0 R^2}$$

or, since $q^*r = \mu_\mathrm{a}^*$,

$$E(M^+, \mu_\mathrm{a}^*) = \frac{-e\mu_\mathrm{a}^*}{4\pi\varepsilon_0 R^2}$$

Using (5.31a) to eliminate μ_a^* we obtain

$$E(\mathrm{M}^+, \mu_\mathrm{a}^*) = \frac{-\alpha_\mathrm{a} e^2}{(4\pi\varepsilon_0 R^2)^2} \qquad (5.35a)$$

Similarly the energy of interaction between the net negative charge on the anion and the induced dipole moment on the cation is given by

$$E(\mathrm{M}^-, \mu_\mathrm{c}^*) = \frac{-e\mu_\mathrm{c}^*}{4\pi\varepsilon_0 R^2}$$

$$E(M^-, \mu_\mathrm{c}^*) = \frac{-\alpha_\mathrm{c} e^2}{(4\pi\varepsilon_0 R^2)^2} \qquad (5.35b)$$

Comparison of equations (5.35a) and (5.35b) with equations (5.33a) and (5.33b) shows that *the monopole/dipole Coulomb energy released when the polarized ions are brought together is exactly twice as large as the energy required to polarize the ions at infinite distance from each other.*

Problem 5.9 Show that the energy of interaction between two dipoles pointing in the same direction, like μ_a^* and μ_c^* in Fig. 5.2, is given by

$$E(\mu_\mathrm{a}^*, \mu_\mathrm{c}^*) = \frac{-2\mu_\mathrm{a}^*\mu_\mathrm{c}^*}{(4\pi\varepsilon_0 R^3)} = \frac{-2\alpha_\mathrm{a}\alpha_\mathrm{c} e^2}{(4\pi\varepsilon_0)^3 R^7} \qquad (5.36)$$

Hint: Replace μ_c^* by two charges $\pm q^*$ separated by a small distance r, and calculate the energy of Coulomb interaction of these charges with the dipole μ_a^*.

The magnitudes of the Coulomb interaction energies decrease strongly along the series

$$E(\mathrm{M}^+, \mathrm{X}^-) \gg E(\mathrm{M}^+, \mu_\mathrm{a}^*) > E(M^-, \mu_\mathrm{c}^*) \gg E(\mu_\mathrm{a}^*, \mu_\mathrm{c}^*)$$

and the last term is often neglected.

Collection of terms gives the total energy of the ion pair

$$V(R) = \mathrm{IE(M)} - \mathrm{EA(X)} + E_{\mathrm{pol,c}} + E_{\mathrm{pol,a}} - \frac{e^2}{4\pi\varepsilon_0 R} + E(\mathrm{M}^+, \mu_{\mathrm{a}}^*) + E(\mathrm{M}^-, \mu_{\mathrm{c}}^*) + E(\mu_{\mathrm{a}}^*, \mu_{\mathrm{c}}^*) + V_{\mathrm{rep}}(R)$$

or

$$V(R) = \mathrm{IE(M)} - \mathrm{EA(X)} - \frac{e^2}{4\pi\varepsilon_0 R} - \frac{1}{2}(\alpha_{\mathrm{a}} + \alpha_{\mathrm{c}}) - \frac{e^2}{(4\pi\varepsilon_0 R^2)^2} - 2\alpha_{\mathrm{a}}\frac{\alpha_{\mathrm{c}} e^2}{(4\pi\varepsilon_0)^3 R^7} + V_{\mathrm{rep}}(R) \quad (5.37)$$

where $V_{\mathrm{rep}}(R)$ represents the Born repulsion energy.

The most recent calculations used a rather complicated expression for the repulsion energy:

$$V_{\mathrm{rep}}(R) = (A/R^3)e^{-bR}$$

The constants A and b were eliminated by demanding that $(\mathrm{d}V(R)/\mathrm{d}R)$ is equal to zero at the experimentally determined bond distance R_{e} and that the second derivative $(\mathrm{d}^2V(R)/\mathrm{d}R^2)$ evaluated at R_{e} is equal to the experimentally determined force constant[5].

Under these conditions the dissociation energies D_0 calculated for the polarizable ion model reproduce the experimental values with an average error of about 6% as compared to the 9% obtained for the spherical ion model.

We have seen that the electric dipole moments of the gaseous alkali metal halides are strongly dependent on the shape of the ions and can only be estimated with accuracy if the model includes polarization effects. The dissociation energies, on the other hand, depend mainly on the Coulomb attraction between the net charges on the ions and may therefore be calculated with reasonable accuracy without including polarization.

5.16 Electrides and alkalides

Ionic compounds with a free electron serving as anion are known as electrides. The crystal structure of a typical electride is shown in Fig. 5.7. The cation consists of a Cs^+ ion sandwiched between two cyclic polyethers, that is two 15-membered rings consisting of 10 CH_2 groups and five O atoms symmetrically placed around the cycle. This polyether is referred to as 15-crown-5: the first number indicates the total number of atoms in the ring, the second the number of oxygen atoms. The Cs^+ ion is bonded to 10 O atoms at an average distance of 315 pm. The electride anions (electrons) are moving around in cavities with radii of about 220 pm which are found between the large cations. The thermodynamic stability of the electrides show that the energy required to remove the electron from the metal atom, i.e. the ionization energy, has been compensated by the energy released during formation of 10 electron donor–acceptor bonds between the cation and the O atoms. We shall return to a discussion of electron donor–acceptor bonds in Chapter 16.

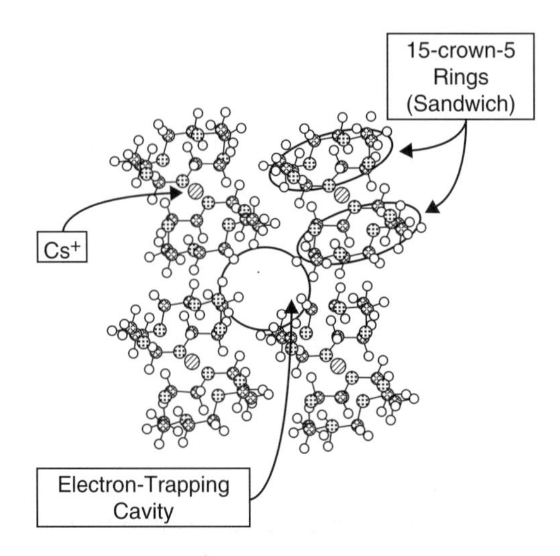

Fig. 5.7. The crystal structure of $[Cs^+(15\text{-crown-}5)_2]e^-$. Reproduced with permission from J. L. Dye, *Inorg. Chem.*, 36 (1997) 3816. Copyright (1997) American Chemical Society.

Ionic compounds containing alkali metal anions, M^-, are known as alkalides. More than 30 such compounds with Cs, Rb, K or Na anions are known and the structures of a few have been determined by X-ray crystallography. The cation in $[Cs^+(18\text{-crown-}6)_2]Cs^-$ is similar to the $[Cs^+(15\text{-crown-}5)_2]$ cation in the electride, but the radius of the cavity containing the anion has been increased to about 310 pm[6].

Formation of the alkalides is favored both by the formation of donor–acceptor bonds between the cation and the crown ethers and by the non-negligible electron affinity of the metal atoms. See Table 3.3.

Problem 5.10 It has been said that the Cs^- ion is "the largest monatomic ion known to man." Does this strike you as a reasonable claim? Do you see any hope of finding a larger monatomic ion in the future? Give reasons for your answers.

References

[1] Electric dipole moments, bond distances, vibrational frequencies and dissociation energies from K. P. Huber and G. Herzberg, *Molecular Spectra and Molecular Structure*, Vol. IV, *Constants of Diatomic Molecules*, Van Nostrand, New York 1979. The dissociation energy of NaF has, however, been taken from P. Brunner and M. Karplus, *J. Chem. Phys.*, 58 (1973) 3903.

[2] B. Trembley, L. Manceron, P. Roy, A.-M. LeQuére and D. Roy, *Chem. Phys. Lett.*, 228 (1994) 410.

[3] J. Modisette, L. Lou and P. Nordlander, *J. Chem. Phys.*, 101 (1994) 8903.

[4] The ionic radii in Table 5.3 have been taken from R. D. Shannon, *Acta Cryst.*, A32 (1976) 751.

[5] S. Hati, B. Datta and D. Datta., *J. Phys. Chem.*, 100 (1996) 19808.

[6] R. H. Huang, D. L. Ward, M. E. Kuchenmeister and J. L. Dye, *J. Am. Chem. Soc.*, 109 (1987) 5561.

Chapter 6

Other heteronuclear diatomic molecules: polar covalent bonds

Introduction

The electric dipole moments, dissociation energies at zero K, equilibrium bond distances, vibrational frequencies and force constants of the gaseous monomeric hydrogen halides, HX(g), are listed in Table 6.1 along with the corresponding parameters for the diatomic interhalogen compounds, XX′(g). Our knowledge of the gaseous alkali metal hydrides, MH(g), is less complete: dipole moments are unknown for all of them except LiH(g). Information about the 12 gaseous diatomic inter-alkali metal compounds, MM′(g), is even more spotty: the equilibrium bond distances and dissociation energies of RbLi(g), CsLi(g) and CsK(g) are all unknown.

We shall define a single bond in a molecule as homolytic if minimum-energy rupture in the gas phase yields neutral, paramagnetic species (atoms or molecules with one or more unpaired electrons). As already pointed out, minimum-energy dissociation of diatomic molecules in the gas phase will always yield neutral atoms. Since the hydrogen, halogen and alkali metal atoms all contain an odd number of electrons, the molecules in Tables 5.1 and 6.1 provide us with a sample of 45 heteronuclear diatomic molecules containing single homolytic bonds.

The polarity of the bonds may be assessed by using the experimental electric dipole moments and equilibrium bond distances to calculate their so-called "ionic character:" this ionic character is found to increase with increasing difference between the electronegativity coefficients of the two elements. We then go on to show that dissociation energies increase and bond distances decrease with increasing bond polarity. Finally we introduce a formula for the calculation of single bond distances between p-block elements from their bonding radii and electronegativity coefficients.

6.1 Electric dipole moments and ionic characters

All heteronuclear diatomic molecules AB have non-zero dipole moments. In the following we shall denote the atom carrying positive charge as A. The magnitude of the positive charge is expected to increase with increasing electronegativity of the bonding partner B, and with decreasing ability of A to hold on to its own electrons, i.e. with decreasing electronegativity of A.

As we have seen in the preceding chapter, the properties of the gaseous monomeric alkali metal halides are in very good agreement with those predicted by the polarizable ion model. Their experimental dipole moments are all very large, ranging from 6.28 D in LiF to 11.69 D in CsI. The electric dipole moment of LiH, 5.88 D, is nearly as large as

Table 6.1. Gaseous heteronuclear diatomic molecules AB(g): hydrogen halides HX; interhalogen compounds XX′; alkali metal hydrides MH, and inter-alkali-metal compounds MM′. Electric dipole moments, μ_{el}; ionic characters, q_{IC}; equilibrium bond distances, R_e; vibrational wavenumbers, ω; dissociation energies at zero K, D_0; reduced masses of the predominant isotopomers, μ_M; and force constants, f_R.

AB(g)	μ_{el} (D)	q_{IC} (a.u.)	R_e (pm)	ω(cm^{-1})	D_0 (kJ mol^{-1})	μ_M (a.u.)	f_r(Nm^{-1})
H–F	1.83	0.41	91.7	4138	566	0.957	966
H–Cl	1.11	0.18	127.5	2991	428	0.980	516
H–Br	0.83	0.12	141.4	2649	363	0.995	411
H–I	0.45	0.06	160.9	2309	295	1.000	314
Cl–F	0.89	0.11	162.8	786	253	12.31	448
Br–F	1.29	0.15	175.9	671	249	15.31	406
I–F	1.95	0.21	191.0	610	278	16.52	362
Br–Cl	0.52	0.05	213.6	444	216	24.23	282
I–Cl	1.24	0.11	232.1	384	208	27.41	239
I–Br	0.74	0.06	246.9	269	175	48.66	207
Li–H	5.88	0.77	159.6	1406	234	0.881	103
Na–H	–	–	188.7	1172	181	0.965	78
K–H	–	–	224.2	984	168	0.982	56
Rb–H	–	–	236.8	937	165	0.996	52
Cs–H	–	–	249.2	891	172	1.000	47
Na–Li	0.46	0.03	288.5	257	83	5.376	21
K–Li	3.51	0.24	331.9	210	73	5.945	15
Rb–Li	4.05	–	–	–	–	6.481	–
Cs–Li	6.30	–	–	–	–	6.667	–
K–Na	2.69	0.20	350.0	124	62	14.46	13
Rb–Na	3.10	0.20	364.3	106	60	18.09	12
Cs–Na	4.75	0.32	385.0	99	57	19.62	11
Rb–K	0.20	0.01	406.9	76	50	26.71	9
Cs–K	2.58	–	–	–	–	30.18	–
Cs–Rb	2.39	0.11	437.0	50	45	51.96	8

that of LiF(g): LiH and all the other alkali metal hydrides are believed to be as ionic as the alkali metal halides. The electric dipole moments of the hydrogen halides range from 0.45 D in HI to 1.83 D in HF, and those of the interhalogen compounds from 0.52 D in BrCl to 1.95 D in IF. These compounds are all gaseous at normal temperature and pressure and are commonly described as *polar covalent*. The dipole moments of the inter-alkali metal compounds range from 0.20 D in RbK to 6.30 D in CsLi: though the dipole moment of the latter is unexpectedly large, other molecular properties show that the bonding is polar covalent rather than ionic.

Figure 6.1 A shows the observed dipole moments of the molecules in Tables 5.1 and 6.1 as a function of the difference between the electronegativity coefficients of the two atoms, $\Delta\chi = \chi_B - \chi_A$. The graph indicates that there is a general tendency towards higher dipole moments with increasing electronegativity difference, but the correlation is not impressive: the linear correlation coefficient which would be equal to unity if all the points were lying on a straight line, is only 0.65.

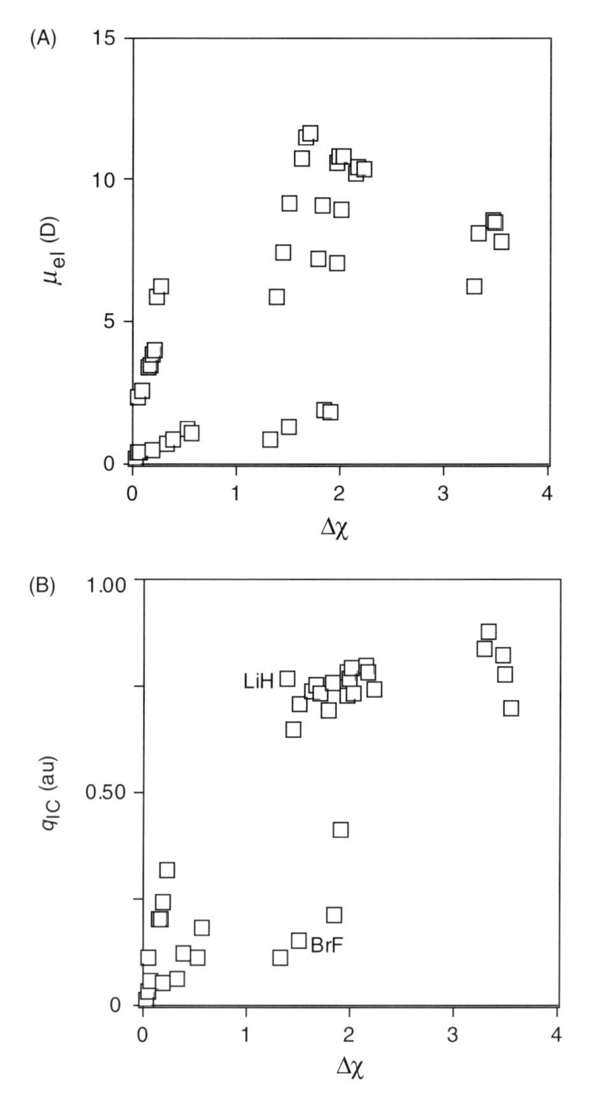

Fig. 6.1. A: The electric dipole moments (in Debye) of 41 heteronuclear diatomic molecules AB(g), A and B = hydrogen, halogen, or alkali metal as a function of the electronegativity difference $\Delta\chi = \chi_B - \chi\chi_A$. B: The ionic characters of the same molecules as a function of $\Delta\chi$.

If we assume the molecules to consist of spherical, non-penetrating atoms, the magnitude of the dipole moment is given by

$$|\mu_{el}| = Q_A R$$

where Q_A is the net positive charge on atom A. See equation (5.6). A large dipole moment may therefore reflect a large bond distance rather than a large net atomic charge. A rough estimate of the net atomic charge in a diatomic molecule may be obtained by dividing the observed dipole moment by the bond distance:

$$Q_A = \frac{|\mu_{el}|}{R}$$

Further division by the elementary charge yields the net charge in atomic units. This charge estimate has traditionally been referred to as the "ionic character" of the molecule and we denote it therefore by q_{IC}:

$$q_{IC} = \frac{|\mu_{el}|}{eR} \tag{6.1}$$

The ionic characters of the gaseous alkali metal halides range from 0.65 to 0.84, while the ionic character of LiH is 0.77. The ionic charges of the remaining molecules in Table 6.1 vary from 0.01 in RbK to 0.41 in HF. Figure 6.1B shows the ionic characters as a function of $\Delta\chi$. The correlation is much improved, linear regression analysis yields a correlation coefficient of 0.82, but is still far from perfect. Thus LiH with $\Delta\chi = 1.39$ has ionic character = 0.77, while BrF with a slightly larger electronegativity difference $\Delta\chi = 1.51$ has an ionic character of only 0.15!

For the lack of something better, we shall nevertheless regard the difference between the electronegativity coefficients as an indicator of bond polarity.

6.2 Bond strength and bond polarity

What determines the strength of a chemical bond? Does each atom possess a certain bonding capacity so that the strength of a bond is determined by the sum – or perhaps the product – of the bonding capacities of the two atoms? In the first case the dissociation energy of the heteronuclear diatomic molecule AB(g) should be equal to the arithmetic mean of the bond dissociation energies of A_2(g) and B_2(g). In the second case the bond energy of AB should be equal to the geometric mean of the bond energies of A_2 and B_2.

In Fig. 6.2A we plot the observed bond dissociation energies D_0 of 42 heteronuclear diatomic molecules AB(g) from Tables 5.1 and 6.1 as a function of the arithmetic mean of the dissociation energies of A_2 and B_2 taken from Table 4.1. If the dissociation energy of A–B were equal to the arithmetic mean, all points would fall on a straight line with slope equal to unity. The bond energies of the five alkali metal hydrides MH are, however, significantly *smaller* than estimated from the arithmetic mean, the discrepancies range from 30 to 75 kJ mol^{-1}. The bond energies of the inter-alkali metal compounds MM$'$ are found to estimated to within 3 kJ mol^{-1}, while the rest of the molecules have bonds which are much stronger than estimated.

In Fig. 6.2B we plot the observed bond dissociation energies in AB as a function of the geometrical mean of the dissociation energies of A_2 and B_2, that is the square root of the product $D_0(A_2) \times D_0(B_2)$. The bond energies of six molecules are estimated to within $\pm$3 kJ mol^{-1}, namely LiNa, LiK, NaK, NaCs, KRb, RbCs, and ClBr. *All the other bonds are stronger than estimated from the geometric mean.*

The strongest bonds listed in Tables 5.1 and 6.1 are those of LiF, $D_0 = 570$ kJ mol^{-1}, and HF, $D_0 = 566$ kJ mol^{-1}. The H–F bond is twice as strong as calculated from the geometric mean, the Li–F bond is four times stronger!

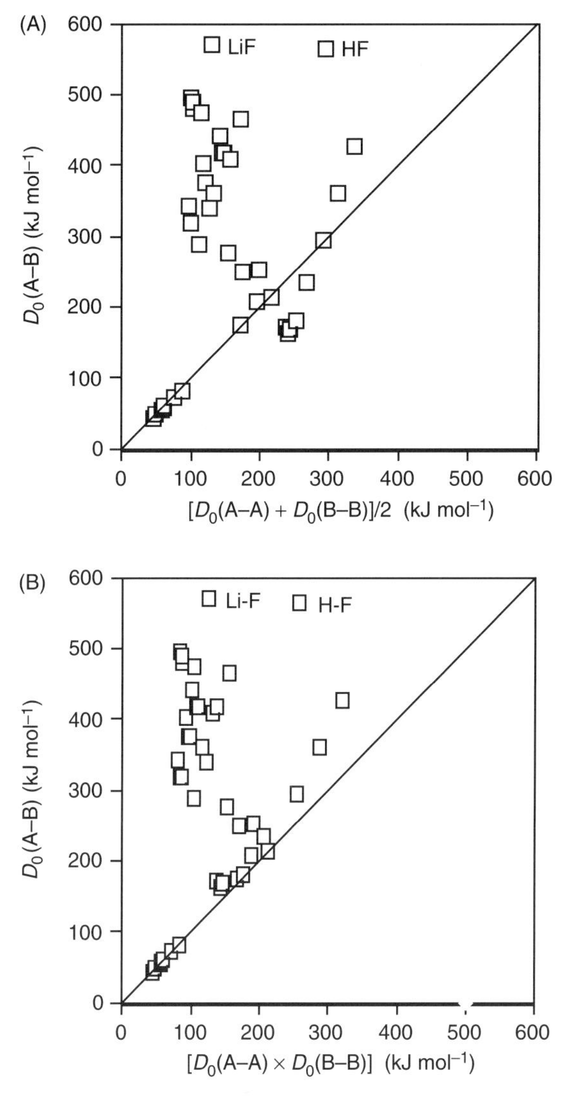

Fig. 6.2. A: The dissociation energies of 42 heteronuclear diatomic molecules AB(g) as a function of the *arithmetic mean* of the dissociation energies of A_2(g) and B_2(g). B: Dissociation energies of heteronuclear diatomic molecules AB(g) as a function of the *geometric mean* of the dissociation energies of A_2(g) and B_2(g).

Comparison of the experimental dissociation energies of the heteronuclear molecules with the *geometric* mean of the dissociation energies of the homonuclear molecules gives the simpler result since all A–B dissociation energies are either equal to or larger than estimated. In the following we shall base our discussion on the assumption that the geometric mean of the dissociation energies of A_2(g) and B_2(g) is more meaningful than the arithmetic. We define the *excess bond strength* in a diatomic molecule AB(g) as the difference between the

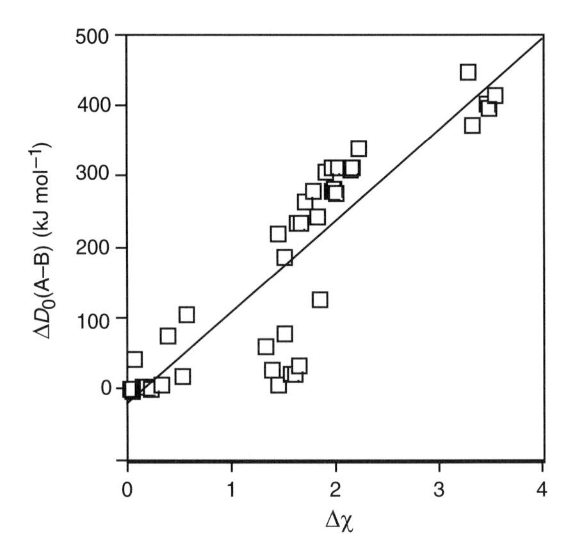

Fig. 6.3. Excess bond strength, ΔD_0(A–B), as a function of the electronegativity difference $\Delta\chi$.

observed dissociation energy of AB(g) and the geometric mean of the dissociation energies in A_2(g) and B_2(g):

$$\Delta D_0(\text{A–B}) = D_0(\text{A–B}) - \sqrt{[D_0(\text{A–A}) \times D_0(\text{B–B})]}$$

Figure 6.3 shows the excess bond strength of the heteronuclear molecules in Tables 5.1 and 6.1 as a function of the difference between the electronegativity coefficients. Linear regression analysis yields a correlation coefficient of 0.87. A plot of the excess bond strength as a function of the ionic character q_{IC} yields a linear correlation coefficient of 0.85. *There is a clear tendency for the bond strength to increase with increasing bond polarity.* As indicated by the examples of HF and LiF, the polarity effect may in fact be as large – or larger than – the bond energy estimated from the geometric mean.

6.3 Bond distances and bond polarities

Is the length of a chemical bond determined simply by the sum of the bonding radii of the two atoms? If this is the case, the bond distance in a heteronuclear diatomic AB should be equal to the arithmetic mean of the bond distances in A_2 and B_2.

$$R(\text{A–B}) = 1/2[R(\text{A–A}) + R(\text{B–B})]$$

In Fig. 6.4 we plot the observed bond distances in 42 heteronuclear diatomic molecules as a function of the arithmetic mean of the bond distances in A_2(g) and B_2(g). If the hypothesis is valid, the points should fall on a straight line with slope equal to unity. It is found that the bond distances of nine molecules are estimated to within 3 pm by calculating the arithmetic mean. *The bond distances in the remaining molecules are all significantly shorter than*

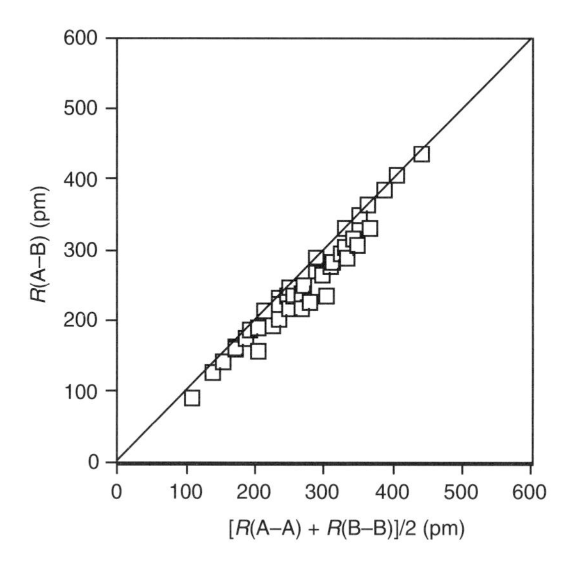

Fig. 6.4. Bond distances in 42 heteronuclear diatomic molecules AB(g) as function of the arithmetic mean of the bond distances in A_2(g) and B_2(g).

estimated from the arithmetic mean. For LiH, CsH, HF, HBr and IF the discrepancies are of the order of 10 pm or larger, in the alkali metal halides the discrepancies range from 15 to 70 pm. We define the *bond shortening* in a heteronuclear diatomic molecule as the difference between the arithmetic mean of the bond distances in A_2(g) and B_2(g) and the observed bond distance in AB(g):

$$-\Delta R(\text{A–B}) = \frac{1}{2}[R(\text{A–A}) + R(\text{B–B})] - R(\text{A–B})$$

Figure 6.5 shows the bond shortening a function of the difference between the electronegativity coefficients. Linear regression analysis indicates the relationship

$$-\Delta R(\text{A–B}) = \Delta\chi \times 15\text{pm}$$

and yields a correlation coefficient of 0.89. *There is a clear tendency for the bond to become shorter with increasing electronegativity difference, i.e. with increasing bond polarity.*

Statistical correlation between variables does not prove that there is a causal connection between them. We shall nevertheless assume that there is such a connection: that the bonds are strong because they are polar, and short because they are strong. According to the spherical ion model, the exceptional strength of the MX bonds in the monomeric alkali metal halide molecules is due to Coulomb attraction between the net ionic charges. Similarly the enhanced stability of polar covalent molecules might be interpreted as due to Coulomb attraction between the (non-integer) net atomic charges. Be that as it may, we shall assume that the stability of a polar covalent bond will increase with increasing negative values of the product of the net charges on the two atoms.

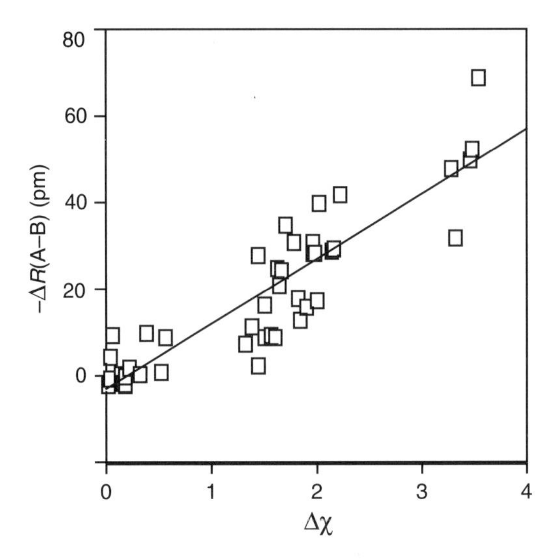

Fig. 6.5. Bond shortening, $-\Delta R$(A–B), as a function of the electronegativity difference $\Delta\chi$.

6.4 Single bond distances in polyatomic molecules

As long ago as in 1941, Schomaker and Stevenson suggested that single bond distances should be given by

$$R(\text{A–B}) = r_A + r_B - c|\chi_A - \chi_B|$$

where r_A and r_B were the "covalent radii" of the two elements, χ_A and χ_B their electronegativity coefficients, and the constant $c = 9$ pm. Since then it has become abundantly clear that the length of a single bond in a polyatomic molecule, R(A–B), does not only depend on the bonding radii and electronegativities of the two atoms but on the entire molecular environment. Among the significant variables are the following:

(i) The valences of the atoms: Sn-Cl bond distances in $SnCl_4$ and $SnCl_2$ differ by 7 pm.
(ii) Multiple bonding to other atoms: the C–C single bond distances in $H_3C–CH_3$ and $H_3C–C \equiv C–CH_3$ differ by 14 pm.
(iii) Steric effects: due to repulsion between Br atoms the C–Ge bond in $C(GeBr_3)_4$ are 7 pm longer than in $C(GeH_3)_4$.
(iv) Electronic effects: replacement of the two H atoms in hydrogen peroxide by F atoms reduces the O–O bond distance by 25 pm.
(v) Intermolecular interactions in condensed phases: the BiCl bond distance in crystalline $BiCl_3$ is 8 pm longer than in the gas phase.

6.5 A modified Schomaker–Stevenson rule for the prediction of single bond distances between *p*-block elements

In the following we shall confine our discussion to bond distances between "Lewis-valent" atoms, i.e. trivalent atoms in Group 13, tetravalent atoms in Group 14, trivalent in Group

15, divalent in Group 16, and monovalent in Group 17. The 16 *p*-block elements within the square in the periodic table defined by the elements C, F, Sn and I plus hydrogen may be combined to form 153 types of single bonds, 17 homonuclear and 136 heteronuclear. Table 6.2 contains a list of 100 reference bond distances taken from accurate gas-phase structures of simple molecules, that is molecules without double or triple bonds, without two or more bulky ligands bonded to the same atom, and – in order to avoid electronic effects – without two or more fluorine atoms[1]. No satisfactory reference bond distance could be found for the remaining 53 bonding combinations.

With the passing years it has become clear that the original Schomaker–Stevensen rule underestimates the bond shortening for large electronegativity differences, say $|\chi_A - \chi_B| > 1$. This difficulty may be overcome by a simple modification:

$$R(\text{A–B}) = r_A + r_B - c|\chi_A - \chi_B|^{1.5} \tag{6.2}$$

The constant c and the bonding radii of the 17 elements H, C, Si, Ge, Sn, N, P, As, Sb, O, S, Se, Te, F, Cl, Br and I have been adjusted to yield the best possible agreement with the 100 reference bond distances in Table 6.2.

The best fit is obtained with the constant c = 8.3 pm and the "normal bond radii" listed in Table 6.3. These radii and the modified Schomaker–Stevenson (MSS) relationship,

Table 6.2. Reference bond distances (in pm) for single, homolytic bonds between the 17 elements, H, C, Si, Ge, Sn, N, P, As, Sb, O, S, Se, Te, F, Cl, Br and I. Bond distances are listed only once. Thus the prototype C–Si bond distance is listed under C and not repeated under Si.[a]

C: 17 bond distances		P: 9 bond distances	
C–H in CH_4	109.4	P–H in PH_3	142.0
C–C in $C(CH_3)_4$	153.9	P–P in P_2H_4	221.9
C–Si in $Si(CH_3)_4$	187.5	P–S in Me_2PSMe	211.1
C–Ge in $Ge(CH_3)_4$	195.8	P–Cl in PCl_3	204.3
C–Sn in $Sn(CH_3)_4$	214.4	P–Br in PBr_3	222.0
C–N in $N(CH_3)_3$	145.4		
C–P in $P(CH_3)_3$	184.7	As: 7 bond distances	
C–As in $As(CH_3)_3$	196.2	As–H in AsH_3	151.1
C–Sb in $Sb(CH_3)_3$	216.3	As–As in As_2Me_4	243.3
C–O in $O(CH_3)_2$	141.6	As–Cl in $AsCl_3$	216.5
C–S in $S(CH_3)_2$	180.5	As–Br in $AsBr_3$	232.9
C–Se in $Se(CH_3)_2$	194.3	As–I in AsI_3	255.7
C–Te in $Te(CH_3)_2$	214.2		
C–F in FCH_3	138.3	Sb: 10 bond distances	
C–Cl in $ClCH_3$	178.5	Sb–H in SbH_3	170.4
C–Br in $BrCH_3$	193.3	Sb–Sb in Sb_2Me_4	281.8
C–I in ICH_3	213.2	Sb–S in $S(SbMe_2)_2$	241.4
		Sb–Se in $Me_2SbSeMe$	255.5
Si: 15 bond distances		Sb–Te in $Te(SbMe_2)_2$	278.1
Si–H in SiH_4	148.1	Sb–Cl in $SbCl_3$	232.3
Si–Si in Si_2H_6	233.1	Sb–Br in $SbBr_3$	249.0
Si–Ge in H_3SiGeH_3	235.7	Sb–I in SbI_3	271.9

Table 6.2. (*Continued*)

Si–N in $N(SiH_3)_3$	173.4		
Si–P in $P(SiH_3)_3$	224.8	O: 10 bond distances	
Si–As in $As(SiH_3)_3$	235.5	O–H in H_2O	95.8
Si–Sb in $Sb(SiH_3)_3$	255.7	O–O in HOOH	147.5
Si–O in $O(SiH_3)_2$	163.4	O–S in HOSH	166.2
Si–S in $S(SiH_3)_2$	213.6	O–F in FOH	144.2
Si–Se in $Se(SiH_3)_2$	227.4	O–Cl in Cl_2O	167.3
Si–F in $FSiH_3$	159.1		
Si–Cl in $ClSiH_3$	204.8	S: 10 bond distances	
Si–Br in $BrSiH_3$	221.0	S–H in SH_2	133.6
Si–I in $ISiH_3$	243.7	S–S in S_2H_2	205.5
		S–Cl in $ClSCH_3$	203.7
Ge: 13 bond distances			
Ge–H in GeH_4	152.5	Se: 7 bond distances	
Ge–Ge in Ge_2H_6	240.3	Se–H in SeH_2	146.0
Ge–N in $N(GcH_3)_3$	183.6	Se–Se in Se_2Me_2	232.6
Ge–P in $P(GeH_3)_3$	230.8	Se–Cl in $SeCl_2$	215.7
Ge–O in $O(GeH_3)_2$	176.6		
Ge–S in $S(GeH_3)_2$	220.9	Te; 5 bond distances	
Ge–Se in $Se(GeH_3)_2$	234.4	Te–H in TeH_2	165.8
Ge–F in $FGeH_3$	173.2	Te–Te in Te_2Me_2	268.6
Ge–Cl in $ClGeH_3$	215.0	Te–Cl in $TeCl_2$	232.9
Ge–Br in $BrGeH_3$	229.7		
Ge–I in $IGeH_3$	250.8	F; 9 bond distances	
		F–H in HF	91.7
Sn: 8 bond distances		F–Cl in ClF	162.8
Sn–H in SnH_4	171.1	F–Br in BrF	175.9
Sn–Sn in Sn_2Me_6	277.6	F–I in IF	190.9
Sn–N in $N(SnMe_3)_3$	204.1		
Sn–O in $O(SnMe_3)_2$	194.0	Cl; 17 bond distances	
Sn–Cl in $ClSnH_3$	232.7	Cl–H in HCl	127.4
Sn–Br in $BrSnH_3$	246.9	Cl–Cl in Cl_2	198.9
Sn–I in $ISnH_3$	267.4	Cl–Br in BrCl	213.6
		Cl–I in ICl	232.1
N: 10 bond distances			
N–H in NH_3	101.2	Br: 13 bond distances	
N–N in N_2H_4	144.9	Br–H in HBr	141.4
N–O in H_2NOH	145.3	Br–Br in Br_2	228.4
N–S in $S(NMe_2)_2$	168.8	Br–I in IBr	246.9
N–F in FNH_2	143.3		
N–Cl in NCl_3	175.9	I; 11 bond distances	
		I–H in HI	160.9
		I–I in I_2	266.7

[a] Me = CH_3

Table 6.3. Bonding radii (r in pm) and Allen electronegativity coefficients (χ) of p-block elements for calculation of single bond distances with the MSS rule:[a]

$$R(\text{A–B}) = r_A + r_B - (8.3\ \text{pm})|\chi_A - \chi_B|^{1.5}$$

Carbon-based radii are indicated by asterisks.

B	*C*	*N*	*O*	*F*
82*; 2.05	78.4; 2.54	71.9; 3.07	73.0; 3.61	77.2; 4.19
Al	Si	P	S	Cl
125*; 1.61	114.1; 1.92	109.5; 2.25	102.8; 2.59	100.5; 2.87
Ga	Ge	As	Se	Br
124*; 1.76	121.1; 1.99	121.2; 2.21	116.3; 2.42	113.9; 2.69
In	Sn	Sb	Te	I
145*; 1.66	140.3; 1.82	140.6; 1.98	135.6; 2.16	133.1; 2.36
Tl	Pb	Bi		
148*; 1.79	150*; 1.85	151*; 2.01		

[a] For hydrogen $r = 32.3$ pm and $\chi = 2.30$.

equation (6.2), reproduce the 100 reference bond distances with an average deviation of 2.0 pm and a maximum deviation of 5.6 pm in $O(SiH_3)_2$. The 17 bond distances involving carbon atoms were reproduced with average and maximum deviations of 1.2 and 2.2 pm respectively.

6.6 Carbon-based bonding radii

The bonding radii of the 17 elements discussed in the preceding section were obtained by adjustment to at least five experimental reference bond distances for each for element. Much less information is available about bond distances to the sixth-period elements Pb and Bi or to the Group 13 elements A = B, Al, Ga, In and Tl. Bonding radii for these elements were therefore calculated from the observed bond distances in $Pb(CH_3)_4$, $Bi(CH_3)_3$ and trimethyl derivatives of the Group 13 elements using the MSS rule. The resulting "carbon-based" bonding radii are listed in Table 6.2.

References

[1] The standard reference for gas phase structures of polyatomic molecules is Landolt-Börnstein, *Numerical Data and Functional Relationships in Science and Technology*, New Series, Volume II/7, J. H. Callomon, E. Hirota, K. Kuchitsu, W. J. Lafferty, A. G. Maki, and C. S. Pote, Eds.; Volume II/15, J. H. Callomon, E. Hirota, T. Iijima, K. Kuchitsu, and W. J. Lafferty, Eds.; Volume II/21, E. Hirota, T. Iijima, K. Kuchitsu, W. J. Lafferty, and D. A. Ramsay, Eds.; and Volume II/23, K. Kuchitsu, Ed., Springer Verlag, Berlin.

Chapter 7

Lewis' "cubical atom" model. Molecular orbital calculations on the one-electron molecule H_2^+ and the two-electron molecule H_2

Introduction

The modern theory of chemical bonding begins with the article "The Atom and the Molecule" published by the American chemist G. N. Lewis in 1916 [1]. In this article, which is still well worth reading, Lewis for the first time associates a single chemical bond with "one pair of electrons held in common by the two atoms." After a brief review of Lewis' model we turn to a quantum-mechanical description of the simplest of all molecules, viz. the hydrogen molecule ion H_2^+. Since this molecule contains only one electron, the Schrödinger equation can be solved exactly once the distance between the nuclei has been fixed. We shall not write down these solutions since they require the use of a rather exotic coordinate system. Instead we shall show how approximate wavefunctions can be written as linear combinations of atomic orbitals of the two atoms. Finally we shall discuss so-called molecular orbital calculations on the simplest two-electron atom, viz. the hydrogen molecule.

7.1 The electron octet, the "cubical atom," and the electron pair bond

In 1916 Mendeleev's periodic table of the elements had been known for more than 40 years. The atoms had been shown to contain a minute, positively charged nucleus that contains most of the atomic mass. The relationship between the atomic number of an element and the charge on the nucleus had been established in 1913. The particle called an electron had been characterized, and was known to be a part of atoms and molecules.

The Bohr model of the H atom had been proposed some five years earlier, but Lewis was not impressed: One way "in which a body may hold another is that in which the planets are held by the sun, and this is the way that in some current theories of atomic structure the electrons are supposed to be held by the atom. Such an assumption seems inadequate to explain even the simplest chemical properties of the atom, and I imagine that it has been introduced only for the sake of maintaining the laws of electromagnetics known to be valid at large distances. The fact is, however, that in the more prominent of these theories even this questionable advantage disappears, for the common laws of electricity are not preserved. The most interesting and suggestive of these theories is the one proposed by Bohr and based on Planck's quantum theory. Planck in his elementary oscillator which maintains its motion even at the absolute zero, and Bohr in his electron which is moving in a fixed orbit, have invented systems containing electrons of which the motion produces no effect upon external charges. Now this is not only inconsistent with the accepted laws of electromagnetics, but

I may add, is logically objectionable, for a state of motion which produces no physical effect whatsoever may better be called a state of rest."

Lewis then proceeded to build his model of "The Cubical Atom" based on chemical evidence. This evidence may be summarized as follows:

(i) The noble gases are chemically inert.
(ii) The alkali metal cations, the halide anions and the doubly charged oxide anion all have the same number of electrons as the nearest noble gas.
(iii) Out of the "tens of thousands of known compounds" only a handful contain an odd number of electrons, viz. NO, NO_2, ClO_2 and "$(C_6H_5)_3C$ as well as other tri-aryl methyls."
(iv) The only known molecule of composition CCl_n is carbon tetrachloride; CCl, CCl_2, CCl_3 or CCl_5 do not exist as stable molecules. Similarly the only known molecule of composition NF_n was nitrogen trifluoride, of composition OF_n oxygen difluoride, and of composition H_nF hydrogen fluoride, HF.
(v) Many compounds like the hydrogen halides that do not contain ions when pure, produce ions when dissolved in water.

On the basis of this evidence Lewis suggested that every atom consists of a "*kernel* which remains essentially unaltered during ordinary chemical changes and which possesses an excess of positive charges" corresponding to 1, 2, 3, 4, 5, 6, 7 or 8 in Groups 1, 2, 13, 14, 15, 16, 17, and 18 respectively. Using today's terminology we would equate the kernel with the nucleus and the electrons in inner shells.

"The atom is composed of the kernel and the outer atom or shell, which in the case of the neutral atom, contains negative electrons in equal numbers to the excess of positive charges of the kernel." Today we refer to the electrons in the outer shell as valence electrons. Lewis suggested the use of the atomic symbol in boldface to represent the kernel of the atom and that valence electrons should be indicated by dots. Figure 7.1 shows the electronic structures of the second-period elements from Li to F as suggested by Lewis.

After having constructed his model of the atom, Lewis goes on to describe the formation of chemical bonds between the cubical atoms as due to sharing of edges or square faces: all halogen atoms are represented by a cube with one electron at each of seven corners, all dihalogen molecules by two cubes sharing an edge. See Fig. 7.1. The three-dimensional

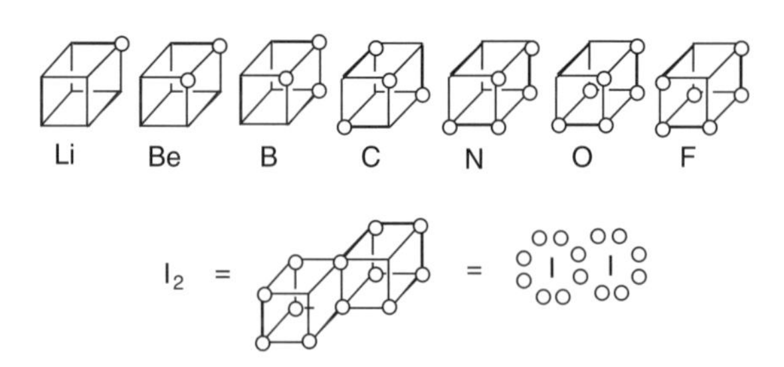

Fig. 7.1. Above: the electronic structures of the second-period elements from Li to F according to Lewis [1]. Below: the electronic structure of the iodine molecule.

figure of the two cubes might also be represented by a formula in which the atomic kernels were represented by the atomic symbol in boldface, all valence electrons are indicated by dots, and the two electrons on the shared edge are drawn between the symbols for the kernels.

The two electrons on the shared edge should be counted as valence electrons of both atoms, both of them are thus surrounded by eight valence electrons. This is also the case for the noble gases and presumably represents a particularly stable arrangement.

In general an atom in Groups 14, 15 or 16 would go on sharing edges till all corners were occupied by electrons: "The atom tends to hold an even number of electrons in the [outer] shell, and especially to hold eight electrons which are normally arranged symmetrically at the eight corners of a cube." Thus O and the other elements in Group 16 would combine with two halogen atoms, N and the other Group 15 elements would share edges with three cubical halogen atoms, C and the other Group 14 elements would share edges with four halogen atoms. Lewis does not discuss any covalent compounds of the elements in Groups 1, 2 or 13, but the electron pair model for the chemical bond and his emphasis on the stability of the electron octet suggest that the elements in these groups would use all their electrons for bond formation.

In the following we shall refer to atoms in the Groups 1, 2, 13 or 14 that form one, two, three or four single bonds respectively, and to atoms in the Groups 15, 16, or 17 that form three, two or one bond respectively, as *Lewis-valent*. Atoms that form fewer bonds than expected from Lewis' model will be described as *subvalent*, and atoms that form more bonds than expected from his model will be described as *hypervalent*.

The hydrogen atom was a special case, since it tends to hold two (or zero) rather than eight electrons in the valence shell.

Double bonds, as in O_2, could be represented by two cubic atoms sharing a face or by an electron dot formula with two electron pairs between the kernels. Triple bonds, as in N_2, would require a rearrangement of the cubical to a tetrahedral atom in which pairs of valence electrons occupy positions on the corners of a tetrahedron. The N atom would consist of an atomic kernel surrounded by one electron pair and three single electrons at the four corners of the tetrahedron, and the molecule by two tetrahedral atoms sharing a face or by an electron dot formula with three electron pairs between the kernels.

Lewis' model represented a great step forwards: It identified a single bond with an electron pair shared between two atoms, a double bond with two pairs and a triple bond with three electron pairs. It provided an easy rationalization of the stoichiometry of many chemical compounds of main group elements and of the connectivity of the constituent atoms. The model was wrong, however, in assigning more or less fixed positions to the electrons in the valence shell.

A more complete description of atoms and molecules required the discovery of electron spin, the development of quantum mechanics and the accumulation of information of the geometric structure of molecules made possible by new and improved methods of structure determination. With hindsight, one might modernize Lewis' model and describe the atoms as tetrahedral rather than cubical: The valence electrons would occupy positions at the corners of a tetrahedron, each site may accommodate one or two electrons. If two electrons occupy the same site they must have opposite spin. The electrons are not fixed to one position, but move about in the vicinity of the site. The part of space containing one or two such electrons is sometimes referred to as an electron domain.

A single bond would then be described as two atoms sharing a corner, a double bond as two atoms sharing an edge, and the triple bond as two atoms sharing a face (as Lewis indeed suggested for the N_2 molecule).

Problem 7.1 Use the cubical and tetrahedral models of the atom to suggest molecular structures for OF_2, NCl_3, CCl_4 and tetrachloroethene, C_2Cl_4. Try to reproduce the angular structure of OF_2, the pyramidal structure of NCl_3, the tetrahedral structure of CCl_4 and the planar structure of C_2Cl_4 found by experiment.

7.2 Molecular orbitals: the hydrogen molecule ion, H_2^+

We shall now use quantum mechanics to describe chemical bonds and begin with the simplest of all molecules, H_2^+. The hydrogen molecule ion has not been found in solids or melts, but is easily formed by electric discharge through hydrogen gas. It is also one of the most common molecules in interstellar space. The properties are well known from experimental studies, the equilibrium bond distance is $R_e = 106.0$ pm and the dissociation energy $D_e = 269$ kJ mol^{-1}. Comparison with the H_2 molecule, $R_e = 74.1$ pm and $D_e =$ 455 kJ mol^{-1}, shows that the one-electron bond in the ion is 43% longer and 41% weaker than the two-electron bond in the neutral molecule.

The importance of the hydrogen molecule ion for the theory of diatomic molecules is similar to the importance of the hydrogen atom for our understanding of atoms: both H and H_2^+ are one-electron systems for which the Schrödinger equation can be solved exactly. The exact solution of the one-electron species is then used as a starting point for the discussion of polyelectron species for which exact solutions of the Schrödinger are unavailable.

In our discussion of the H_2^+ ion we shall use several coordinate systems, sometimes we shall even mix different coordinate systems in the same expression. We begin by defining a right-handed Cartesian coordinate system with the z-axis running through both nuclei and the origin at the midpoint between them. See Fig. 7.2. We shall also use Cartesian coordinate systems with their origins at one of the nuclei, A or B. The z_A and z_B axes coincide with the

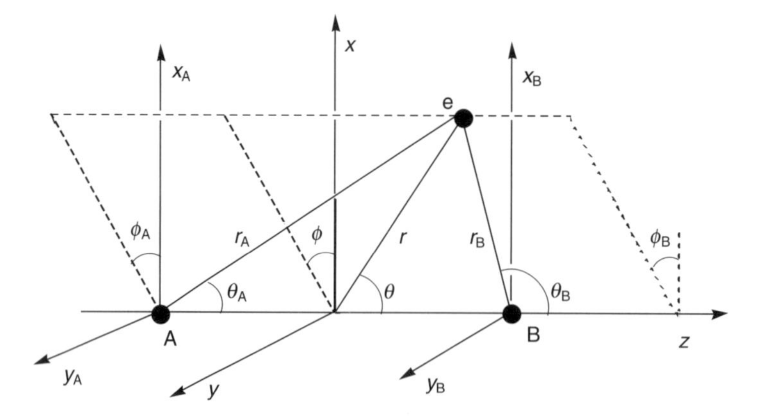

Fig. 7.2. Polar and Cartesian coordinate systems used to describe the position of an electron in the hydrogen molecule ion. The stippled lines indicate a plane containing the z-axis and the electron.

original z-axis, the x_{A} and x_{B} axes are parallel to the original x axis while the y_{A} and y_{B} axes are parallel to the original y axis. Alternatively we might use polar coordinate systems with the origin at the bond midpoint (r, θ and ϕ_{p}), or with the origin at one of the two nuclei A or B. See Fig. 7.2.

The polar coordinates r_A and r_B are given by

$$r_{\mathrm{A}} = \sqrt{x^2 + y^2 + (z + R/2)^2} \quad \text{and} \quad r_{\mathrm{B}} = \sqrt{x^2 + y^2 + (z - R/2)^2}$$

where R is the internuclear distance. Note that the collinearity of the three z-axes and the parallelism of the three x- or y-axes imply that the three azimutal angles are equal: $\phi_{\mathrm{A}} = \phi_{\mathrm{B}} = \phi$.

In Section 4.2 we defined the "electronic energy" of the molecule as the sum of the kinetic energy of the electron and the Coulomb energies due to attraction between the electron and each of the two nuclei and due repulsion between the nuclei:

$$E = \frac{1}{2}m_{\mathrm{e}}v^2 + \frac{e^2}{4\pi\varepsilon_0}\left(-\frac{1}{r_{\mathrm{A}}} - \frac{1}{r_{\mathrm{B}}} + \frac{1}{R}\right)$$

This classical expression for the energy may be transformed into the Hamiltonian operator of the system:

$$\mathcal{H} = -\frac{h^2}{8\pi^2 m_{\mathrm{e}}}\nabla^2 + \frac{e^2}{4\pi\varepsilon_0}\left(-\frac{1}{r_{\mathrm{A}}} - \frac{1}{r_{\mathrm{B}}} + \frac{1}{\mathrm{R}}\right) \tag{7.1}$$

According to the Born–Oppenheimer approximation the electronic energy and the wavefunction of the electron can be determined for any instantaneous value of R by solving the Schrödinger equation

$$\mathcal{H}\Psi(x, y, z) = E\Psi(x, y, z)$$

Since the magnitude of the last three terms in the Hamiltonian operator depend on the positions of the nuclei, i.e. on R, so will the energy: $E(R)$, and the wavefunction: $\Psi(x, y, z; R)$.

Like the Schrödinger equation of the H atom, the Schrödinger equation of the hydrogen molecular ion has infinitely many solutions for each value of R. We shall, however, only discuss the ground state and the first excited state. The electronic energies of these states as functions of R are shown in Fig. 7.3.

When the two nuclei are far apart, the two states are degenerate, i.e. they have the same energy. This energy is equal to $-1\mathcal{R} = -1313\,\mathrm{kJ\,mol^{-1}}$, which is equal to the energy of a hydrogen atom at a large distance from a proton. When the distance between the nuclei is reduced, the degeneracy is lifted. The energy associated with the electronic state Ψ_{b} decreases, reaches a minimum at $R_{\mathrm{e}} = 106.0$ pm, and increases again when the internuclear distance is further reduced. See Fig. 7.3. The probability density calculated from the wavefunction Ψ_{b} is particularly high in the regions around and between the two nuclei, i.e. in those regions of space where the Coulomb energy of the electron is low. A wavefunction that describes *one* electron moving in the space around and between two or more nuclei is called a *molecular orbital*, MO. An MO which describes an electron which moves in the space between and around *two* nuclei, is described as a two-center, or $2c$, MO.

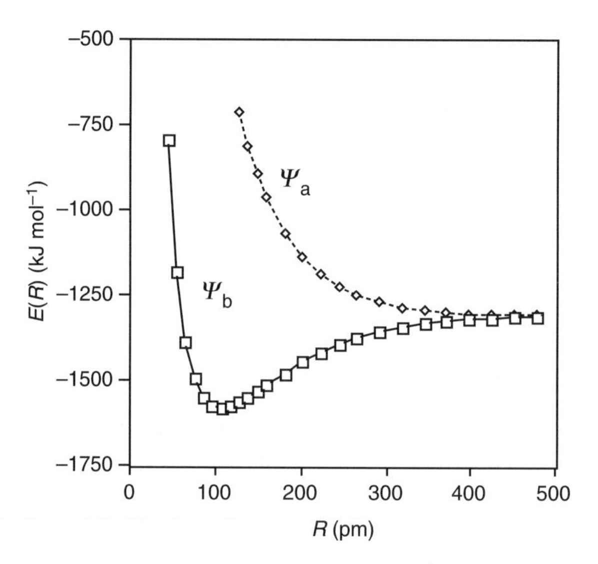

Fig. 7.3. The electronic energies of the two lowest states of the H_2^+ ion as a function of the internuclear distance R : Ψ_b (b for bonding) and Ψ_a (a for antibonding). Based on data from refrence [2].

Since Ψ_b describes a molecule which has lower energy than the separated atoms (H and H^+), it is described as a *bonding* MO, hence the subscript: Ψ_b.

The energy associated with the second wavefunction Ψ_a increases monotonically as the internuclear distance is decreased. This means that if we somehow have succeeded in bringing the two nuclei close together, they will spontaneously move apart and accelerate as they do so. The Ψ_a state is therefore described as *repulsive* or *antibonding*. The probability density calculated from Ψ_a has maxima at the two nuclei, but the density between them is smaller than calculated by supposition of two half hydrogen atoms with the electrons in 1*s* orbitals. This means that an electron in Ψ_a is less likely to be found in the region of space where the Coulomb energy is low.

We begin our discussion of the wavefunctions by noting that when the nuclei are infinitely far apart, they are given by

$$\Psi_b(x, y, z; R = \infty) = [\Phi_A(x, y, z) + \Phi_B(x, y, z)]/\sqrt{2}$$
$$\Psi_a(x, y, z; R = \infty) = [\Phi_A(x, y, z) - \Phi_B(x, y, z)]/\sqrt{2} \tag{7.2}$$

where $\Phi_A(x, y, z)$ and $\Phi_B(x, y, z)$ are given by

$$\Phi_A(x, y, z) = (\pi a_0^3)^{-1/2} \exp(-r_A/a_0)$$
$$\Phi_B(x, y, z) = (\pi a_0^3)^{-1/2} \exp(-r_B/a_0) \tag{7.3}$$

Note that Φ_A and Φ_B are identical to the wavefunctions of a 1*s* electron on nucleus A or B respectively.

The probability densities of the two states are given by

$$\Psi^2(x, y, z; R = \infty) = \left(\frac{1}{2}\right)(\pi a_0^3)^{-1}[\exp(-2r_A/a_0)$$
$$\pm 2\exp[-(r_A + r_B/a_0)] + \exp(-2r_B/a_0)] \tag{7.4}$$

where the plus sign in front of the second term in the brackets pertains to Ψ_B and the minus sign to Ψ_A. Since the nuclei are infinitely far apart, $(r_A + r_B)$ will be infinite for all values of x, y and z. The second term in the brackets vanishes, and the probability density of both states is given by

$$\Psi^2(x, y, z; R = \infty) = (1/2)(\pi a_0^3)^{-1}[\exp(-2r_A/a_0) + \exp(-2r_B/a_0)] \tag{7.5}$$

The electron density thus calculated is equal to *half* the probability density of a $1s$ electron on nucleus A plus *half* the probability density of a $1s$ electron on nucleus B: the electron occupies a $1s$ orbital at one of the nuclei, but we do not know which one!

7.3 Approximate molecular orbitals obtained by linear combinations of atomic orbitals

We have seen that when the distance between the two nuclei is infinite, the ground state of the H_2^+ ion is doubly degenerate and that the two wavefunctions are exactly equal to the sum or the difference of the two $1s$ atomic orbitals.

The shape of the exact wavefunction, $\Psi_b(x, y, z)$, for the ground state of H_2^+ when R is approximately equal to the equilibrium bond distance, suggests that it may be *approximated* by the sum of the $1s$ atomic orbitals of the two atoms:

$$\Psi_+(x, y, z) = k_+[\Phi_A(x, y, z) + \Phi_B(x, y, z)] \tag{7.6}$$

The constant k_+ is fixed by requiring that the MO is normalized:

$$\int \Psi_+^2(x, y, z)\mathrm{d}\tau = k_+^2 \int [\Phi_A(x, y, z) + \Phi_B(x, y, z\, z)]^2 \mathrm{d}\tau$$

$$= k_+^2 \left[\int \Phi_A^2(x, y, z)\mathrm{d}\tau + 2\int \Phi_A(x, y, z)\Phi_B(x, y, z)\mathrm{d}\tau + \int \Phi_B^2(x, y, z)\mathrm{d}\tau\right] = 1$$

If the AOs have been normalized, the first and the third integrals are both equal to unity. The second integral, $S_{AB} = \int \Phi_A(x, y, z)\Phi_B(x, y, z)\mathrm{d}\tau$, is referred to as the *overlap integral*. The overlap integral is a dimensionless number. If the AOs have been normalized, S will have a value between zero and one depending on the distance between the nuclei.

$$\int \Psi_+^2(x, y, z)\mathrm{d}\tau = k_+^2(2 + 2S_{AB}) = 1$$

or

$$k_+ = \frac{1}{\sqrt{2 + 2S_{AB}}} \tag{7.7}$$

The energy corresponding to such an approximate solution of the Schrödinger equation may then be calculated from

$$E_+ = \int \Psi_+(x, y, z)\mathcal{H}\Psi_+(x, y, z)\mathrm{d}\tau \tag{7.8}$$

where $\mathcal{H}$ is the Hamiltonian operator given by equation (7.1).

We now introduce the four integrals

$$H_{\mathrm{AA}} = \int \Phi_{\mathrm{A}}(x, y, z)\mathcal{H}\Phi_{\mathrm{A}}(x, y, z)\,\mathrm{d}\tau \quad \text{and} \quad H_{BB} = \int \Phi_{\mathrm{B}}(x, y, z)\mathcal{H}\Phi_{\mathrm{B}}(x, y, z)\mathrm{d}\tau$$

$$H_{\mathrm{AB}} = \int \Phi_{\mathrm{A}}(x, y, z)\mathcal{H}\Phi_{\mathrm{B}}(x, y, z)\mathrm{d}\tau \quad \text{and} \quad H_{\mathrm{BA}} = \int \Phi_{\mathrm{B}}(x, y, z)\mathcal{H}\Phi_{\mathrm{A}}(x, y, z)\mathrm{d}\tau \tag{7.9}$$

All these integrals have the dimension energy. Their magnitude depends on the internuclear distance R, but due to the two negative terms in the Hamiltonian operator representing the energy of attraction between the electron and the two nuclei, they will always be negative. Inserting equations (7.6) in (7.8) and using (7.7) and (7.9) we obtain:

$$E_+ = \frac{H_{\mathrm{AA}} + H_{\mathrm{AB}} + H_{\mathrm{BA}} + H_{\mathrm{BB}}}{2 + 2S_{\mathrm{AB}}}$$

or, accepting that H_{AA} is equal to H_{BB}, and that H_{AB} is equal to H_{BA}:

$$E_+ = \frac{H_{\mathrm{AA}} + H_{\mathrm{AB}}}{1 + S_{\mathrm{AB}}} \tag{7.10}$$

The energy corresponding to the approximate wavefunction Ψ_+ for any particular internuclear distance may be calculated from (7.10) and plotted as a function of R. The resulting potential energy curve has the same general shape as the exact curve in Fig. 7.3. The asymptotic value for large internuclear distances is $E = -1313$ kJ mol^{-1}, corresponding to a hydrogen atom at a large distance from a proton. The minimum in the potential energy curve is found for $R_e = 132$ pm; the corresponding energy is -1483 kJ mol^{-1}. The dissociation energy is calculated as

$$D_{\mathrm{e}} = E_+(R = \infty) - E_+(R = R_{\mathrm{e}}) = 170\,\mathrm{kJ\,mol^{-1}}.$$

Comparison with the experimental values, $R_{\mathrm{e}} = 106.0$ pm and $D_{\mathrm{e}} = 269.3$ kJ mol^{-1}shows that these calculations overestimate the bond distance by 25% and underestimate the dissociation energy by 37%. The wavefunction Ψ_+ is clearly not very good!

In order to find out how the wavefunction might be improved, we calculate the probability density for comparison with that obtained from the exact solution Ψ_{B}.

The probability density calculated for our approximate wavefunction is

$$\Psi_+^2(x, y, z) = k_+^2[\Phi_{\mathrm{A}}^2(x, y, z) + 2\Phi_{\mathrm{A}}(x, y, z)\Phi_{\mathrm{B}}(x, y, z) + \Phi_{\mathrm{B}}^2(x, y, z)] \tag{7.11}$$

Since Φ_{A} and Φ_{B} are 1s AOs at atoms A and B respectively, see equation (7.3):

$$\Psi_+^2(x, y, z)) = C_+^2[\exp(-2r_{\mathrm{A}}/a_0) + 2\exp[-(r_{\mathrm{A}} + r_{\mathrm{B}})/a_0] + \exp(-2r_{\mathrm{B}}/a_0)] \tag{7.12}$$

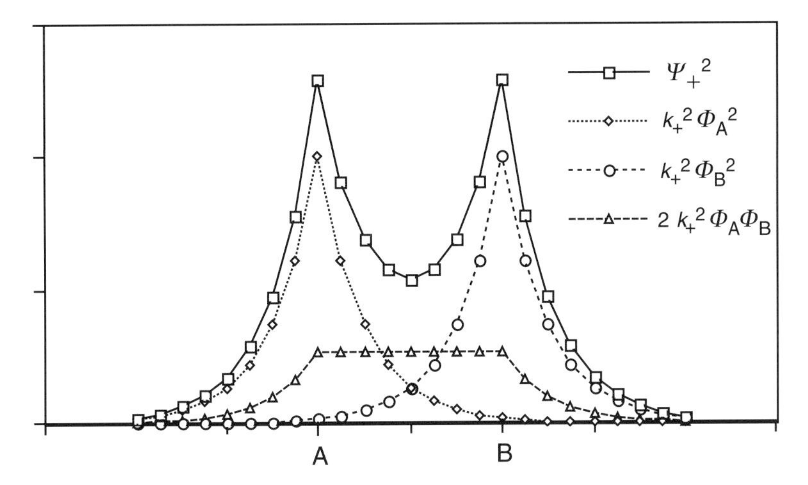

Fig. 7.4. The variation of the probability density along the z-axis in H_2^+ in the ground state calculated from the approximate wavefunction Ψ_+, equation (7.12): $k_+^2\Phi_A^2(r_A)$, $k_+^2\Phi_B^2(r_B)$, $2k_+^2\Phi_A(r_A)\Phi_B(r_B)$, and the total probability density Ψ_+^2.

Here the constant C_+^2 is the product of the squared normalization constant of the 1s atomic orbital, $(\pi a_0^3)^{-1}$, and the squared normalization constant of the MO, $k_+^2 = 1/(2 + 2S_{AB})$.

The probability density of equation (7.11) thus consists of three parts. The first is a cloud with the shape of the 1s probability density around nucleus A, and the third corresponds to the 1s density around nucleus B, both are reduced by the factor k_+^2. The second component of the probability density has its maximum value for all points on the line connecting the two nuclei, where $r_A + r_B = R$. It is equal for all points where $r_A + r_B$ are equal, i.e. for points on ellipsoids with the nuclei A and B in the focal points. This component is sometimes referred to as the overlap density or overlap cloud.

The variation of the densities of the three components along a line connecting the nuclei is shown in Fig. 7.4.

Problem 7.2 Sketch constant probability density contours of the three components in a plane containing the two nuclei.

We now turn our attention to the antibonding state. Examination of the exact solution Ψ_a suggests that it may be approximated by

$$\Psi_-(x, y, z) = k_-[\Phi_A(x, y, z)) - \Phi_B(x, y, z)] \tag{7.13}$$

Problem 7.3 Show that the antibonding MO is normalized if

$$k_- = \frac{1}{\sqrt{2 - 2S_{AB}}} \tag{7.14}$$

and that the energy is given by

$$E_- = \frac{H_{AA} - H_{AB}}{1 - S_{AB}} \tag{7.15}$$

The energy E_- may be calculated for any instantaneous internuclear distance from equation (7.15) and plotted as a function of R. It is found to increase monotonically with decreasing R.

The probability density calculated for an electron in Ψ_- is

$$\Psi_-^2(x, y, z) = C_-^2[\exp(-r_A/a_0) - 2\exp[-(r_A + r_B)/a_0] + \exp(-r_B/a_0)] \quad (7.16)$$

Here the constant C_-^2 is the product of the squared normalization constant of the 1s functions, $(\pi a_0^3)^{-1}$, and $k_-^2 = 1/(2 - 2S_{AB})$.

The probability density in (7.16) consists of three parts. The first is a cloud of the shape of a 1s electron cloud around nucleus A reduced by the factor k_-^2. The third term corresponds to *a* 1s electron density around nucleus B reduced by the same factor. The second term in (7.16) has its maximum absolute value at the line connecting the two nuclei, and has a constant value at points on an ellipsoid with the two nuclei in the focal points. This term is, however, to be *subtracted* from the two others. The effect is to *reduce* the electron density between the two nuclei, i.e. in the region where the potential energy of the electron is particularly low.

Ψ_+ and Ψ_- represent first approximations to the true wavefunctions of the ground state and the first excited state of H_2^+, Ψ_B and Ψ_A respectively. Molecular orbitals that have been formed in this manner, are referred to as LCAO (linear combination of atomic orbitals) MOs. The AOs used are referred to as the basis (for the calculation).

7.4 Improvement of the LCAO MO

As we have seen the MO calculations of the last paragraph yielded a bond distance which was 25% too long and a dissociation energy which was 37% too low. How could the accuracy be improved?

Comparison of the probability densities calculated for the approximate wavefunction Ψ_+ with those calculated from the exact wavefunction Ψ_b shows that the probability density in the low-energy region near and in between the two nuclei is underestimated by some 40%. No wonder then that the calculations fail to reproduce the true strength of the bond!

The simplest way to increase the probability density near the nuclei is to modify the 1s orbitals:

$$\Phi_A(x, y, z) = \left(\frac{\zeta^3}{\pi a_0^3}\right)^{1/2} \exp\left(-\frac{\zeta r_A}{a_0}\right)$$

$$\Phi_B(x, y, z) = \left(\frac{\zeta^3}{-\pi a_0^3}\right)^{1/2} \exp\left(-\frac{\zeta r_B}{a_0}\right) \quad (7.17)$$

where ζ ("zeta") is an adjustable constant much like the effective nuclear charge which we used in the calculations on atoms with more than one electron.

When the two nuclei are far apart, the best value for ζ is obviously equal to 1.0. In the other limiting situation, i.e. when the internuclear distance approaches zero, the H_2^+ molecule is transformed into a He^+ ion, and then the best value for $\zeta = 2$: the best value for ζ then, has to be determined for each value of R.

The calculations are carried out as outlined in the preceding section. Thus E is calculated from equation (7.10):

$$E_1 = \frac{H_{AA} + H_{AB}}{1 + S_{AB}}$$

and ζ varied to give the lowest possible energy for each value of R. The potential energy curve obtained in this manner has its minimum at $R_e = 106$ pm (i.e. equal to the experimental value!) and the dissociation energy becomes 217 kJ mol^{-1} as compared to an experimental value of 260 kJ mol^{-1}. (The optimal value of ζ at $R = 106$ pm is 1.23.)

The next step in improvement of the calculations is to write the MO orbital as a linear combination of *four* atomic orbitals, one $1s$ type function and one $2p_z$ type function (since the coordinate system is defined in such a way that the two nuclei are lying on the z-axis) for each atom [3]. The ζ parameters of the $1s$- and $2p_z$-type AOs are adjusted separately. Such calculations reduce the discrepancy between experimental and calculated dissociation energies to 5 kJ mol^{-1}.

The best wavefunction obtained for $R = R_e = 106$ pm may be written on the form

$$\Psi(x, y, z) = C[(1s)_A + (1s)_B + c(2p_z)_A - c(2p_z)_B]$$

with $c = 0.15$. (We have assumed that the z-coordinate of nucleus A is smaller than the z-coordinate of nucleus B. The negative coefficient in front of the function $(2p_z)_B$ then assures that the positive lobe of the orbital is pointing towards nucleus A.)

What does this result mean? Are electrons in bonding molecular orbitals found in atomic orbitals *outside* the valence shell? Most theoretical chemists would answer no, the *shape* is that of a $2p_z$ orbital, but the *size* of the orbital is much smaller than the $2p_z$ orbital of an isolated H atom: the form of the orbitals after adjustment of ζ, is given by

$$(2p_z) = z\exp(-2.87r/2a_0)$$

This, in fact, is close to the form of the $2p_z$ orbital of a Li^{2+} cation,

$$(2p_z) = z\exp(-3.00r/2a_0)$$

with an average electron-nucleus distance which is only 1/3 of that of a hydrogen atom $2p_z$ orbital. The best way to view the matter is to assume that the shape of the $1s$ orbital of the isolated atom has to be modified in order to reproduce the real electron distribution in the H_2^+ molecule. The effect of the admixture of the $2p_z$ type orbitals is to increase the electron density between the two nuclei. We say that the H atom $1s$ orbitals have been polarized and refer to the $2p_z$ type orbitals as *polarization functions*.

7.5 The hydrogen molecule and the molecular orbital approximation

Since the hydrogen molecule ion H_2^+ contains only one electron, the Schrödinger equation can be solved exactly (once we have accepted the Born–Oppenheimer approximation). In the preceding section we have shown how *approximate* solutions may be obtained by linear combinations of atomic orbitals on each atom. In this context, the term "atomic orbitals" should *not* be taken to mean "solutions of the Schrödinger equation for the atom" or the "best possible AOs for the atom." The term only implies that they resemble the AOs for the hydrogen-like atoms, in particular that the angular functions correspond to H atom s, p, or d orbitals. The LCAO-MO functions of H_2^+ can be brought into very good agreement with the exact solution, the discrepancy can in fact be reduced below any limit by using a sufficiently large number of basis functions. We now turn our attention to the simplest two-electron molecule, viz. the hydrogen molecule.

The electronic energy of an H_2 molecule contains two terms representing the kinetic energy of the electrons, four terms representing the energies due to Coulomb attraction between two electrons and two nuclei, one term representing the energy due to repulsion between the two electrons and a last term representing the energy due to repulsion between the two nuclei.

Problem 7.4 Place the H_2 molecule in a Cartesian coordinate system in such a manner that the origin is at the midpoint of the H–H bond and the z-axis runs through both nuclei. Write down an expression for the potential energy of the molecule. Which terms depend on the internuclear distance R? What state corresponds to zero energy? Write down the Hamiltonian operator $\mathcal{H}$.

The hydrogen molecule, like the He atom, poses a real problem: the Hamiltonian operator contains a term representing the repulsion between the electrons, and the presence of this term makes an exact solution of the Schrödinger equation impossible. In the case of the helium atom we turned to the hydrogen atom for guidance in the choice of approximate wavefunctions. In the case of the hydrogen molecule we turn to the H_2^+ ion and assume that the wavefunction may be approximated by the product of two molecular orbitals

$$\Psi(\mathbf{r}_1, \mathbf{r}_2) = \Psi_1(\mathbf{r}_1)\Psi_2\mathbf{r}_2) \tag{7.18}$$

where Ψ_1 and Ψ_2 are molecular orbitals that resemble the solutions for the H_2^+ ion and $\mathbf{r}_1$ and $\mathbf{r}_2$ represent the coordinates of the first and second electron respectively.

As a first approximation for the ground state of H_2 we use MOs of the form

$$\Psi(x, y, z) = \frac{\Phi_A(x, y, z) + \Phi_B(x, y, z)}{\sqrt{2 + 2S_{AB}}} \tag{7.19}$$

where $\Phi_A(x, y, z)$ and $\Phi_B(x, y, z)$ are hydrogen atom 1s functions centered on nuclei A or B respectively.

When discussing the He atom, we stressed that a wavefunction written as the product of two atomic orbitals is inherently wrong since it fails to reflect the fact that repulsion between the electrons tends to keep them far apart. The energy calculated from such a wavefunction will therefore be higher than that of the real atom. What was true for the two-electron atom, is equally true for a two-electron molecule. Like atomic orbital calculations, molecular orbital calculations on molecules containing two or more electrons are inherently wrong.

The electronic energy associated with the wavefunction given by (7.18) and (7.19) may be calculated from

$$E(R) = \int \Psi(\mathbf{r}_1, \mathbf{r}_2)\mathcal{H}\Psi(\mathbf{r}_1, \mathbf{r}_2)\mathrm{d}\tau_1\,\mathrm{d}\tau_2 \tag{7.20}$$

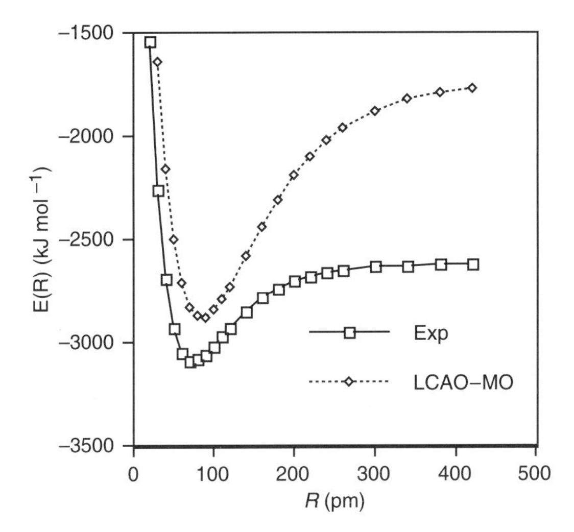

Fig. 7.5. Schematic representation of the electronic energy of the H_2 molecule calculated from the simple LCAO wavefunction, equations (7.18) and (7.19), and the experimentally determined counterpart.

where the integral extends over all values for the coordinates of both electrons. See Fig. 7.5.

The electronic energy thus obtained has a minimum at $R_{\text{min}} = 85$ pm (as compared to an experimental bond distance of 74.1 pm) and an energy

$$E(R_{\text{min}}) = -2885\,\text{kJ mol}^{-1} \tag{7.21}$$

which is 196 kJ mol^{-1} above the energy at the minimum of the experimental electronic energy curve. In order to calculate the dissociation energy D_{e} we need to compare with the energy calculated when the two nuclei are infinitely far apart:

$$D_{\text{e}} = E(R = \infty) - E(R_{\text{min}}) \tag{7.22}$$

The molecular orbital calculations yield

$$E(R = \infty) = -1745\,\text{kJ mol}^{-1} \tag{7.23}$$

This result is obviously wrong: when the internuclear distance is very large, the ground state of the system must correspond to two separate H atoms, each in the 1s ground state, and the true energy is thus $E(R = \infty) = -2\mathcal{R} = -2626\,\text{kJ mol}^{-1}$. *Molecular orbital calculations on diatomic molecules with more than one electron yield energies which are higher than the true, experimental energies for all values of R. In fact, the error is larger when nuclei are far apart than when they are close together.*

Combination of equations (7.21), (7.22), and (7.23) yields an estimated dissociation energy of

$$D_{\text{e}} = E(R = \infty) - E(R_{\text{min}}) = -1754\,\text{kJ mol}^{-1} - (2885\,\text{kJ mol}^{-1}) = 1131\,\text{kJ mol}^{-1}$$

which is more than twice as large as the experimental value.

The estimated dissociation energy is much improved if we combine the calculated energy at the minimum of the potential curve with the sum of the energies obtained by calculations on the two atoms one at a time:

$$D_e = -2\mathcal{R} - E(R_{\min}) = -2626 \text{ kJ mol}^{-1} - (-2885 \text{ kJ mol}^{-1}) = 259 \text{ kJ mol}^{-1}$$

as compared to an experimental value of 455 kJ mol^{-1}. The calculations based on the simple LCAO wavefunction given by equations (7.18) and (719) thus overestimate the bond distance by some 15% and underestimate the dissociation energy by more than 40%.

One way to improve the result is obviously to modify the $1s$-like basis functions to include an adjustable ζ parameter as indicated in (7.17): When the energy is calculated and ζ varied to give the lowest energy for each R, one obtains an electronic energy curve with a minimum at 73 pm (1 pm *shorter* than the experimental bond distance) and a dissociation energy of 335 kJ mol^{-1}.

Somewhat better agreement between calculated and experimental bond dissociation energies are obtained by expanding the basis, for instance by including $2p_z$ type functions. However, the highest and best estimate for the dissociation energy that can be obtained under the molecular orbital approximation is 351 kJ mol^{-1}, which still represents an error of 23%. This error is inherent in the form of the wavefunction (7.18). The difference between the experimental dissociation energy and the best value that can be obtained under the molecular orbital approximation is referred to as the "electron correlation energy."

Molecular orbital calculations on molecules with three or more electrons are based on Slater-determinant wavefunctions. If the basis sets are sufficiently large, such Hartree–Fock or HF calculations are generally found to reproduce experimental bond distances to within 3 or 4 pm, and experimental valence angles to within 3 or 4 degrees. The dissociation energies obtained by such calculations are, however, too inaccurate to be useful.

Algorithms and programs that allow calculation of the correlation energy – or at least a part of it – have been developed. We shall refer to such calculations in Section 20.6.

7.6 The electric dipole moment of HD: failure of the Born–Oppenheimer (adiabatic) approximation

If we accept the Born–Oppenheimer approximation, the electron distribution in the hydrogen molecule or the isotopomer HD would always be symmetrical, in the sense that the center of gravity of negative charge would fall at the midpoint of the line between the nuclei. The center of positive and negative charges would thus coincide, and the electric dipole moment would be identically equal to zero. Careful investigation by rotational spectroscopy shows, however, that HD has a dipole moment of 0.0006 D [4].

This observation can only be explained as a failure of the Born–Oppenheimer approximation. Instead of assuming that the shape of the electron cloud always is perfectly adjusted to the instantaneous internuclear distance, let us go to the opposite extreme and assume that the electron cloud remains unchanged while the nuclei vibrate. See Fig. 7.6. During the vibration the center of gravity of the molecule must remain at rest. This means that if at one instant the proton has moved a certain distance from the equilibrium position in such a direction that R has increased, the deuteron has moved half as far in the opposite direction,

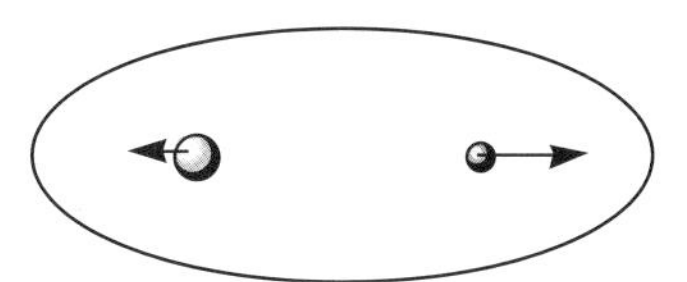

Fig. 7.6. Failure of the Born-Oppenheimer approximation for HD. The shaded circle on the left represents the deuterium nucleus, the shaded circle on the right the proton. The frozen electron distribution is indicated by an ellipse.

and the center of gravity of the positive charge no longer coincides with the center of the negative. At this instant the molecule will have a dipole moment with the positive end at the proton. If the shape of the electron cloud follows the nuclear motion, but with a certain time lag, there will be an instantaneous dipole moment in the same direction, but smaller.

What will be the situation when, a little bit later, the proton is moving towards the deuteron? When the proton has moved a certain distance from the equilibrium position, the deuteron will have moved only half as far. Again the molecule will have an instantaneous dipole moment, but the positive pole will now be at the deuteron.

The electron dipole moment that is measured by rotational spectroscopy represents an average over time as the molecule vibrates, the so-called "permanent" dipole moment. From what we have said, one would expect the instantaneous dipole moments of HD to average to zero during the vibrational cycle, but for some unknown reason, perhaps connected with anharmonicity, it does not. In any case, the observation of a "permanent" dipole moment of HD shows that the Born–Oppenheimer approximation has broken down.

References

[1] G. N. Lewis, *J. Am. Chem. Soc.* 38 (1916) 762.
[2] D. R. Bates, K. Ledsham, A. L. Stewart, *Phil. Trans. Roy. Soc.* London, 246 (1953) 215.
[3] B. N. Dickinson, *J. Chem. Phys.*, 1 (1933) 317.
[4] M. Trefler, H. P. Gush, *Phys. Rev. Lett.*, 20 (1968) 703.

Chapter 8

Molecular orbital calculations on heteronuclear diatomic molecules, hybridization, and estimation of net atomic charges from calculated electron densities

Introduction

In this chapter we shall use lithium hydride, LiH, to discuss the application of the molecular orbital model to a heteronuclear diatomic molecule, and begin by outlining a very simple computational procedure that yields an approximate description of the molecular orbital containing the two valence electrons. We then go on to outline the application of Hartree–Fock (HF) calculations based on a wavefuntion for both the two valence and the two inner-shell electrons. The wavefunction obtained by such calculations indicate that the bonding molecular orbital must be written as a linear combination of the H 1*s* with both 2*s* and $2p_\sigma$ atomic orbitals on the Li atom.

This result leads us to the definition of hybrid atomic orbitals obtained by combination of valence shell *s* and *p* AOs of the same atom. The properties of such hybrid orbitals are examined. Even though hybridization of an atom requires energy, this energy may be more than compensated by the formation of a stronger bond.

Finally we describe how net atomic charges may be estimated from the molecular wavefuntion or by integration of calculated electron densities.

8.1 The LiH molecule: approximate molecular orbital calculations

A full description of the LiH molecule requires a wavefunction that is a function of the space and spin coordinates of four electrons and written as a 4×4 Slater determinant. The electronic energy and the Hamiltonian operator of the molecule both contain four terms representing the kinetic energies of the electrons, eight terms representing the energies of Coulomb attraction between electrons and nuclei, six terms representing energies of repulsion between electrons and one term representing the energy of repulsion between the nuclei. We shall begin, therefore, by showing how simple calculations may be used to obtain a very approximate expression for the molecular orbital containing the two valence electrons if the equilibrium bond distance is known – or assumed.

We first assume – like Lewis – that the electrons in inner shells are unaffected by bond formation, i.e. that two electrons continue to occupy the Li 1*s* orbital. Secondly we assume that the two valence electrons occupy bonding molecular orbitals formed by linear combination of the two AOs that contain the valence electrons when the two atoms are far apart, i.e. by combination of the Li 2*s* and the H 1*s* orbitals, which we denote by $\Phi_1(x, y, z)$ and

$\Phi_2(x, y, z)$ respectively:

$$\Psi(x, y, z) = [c_1\Phi_1(x, y, z) + c_2\Phi_2(x, y, z)]/\sqrt{D}. \tag{8.1}$$

There is, of course, no reason to believe that the two coefficients c_1 and c_2 will be equal.

Problem 8.1 Show that the approximate MO, equation (8.1), is normalized if

$$D = c_1^2 + 2c_1c_2S_{1,2} + c_2^2 \tag{8.2}$$

where the overlap integral is

$$S_{1,2} = \int \Phi_1(x, y, z)\, \Phi_2(x, y, z)\mathrm{d}\tau \tag{8.3}$$

Finally we assume that we know the orbital coefficients c_1 and c_2 for the MO containing one of the valence electrons: electron number 3. The orbital energy of the fourth and last electron may then be written as the sum of its kinetic energy, its Coulomb energy of attraction to both nuclei and its energy of repulsion from the other three electrons. Since we have assumed that we know the wavefunctions of the three other electrons, it should, *in principle*, be possible to write down the Hamiltonian operator of fourth electron ($\hat{h}$) and compute the orbital energy from

$$\varepsilon = \int \Psi(x, y, z)\hat{h}\,\Psi(x, y, z)\mathrm{d}\tau$$

We now introduce the integrals

$$H_{1,1} = \int \Phi_1(x, y, z)\hat{h}\,\Phi_1(x, y, z)\mathrm{d}\tau \quad H_{2,2} = \int \Phi_2(x, y, z)\hat{h}\,\Phi_2(x, y, z)\mathrm{d}\tau$$

$$H_{1,2} = \int \Phi_1(x, y, z)\hat{h}\,\Phi_2(x, y, z)\mathrm{d}\tau \quad H_{2,1} = \int \Phi_2(x, y, z)\hat{h}\,\Phi_1(x, y, z)\mathrm{d}\tau \tag{8.4}$$

It may not be obvious, but it can be shown that the integrals $H_{1,2}$ and $H_{2,1}$ are equal.

Using this result, the orbital energy can now be written as

$$\varepsilon = \frac{c_1^2H_{1,1} + 2c_1c_2H_{1,2} + c_2^2H_{2,2}}{c_1^2 + c_1c_2S_{1,2} + c_2^2} \tag{8.5a}$$

$$= \frac{N(c_1, c_2)}{D(c_1, c_2)} \tag{8.5b}$$

Let us assume that we have carried out the calculation of the various integrals in equation (8.4) with R equal to the equilibrium bond distance. According to the variation principle, the calculated (approximate) orbital energy must be higher than the true energy. We now seek those values of the coefficients c_1 and c_2 that yield the *minimum* value of ε by taking the derivative of ε with respect to c_1 and c_2 and setting these derivatives equal to zero.

The equation $(\partial\varepsilon/\partial c_1) = 0$ yields

$$\frac{[(2c_1H_{1,1} + 2c_2H_{1,2})D - (2c_1 + 2c_2S_{1,2})N]}{D^2} = 0$$

and, after multiplication with $D^2/2$

$$(c_1H_{1,1} + c_2H_{1,2})D - (c_1 + c_2S_{1,2})N = 0 \tag{8.6a}$$

Problem 8.2 Show that the equation $(\partial\varepsilon/\partial c_2) = 0$ yields the equation

$$(c_1H_{1,2} + c_2H_{2,2})D - (c_1S_{1,2} + c_2)\,N = 0. \tag{8.6b}$$

Equations (8.6a) and (8.6b) are satisfied by those values of c_1 and c_2 for which the two derivatives are zero. Solution of the two equations would allow us to determine their value. Insertion of these values into equation (8.5a) would then give us a minimum or maximum value for the orbital energy ε.

Equations (8.6a) and (8.6b) are, however, difficult to solve since they contain the unknown coefficients to the third power. (Remember that N and D in (8.5b) both contain the coefficients to the second power.) We therefore choose a more indirect, but quicker approach. Division of (8.6a) and (8.6b) by D yields

$$(c_1H_{1,1} + c_2H_{1,2}) - (c_1 + c_2S_{1,2})N/D = 0 \tag{8.7a}$$

$$(c_1H_{1,2} + c_2H_{2,2}) - (c_1S_{1,2} + c_2)N/D = 0 \tag{8.7b}$$

Now N/D is a minimum or maximum orbital energy. Replacing N/D by ε_{m} and rearranging we obtain

$$(H_{1,1} - \varepsilon_m)c_1 + (H_{1,2} - S_{1,2}\,\varepsilon_m)c_2 = 0 \tag{8.8a}$$

$$(H_{1,2} - S_{1,2}\,\varepsilon_m)c_1 + (H_{2,2} - \varepsilon_m)c_2 = 0 \tag{8.8b}$$

Equations (8.8a) and (8.8b) are referred to as the "secular equations." If we regard the integrals $H_{1,1}, H_{1,2}$ and $H_{2,2}$ as well as the orbital energy ε_{m} as known quantities, (8.8a) and (8.8b) represent two linear equations for the two unknowns c_1 and c_2. (In the following we refer to the parentheses in front of the two unknowns as the "coefficients".) Linear equations where the constant terms are zero are called homogeneous. A set of two linear homogeneous equations for two unknowns will always have the trivial solutions $c_1 = c_2 = 0$. Mathematicians tell us that there will be non-trivial solutions *only if the determinant formed by the coefficients is equal to zero*:

$$\begin{vmatrix} (H_{1,1} - \varepsilon_m) & (H_{1,2} - S_{1,2}\,\varepsilon_m) \\ (H_{1,2} - S_{1,2}\,\varepsilon_m) & (H_{2,2} - \varepsilon_m) \end{vmatrix} = 0 \tag{8.9}$$

or

$$(H_{1,1} - \varepsilon_m)(H_{2,2} - \varepsilon_m) - (H_{1,2} - S_{1,2}\varepsilon_m)^2 = 0$$

or

$$(1 - S_{1,2}^2)\varepsilon_{\mathrm{m}}^2 - (H_{1,1} - 2H_{1,2}S_{1,2} + H_{2,2})\varepsilon_{\mathrm{m}} + (H_{1,1}H_{2,2} - H_{1,2}^2) = 0 \tag{8.10}$$

This quadratic equation is solved to yield the two values of ε_{m} that satisfies equation (8.9). The lower of the two roots, ε_{b}, represents a minimum value of ε_{m} and is an approximation to the energy of a bonding molecular orbital. The higher of the two roots, ε_a, represents a maximum value of ε_m and is an approximation to the energy of the corresponding anti-bonding MO.

The coefficient c_2 for the bonding MO is determined by introducing $\varepsilon_m = \varepsilon_b$ and $c_1 = 1$ in one of the secular equations. Then the wavefunction Ψ_b is normalized with the use of equations (8.1) and (8.2):

$$D = 1 + 2c_2\, S_{1,2} + c_2^2$$

Finally *both* valence electrons are assumed to occupy the bonding MO.

The reader will notice that the reasoning behind the present "calculations" is circular: We began by assuming that we knew the wavefunctions of all electrons except one. Nevertheless, at the end we assign both valence electrons to a MO that was unknown at the outset. How then is it possible to calculate the four integrals in (8.4)? The answer to the objection is that it is *not* possible. Instead we use experimental information about the atoms to make a rough guess at their values: the integral $H_{1,1}$ describes the energy of the fourth electron when it occupies a Li 2*s* orbital and contains terms representing the kinetic energy of the electron, the energy of attraction to the Li nucleus and of repulsion to the two electrons in the Li 1*s* orbital. In addition there are two more terms, one of them represents the energy of attraction of the fourth electron to the nucleus of the H atom, the other represents the energy of repulsion between the fourth and the third electron. If we assume that the two last terms cancel, $H_{1,1}$ is equal to the 2*s* orbital energy of an isolated Li atom. According to Koopman's theorem the orbital energy is equal to minus the ionization energy. The integral $H_{1,1}$ may therefore be approximated by minus the first ionization energy of Li:

$$H_{1,1} = \varepsilon(Li, 2s) = -\mathrm{IE}(Li) = -\mathrm{IE}_1 \tag{8.11a}$$

Similarly the integral $H_{2,2}$ may be approximated by minus the ionization energy of an isolated H atom:

$$H_{2,2} = \varepsilon(H, 1s) = -\mathrm{IE}(H) = -\mathrm{IE}_2 \tag{8.11b}$$

The "resonance integral" $H_{1,2}$ is often approximated by

$$H_{1,2} = -K\, S_{1,2}\sqrt{\mathrm{IE}_1 \times \mathrm{IE}_2} \tag{8.12}$$

where the constant K is selected to yield results which are considered reasonable. The only integral that has to be calculated is then the relatively simple overlap integral.

The probability density due to *one* electron in the bonding orbital is

$$\Psi_b^2(x,y,z) = [c_1^2\, \Phi_1^2(x,y,z) + 2c_1c_2\, \Phi_1(x,y,z)\Phi_2(x,y,z) + c_2^2\, \Phi_2^2(x,y,z)]/D$$

This total probability density may be regarded as the supposition of three terms. The first is centered on the nucleus if atom number 1, the third on the nucleus of atom number 2, and the second term in the region *between* the two nuclei. This expression may be interpreted to mean that the cloud centered around the first nucleus contains c_1^2/D electrons, the cloud centered around the second nucleus contains c_2^2/D electrons, while the "overlap cloud" contains $[2c_1c_2S_{1,2}]/D$ electrons. It can be shown that if – as in he present case – the ionization energy of atom 2 (H) is larger than the ionization energy of atom 1 (Li), i.e. if atom 2 is more electronegative than atom 1, then c_2 will be larger than c_1. The calculations thus indicate that the higher electron density will be found at the more electronegative atom. With increasing electronegativity difference c_2 approaches unity while c_1 approaches zero: in the ionic limit both electrons will occupy an atomic orbital on the anion.

The same method may be applied to calculations on molecules with more than two atoms and more than two basis functions. Calculations of this type are referred to as Hückel calculations. They are obviously very inaccurate, but are sufficient to bring out consequences of the molecular symmetry or electronegativity differences between the constituent atoms.

Problem 8.3 Show that if the two atoms are equal, i.e. if $H_{1,1} = H_{2,2}$, then equation (8.10) has two roots

$$\varepsilon_1 = \frac{H_{1,1} + H_{1,2}}{1 + S_{1,2}} \quad \text{and} \quad \varepsilon_2 = \frac{H_{1,1} - H_{1,2}}{1 - S_{1,2}}$$

Introduce ε_1 into the first of the secular equations (8.8a), set $c_1 = 1$ and find c_2. Write down the corresponding normalized molecular orbital. Find the normalized wavefunction corresponding to ε_2.

8.2 Simplified Hückel calculations

The calculations described in the preceding section still require the use of computer programs for calculation of overlap integrals. Since the results are inaccurate in any case, one often simplifies the calculations still further in order to be able to carry them out using pencil and paper only: The overlap integral $S_{1,2}$ is set equal to zero and the resonance intergral is approximated by

$$H_{1,2} = -K\sqrt{\text{IE}_1 \times \text{IE}_2} \tag{8.13}$$

where K is a constant usually chosen in the range 0.20–0.50. The secular equations are then reduced to

$$(H_{1,1} - \varepsilon_{\text{m}})c_1 + H_{1,2}\,c_2 = 0 \tag{8.14a}$$

$$H_{1,2}\,c_1 + (H_{2,2} - \varepsilon_{\text{m}})c_2 = 0 \tag{8.14b}$$

and have non-trivial solutions for

$$(H_{1,1} - \varepsilon_{\text{m}})(H_{2,2} - \varepsilon_{\text{m}}) - (H_{1,2})^2 = 0 \tag{8.15}$$

which yields

$$\varepsilon_{\text{b}} = \{(H_{1,1} + H_{2,2}) - \sqrt{(H_{1,1} - H_{2,2})^2 + 4(H_{1,2})^2}\}/2 \tag{8.16}$$

$$\varepsilon_{\text{a}} = \{(H_{1,1} + H_{2,2}) + \sqrt{(H_{1,1} - H_{2,2})^2 + 4(H_{1,2})^2}\}/2 \tag{8.17}$$

Note that the solution ε_{b} always will be lower than ε_{a}.

We shall now use the experimental ionization energies of the H and Li atom and equation (8.13) with $K = 0.40$ to estimate the nature of the bonding orbital and the charge distribution in LiH. Insertion of $H_{1,1} = -520\,\text{kJ mol}^{-1}$, $H_{2,2} = -1313\,\text{kJ mol}^{-1}$ and $H_{1,2} = -331\,\text{kJ mol}^{-1}$ in (8.16) yields

$$\varepsilon_\text{b} = (-917 - 517)\,\text{kJ mol}^{-1} = -1434\,\text{kJ mol}^{-1}.$$

Insertion of $\varepsilon_\text{m} = \varepsilon_\text{b}$ and $c_1 = 1$ in the first of the secular equations (8.8a) yields

$$(-520 + 1434)1 - 331c_2 = 0$$

or

$$c_2 = 2.76$$

Since we have assumed that $S_{1,2} = 0$, the normalization constant becomes

$$\sqrt{D} = \sqrt{1 + 2.76^2} = 2.93$$

and the bonding molecular orbital

$$\Psi_\text{b} = (\Phi_1 + 2.76\Phi_2)/2.93 = 0.34\Phi_1 + 0.94\Phi_2 \quad \text{or}$$
$$\Psi_\text{b} = 0.34(2s)_\text{Li} + 0.94(1s)_\text{H}$$

This MO suggest that when the bonding orbital is occupied by two electrons, $2(0.34)^2 = 0.23$ electrons are in a cloud centered around the Li nucleus, while $2(0.94)^2 = 1.77$ electrons are in a cloud centered around the H nucleus. (Since we have carried out our calculations with $S_{1,2} = 0$ the number of electrons in the overlap cloud is also assumed to be zero.)

The nuclear charge on the Li atom is $+3e$. Counting the two electrons in the inner shell $1s$ orbital, the calculations suggest that the Li nucleus is surrounded by a total of 2.23 electrons. This in turn suggests that the Li atom carries a net positive charge of $+0.77e$ while the H atom carries a net negative charge of $-0.77e$. These atomic charges correspond to a dipole moment of $\mu_\text{el} = 0.77eR = 5.9$ Debye which happens to be very close to the experimental value of 5.88 Debye.

The formation of the bonding and antibonding molecular orbitals of LiH by linear combination of one atomic orbital from the valence shell of each atom is represented in Fig. 8.1.

Problem 8.4 Find the coefficients of the antibonding orbital of LiH and discuss the charge distribution and dipole moment.

Problem 8.5 Use the ionization energy of the $1s^22p^1$ configuration of the Li atom, IE $=$ 342 kJ mol^{-1}, and simplified Hückel calculations to determine the coefficients in the best possible MO formed by linear combination of the $1s$ H and $2p_z$ Li atomic orbitals.

Problem 8.6 Place the F atom of a HF molecule at the origin and the H atom at the positive z axis. Use simplified Hückel calculations to form an estimate of the shape of the bonding molecular orbital and the charge distribution.

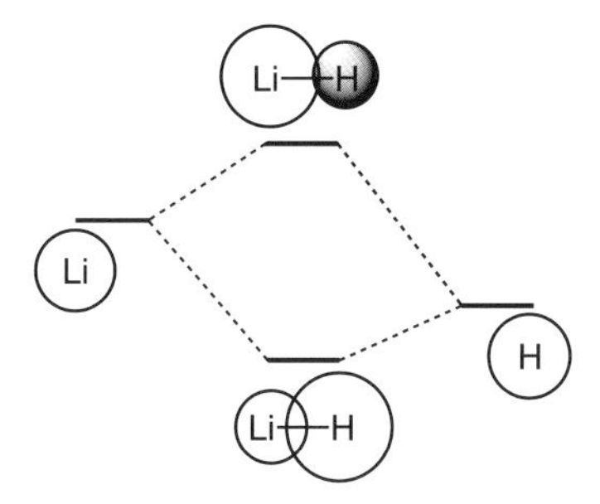

Fig. 8.1. Formation of bonding and antibonding molecular orbitals of LiH by linear combination of one atomic orbital from each atom. Bold lines indicate energy levels, narrow lines the A-B bond axis. On the left the $2s$ AO of the isolated Li atom, on the right the $1s$ AO of the H atom. At the center the bonding and antibonding MOs of LiH. Unshaded circles indicate AOs with positive coefficients, shaded circles AOs with negative coefficient. The relative size of the circles indicates the relative magnitude of the orbital coefficients c_1 and c_2.

8.3 The importance of the "resonance integral", $H_{1,2}$

When using simplified Hückel calculations as a model, we associate the formation of a chemical bond with the formation of a molecular orbital with lower energy than the atomic orbitals containing the valence electrons on the separated atoms. This bonding MO yields particularly high electron density in the space between the two nuclei where the potential energy of the electrons is particularly low.

Let us return to equation (8.15) and assume that the integral $H_{1,2}$ is equal to zero. We then obtain the solutions

$$\varepsilon_1 = H_{1,1} \quad \text{and} \quad \varepsilon_2 = H_{2,2}$$

Inserting $\varepsilon_1 = H_{1,1}$ into (8.14a) we find that $c_2 = 0$. c_1 must therefore be equal to 1. Inserting $\varepsilon_2 = H_{2,2}$ into (8.14b) we obtain $c_1 = 0$ and $c_2 = 1$. Thus, if $H_{1,2}$ is equal to zero, the calculations return the separate AOs with unchanged energy: *no bond has been formed*. The lowering of the energy of the bonding molecular orbital may therefore be associated with the resonance integral $H_{1,2}$.

In Fig. 8.2 we present combinations of atomic orbitals for which overlap and resonance integrals are zero by symmetry: the combinations of σ and π atomic orbitals that do not lead to bonding interactions and are referred to as "forbidden."

In Fig. 8.3 we show the relative orientation atomic orbitals that lead to maximum, positive values of overlap integrals, and to large, negative values for the resonance integrals and hence to maximum bonding interactions. These orientations yield two positive LCAO coefficients for the bonding molecular orbital.

Before going on, we pause to see what would happen if we combine an s and a p_σ atomic orbital oriented in such a way that the overlap integral acquires its maximum *negative* value:

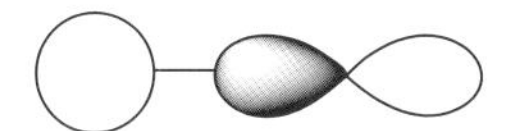

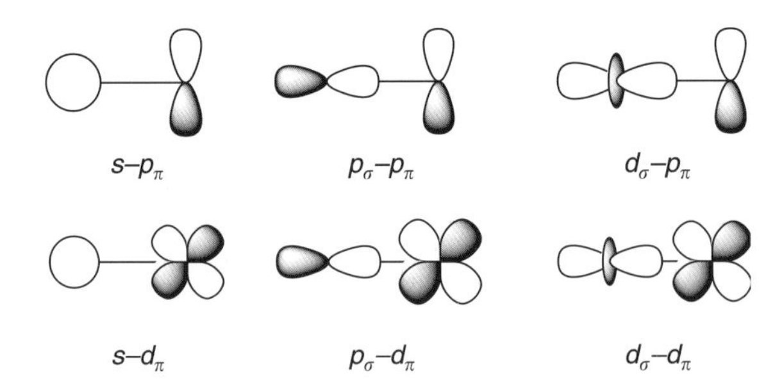

Fig. 8.2. Forbidden combinations of atomic orbitals. As usual, the direction of the z-axes is such that it runs through both nuclei.

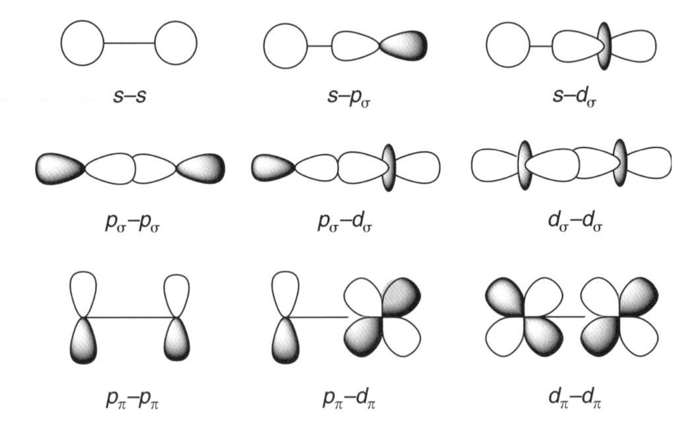

Fig. 8.3. Optimal combinations of atomic orbitals. The orientation of the atomic orbitals has been chosen in such a manner that overlap integrals are positive.

The resonance integral would have same magnitude as for the arrangement shown in Fig. 8.3, but the sign would be now be *positive*. The molecular orbital energies ε calculated for the two arrangements would be equal, but while the orbital coefficients c_1 and c_2 for the bonding MO would come out positive for the arrangement shown in Fig. 8.3, the alternative arrangement would yield a negative value for c_2: the negative coefficient changes the signs of the two lobes of the p_σ orbital.

8.4 A more accurate MO calculation for LiH

MO calculations leading to a full description of the LiH molecule requires the use of an electronic computer. The programs for such calculations are, however, commercially available.

Starting with a basis consisting of $1s$, $2s$ and $2p_\sigma$ atomic orbitals on Li (Φ_1, Φ_2 and Φ_3) and the $1s$ orbital on the H atom (Φ_4), we form *two* molecular orbitals by linear combinations:

$$\Psi_1(x, y, z) = a_1\Phi_1(x, y, z) + a_2\Phi_2(x, y, z) + a_3\Phi_3(x, y, z) + a_4\Phi_4(x, y, z)$$

and

$$\Psi_2(x, y, z) = c_1\, \Phi_1(x, y, z) + c_2\, \Phi_2(x, y, z) + c_3\, \Phi_3(x, y, z) + c_4\, \Phi_4(x, y, z)$$

One molecular orbital, say Ψ_1, would presumably correspond to the $1s$ AO on Li, the coefficient a_1 should be close to unity, a_2, a_3 and a_4 close to zero. The other molecular orbital would presumably be formed from the valence shell orbitals Φ_2, Φ_3 on Li and Φ_4 on H while c_1 would be close to zero.

The coefficients are determined by constructing four spin orbitals, writing down the appropriate Slater determinant, calculating the energy of the molecule from

$$E = \int \Psi\, \mathcal{H}\, \Psi\, d\tau_1\, d\tau_2\, d\tau_3\, d\tau_4\, ds_1\, ds_2\, ds_3\, ds_4$$

and varying the eight coefficients to minimize the energy.

The potential energy curve obtained in this manner has a minimum for $R = 151$ pm as compared to the experimental bond distance of $R_e = 160$ pm [1].

The lowest MO is indeed a pure $1s$ orbital on the Li atom. The second is given by

$$\Psi_2 = 0.43\,(2s)_{\mathrm{Li}} + 0.35\,(2p_\sigma)_{\mathrm{Li}} + 0.54\,(1s)_{\mathrm{H}} \tag{8.18}$$

It is clear that the $2p_\sigma$ atomic orbital on Li is important for a good description of the bonding molecular orbital. This result is hardly surprising, since an electron in a $2p_\sigma$ orbital on Li would yield high electron density on the z-axis, and the energy of the $1s^2 2p^1$ configuration of the Li atom is only 178 kJ mol^{-1} higher than the $1s^2 2s^1$ configuration.

The result nevertheless complicates the simple picture presented in the preceding sections that a normal two-center bond is formed when each atom contributes one valence electron and one valence shell AO to the formation of a bonding MO in which the two valence electrons may be accommodated.

8.5 Hybrid atomic orbitals

Let us consider a function of the type

$$\Phi(x, y, z) = [(2s) + a\,(2p)]/\sqrt{D} \tag{8.19}$$

where the $2s$ and $2p$ atomic orbitals are *centered on the same atom,* and a is a constant. Φ is referred to as a *hybrid* atomic orbital. Since the $2s$ and $2p$ orbitals are orthogonal, the hybrid AO is normalized if $D = (1 + a^2)$.

The hybrid AO yields the electron density

$$\Phi(x, y, z)^2 = \frac{(2s)^2 + 2a\,(2s)(2p) + a^2\,(2p)^2}{1 + a^2} \tag{8.20}$$

The fraction $1/(1 + a^2)$ is referred to as the s-character of the hybrid, and the fraction $a^2/(1 + a^2)$ is referred to as the p-character. Note that the sum of the s- and p-characters is equal to 1. Division of the p-character with the s-character yields the p to s mixing ratio

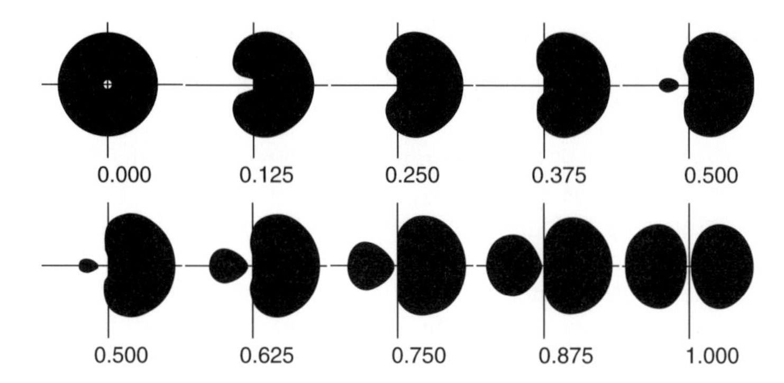

Fig. 8.4. Constant probability density contours of sp_z hybrid atomic orbitals as a function of their p-characters. Top line: p-characters ranging from 0 to 0.500. Bottom line: p-characters ranging from 0.500 to 1.000. Since both s and p_z orbitals have cylinder symmetry about the z-axis, so do the sp_z hybrids. The hybrids have been formed from s and p_z orbitals with identical radial wavefunctions $R(r)$ according to equation (8.20) with positive values for the constant a. The shading covers areas where the electron density is greater than 1/5 of the maximum density for the s orbital.

$\lambda = a^2$. Note also that the mixing ratio has not been restricted to integer values. Finally the hybrid AO is denoted by sp^λ.

Figure 8.4 depicts the probability density contours of sp_z hybrid orbitals in a plane through the z-axis. The $2s$ orbital has spherical symmetry and is positive everywhere, while the $2p_z$ orbital has its highest positive values along the positive z-axis, and equally large negative values along the negative z-axis. The value of the hybrid orbital with p-character equal to 0.125 will therefore be higher than the value of $2s$ orbital for points near the positive z-axis and lower than the $2s$ orbital for points near the negative z-axes. The probability density obtained by squaring the hybrid atomic orbital is thus higher along the positive than along the negative z-axis: the center of gravity of the electron cloud has been moved from the nucleus to some point at the positive z-axis: *the atom has become polarized.*

As the p-character is increased beyond 0.125, the hybrid orbital continues to increase along the positive and to decrease along the negative z-axis. When the p-character reaches 0.250, the hybrid orbital is zero along the negative axis, and when the p-character is further increased the orbital in this part of space becomes increasingly negative. This means that the probability density near the negative z-axis now increases with increasing admixture of the $2p_z$ orbital. When the p-character has reached 0.500, the probability density in this area is sufficiently large to registered by a new lobe on the constant probability density contour diagram. Further increase of p-character leads to further increase of the probability density at negative z, until the two lobes become equal when the p-character reaches 1.000.

Assume that the hybrid orbital

$$\Phi(x, y, z) = \frac{(2s) + a(2p_z)}{\sqrt{1} + a^2} \tag{8.21}$$

contains one electron. We shall calculate the probability of finding the electron at positive z (i.e. above the xy plane):

The probability is given by the integral $\int \Phi^2(x, y, z)\mathrm{d}\tau$ where x and y are integrated from minus to plus infinity but z from zero to plus infinity. If we use polar coordinates, r is integrated from zero to infinity, θ from zero to $\pi/2$, and ϕ from zero to 2π.

$$\int \Phi^2(x, y, z)d\tau = \int \left\{ \frac{[(2s) + a(2p_z)]^2}{1 + a^2} \right\} d\tau$$
$$= \frac{\int (2s)^2 d\tau + 2a \int (2s)(2p_z)d\tau + a^2 \int (2p_z)^2 d\tau}{1 + a^2} \qquad (8.22)$$

Since the $2s$ and $2p_z$ orbitals are normalized, and since both are symmetric with respect to the (xy) plane, the first and the last integral in (8.22) are both equal to $\frac{1}{2}$.

We assume that the $R(r)$ functions of the $2s$ and $2p_z$ orbitals are equal and normalized. The second integral then becomes

$$\int (2s)(2p_z)d\tau = \left[\int R(r)^2 r^2 dr\right]\left[\int (\sqrt{3}/4\pi)\cos\theta\sin\theta \mathrm{d}\theta\right]\left[\int \mathrm{d}\Phi\right] \qquad (8.23)$$

The first of the three integrals in (8.23) is equal to 1, the last is equal to 2π. Hence,

$$\int (2s)(2p_z)\mathrm{d}\tau = \sqrt{\frac{3}{2}} \int \cos\theta \sin\theta \mathrm{d}\theta$$

Integration from $\theta = 0$ to $\pi/2$ yields

$$\int (2s)(2p_z)\mathrm{d}\tau = \sqrt{\frac{3}{4}}$$

The probability of finding the electron at positive z is then

$$\frac{\left[\frac{1}{2} + a\sqrt{\frac{3}{2}} + a^2\frac{1}{2}\right]}{(1 + a^2)} \qquad (8.24)$$

Problem 8.7 Show that the largest probability of finding the electron at positive z is obtained for $a = 1$, i.e. for a hybrid with both s and p characters equal to 0.50. Calculate this probability. Compare with Fig. 8.4.

We now return to the LiH molecule and rewrite the bonding molecular orbital (8.18) as:

$$\Psi(x, y, z) = 0.43[(2s)_{\mathrm{Li}} + 0.81(2p_\sigma)_{\mathrm{Li}}] + 0.54(1s)_{\mathrm{H}} \qquad (8.25)$$

Then we define the hybrid orbital

$$(sp_\sigma)_{\mathrm{Li}} = [(2s) + 0.81(2p_\sigma)]/\sqrt{D}$$

where the p to s mixing ratio is $\lambda = \mathrm{a}^2 = 0.81^2 = 0.66$ and $\sqrt{D} = \sqrt{(1 + \mathrm{a}^2)} = \sqrt{(1 + 0.81^2)} = 1.29$:

$$(sp_\sigma^{0.66})_{\mathrm{Li}} = [(2s) + 0.81(2p_\sigma)]/1.29 \qquad (8.26)$$

Equation (8.25) may now be rewritten as

$$\Psi(x, y, z) = 0.43 \times 1.29(sp_{\sigma}^{0.66})_{\mathrm{Li}} + 0.54(1s)_{\mathrm{H}}$$
$$= 0.55(sp_{\sigma}^{0.66})_{\mathrm{Li}} + 0.54(1s)_{\mathrm{H}}$$

The bonding MO may thus be written as the sum of one atomic orbital from each of the bonded atoms, but the AO on Li is a hybrid.

The effect of the hybridization has been to increase the electron density in the direction of the bond to the H atom. Insertion of $a = 0.81$ into (8.24) yields a probability of 0.92! This means that an electron in the hybrid orbital (8.26) spends more than 90% of the time in a region that brings it close to the H atom. This is presumably favorable for the formation of a short and strong bond.

8.6 Hybridization energies

We may imagine that the bond in LiH is formed in two separate steps: In the first step the Li atom is promoted from the ground state $1s^2 2s^1$ to the "valence state" $1s^2(sp^{\lambda})^1$ electron configuration. Calculations within the atomic orbital approximation indicate that the energy of hybridization is equal to product of the p character of the hybrid orbital and the energy required to excite the Li atom from the $1s^2 2s^1$ to the $1s^2 2p^1$ electron configuration:

$$\Delta E^* = \frac{a^2}{1 + a^2} \Delta E(2s \rightarrow 2p) \tag{8.27}$$

Insertion of $a^2 = 0.66$ and $\Delta E(2s \rightarrow 2p) = 178$ kJmol^{-1} yields $\Delta E^* = 71$ J mol^{-1}.

In the second step the Li-H bond is formed from the Li atom in the valence state (Li*) and the H atom in the ground state. The energy released during this second step is equal to $-BE^* = -\Delta E^* - D_e$. See Fig. 8.5.

Our discussion in this and the preceding section suggests that *the shortest single bond between a given atom pair A–B is formed when the two atoms use hybrid atomic orbitals with s and p characters approximately equal to 0.50 for formation of the two-center molecular*

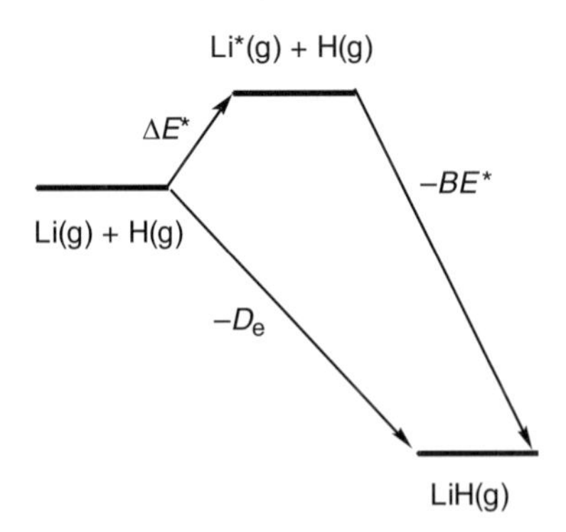

Fig. 8.5. Schematic representation of the formation of LiH(g) from the gaseous Li and H atoms.

orbital. The bond dissociation energy, however, also depends on the energy expended to promote the atom to the valence state.

8.7 Orthogonal hybrid orbitals

Having combined the $2s$ and $2p_z$ atomic orbitals to produce the hybrid orbital

$$(sp_z^{\lambda}) = \frac{(2s) + a\,(2p_z)}{\sqrt{1+a^2}}$$

is it possible to combine the same two atomic orbitals to form a second hybrid which is orthogonal to the first?

Let us write

$$(sp_z^{\kappa}) = \frac{(2s) + b\,(2p_z)}{\sqrt{1+b^2}}$$

where $\kappa = b^2$.

The orthogonality is assured if

$$\int (sp_z^{\lambda})(sp_z^{\kappa})\mathrm{d}\tau = 0 \tag{8.28}$$

Multiplying with $\sqrt{1+a^2}$ and $\sqrt{1+b^2}$ we obtain

$$\int [(2s) + a\,(2p_z)][(2s) + b\,(2p_z)]d\tau = 0$$

$$\int [(2s)^2 + a\,(2s)(2p_z) + b\,(2s)(2p_z) + ab\,(2p_z)^2]\mathrm{d}\tau = 0$$

$$1 + 0 + 0 + ab = 0$$

or

$$b = -\frac{1}{a} \tag{8.29}$$

Since equation (8.29) has only one solution, it is possible to construct no more than two orthogonal hybrid orbitals from the $2s$ and $2p_z$ AOs.

Problem 8.8 Show that both the sum of the s characters of the two hybrids and the sum of their p characters are equal to unity.

Problem 8.9 Assume that the p character of the first hybrid is equal to 0.125. Use the information in Fig. 8.4 to make a sketch indicating the shape and direction of the second hybrid.

Problem 8.10 Consider a hybrid between the $2s$ and $2p_x$ orbitals on Li:

$$(sp_x^{\kappa}) = \frac{(2s) - (1/a)(2p_x)}{1 + (1/a)^2}$$

Is this hybrid orthogonal to the (sp_z) hybrid in equation (8.21)?

8.8 Equivalent hybrid orbitals

Two hybrid AOs, sp^λ and sp^κ, will have the same shape if the p to s mixing ratios are equal. Is it possible to combine the $2s$ and $2p_z$ orbitals on Li in such a way that we obtain two hybrid orbitals that are both orthogonal and have the same shape?

The two hybrids are orthogonal if $b = -1/a$. They will have the same shape if the p to s mixing ratios are equal, i.e. if

$$a^2 = b^2 = (-1/a)^2 = 1/a^2$$

Rearrangement yields

$$a^4 = 1 \quad \text{or} \quad a = \pm 1.$$

If we choose $a = +1$, then $b = -1$, and *vice versa*.

This means that if we want to form two equivalent hybrids, the mixing ratios of both must be equal to unity, and the hybrids are

$$(sp_z^1)_+ = [(2s) + (2p_z)]/\sqrt{2} \quad \text{and} \quad (sp_z^1)_- = [(2s) - (2p_z)]/\sqrt{2}$$

(The superscript 1 is usually deleted, and the two hybrids are referred to simply as *sp* hybrids.)

Our results for the hybridization of one s-orbital with one p-orbital may be generalized as follows:

(i) Beginning with one s orbital and k orthogonal p-orbitals it is possible to construct no more than $1 + k$ orthogonal hybrid orbitals.

(ii) The sum of the s characters of the $k + 1$ orthogonal hybrid orbitals is equal to unity, the sum of their p characters equal to k.

(iii) The orthogonal hybrid orbitals will be equivalent, i.e. have the same shape, if the p to s mixing ratio of each is equal to k.

We now go on to calculate the electron density created by *one electron in each* of the two *sp* hybrid orbitals formed by combination of a $2s$ and a $2p_z$ orbital:

$$\begin{aligned}\rho(x, y, z) &= [(2s) + (2p_z)]^2/2 + [(2s) - (2p_z)]^2/2 \\ &= [(2s)^2 + 2(2s)(2p_z) + (2p_z)^2 + (2s)^2 - 2(2s)(2p_z) + (2p_z)^2]/2 \\ &= [2(2s)^2 + 2(2p_z)^2]/2 \\ &= (2s)^2 + (2p_z)^2\end{aligned}$$

The electron density created by one electron in each of the two equivalent sp_z hybrids is thus identical to the density created by one electron in the $2s$ and one electron in the $2p_z$ orbital.

As long as one works within the atomic and molecular orbital approximations, two wavefunctions that yield the same total electron density can be shown to be mathematically equivalent. Whether we describe an excited Be atom as having the electron configuration $1s^2 2s^1 2p^1$ or as having the one valence electron in each of the two sp_z hybrid orbitals, we are really talking about the same state!

8.9 The estimation of net atomic charges from calculated electron densities

The constant electron density contour diagrams of one homonuclear and five heteronuclear diatomic molecules presented in Fig. 8.6 have been obtained by reasonably accurate quantum chemical calculations. The electric dipole moment of F_2 is zero by symmetry. The electric dipole moments of heteronuclear molecules like LiH, LiF, HF, ClF or CO may be calculated from their electron densities using equation (5.2). These dipole moments, in turn, allow us to calculate the ionic characters: $q_{IC}(\text{calc}) = \mu_{el}(\text{calc})/eR_e(\text{calc})$. In Table 8.1 we compare the calculated ionic characters of 21 heteronuclear diatomic molecules with their experimental counterparts. The agreement between experiment and calculations is good: the average deviation between experimental and calculated values is less than 0.02 a.u, the maximum deviation (in KLi) is 0.05 a.u.

While the electric dipole moment is sufficient to allow us to estimate the atomic charges in a diatomic molecule, this is rarely the case for a polyatomic molecule. Chemists have therefore developed methods to extract estimated atomic charges in such molecules from calculated electron densities. One way to do this is to use the molecular orbitals formed by linear combination of atomic orbitals to distribute the total electron density at any point between the atoms in the molecule, as we have done in Section 8.2 using a simple wavefunction for LiH. Much better results are obtained by using "natural atomic orbitals" NAO. These are hybridized atomic orbitals that are constructed to give the best possible description of the electron density in the molecule under investigation. The total electron density of each atom computed from the natural atomic orbitals extends to infinity in all directions, integration over all space gives the total amount of negative charge

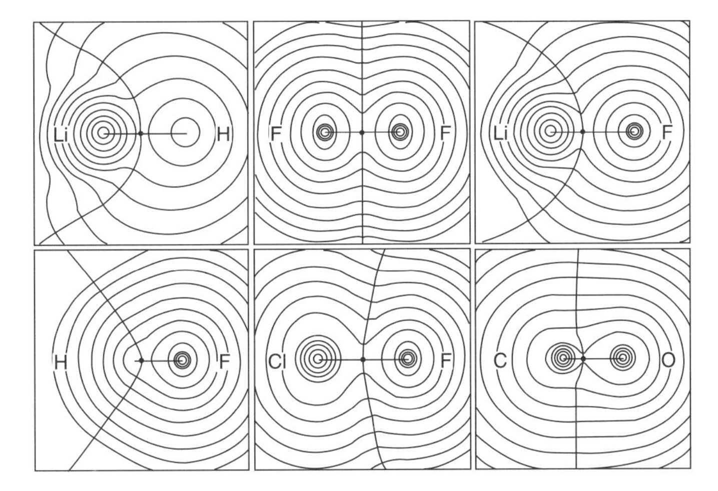

Fig. 8.6. Constant electron density contour diagrams of LiH, F_2, LiF, HF, ClF and CO. Contour lines have been drawn at densities equal to 1.00×10^n and 3.16×10^n atomic units (electrons per a_0^3) with n ranging from -3 to $+2$ in steps of unity (note that $3.16 = \sqrt{10}$). Bonds are indicated by lines between the nuclei, bond critical points (BCPs) by diamonds (♦). Heavy lines indicate boundaries between the atoms according to AIM theory [2].

Table 8.1. Atomic charges in the 21 heteronuclear diatomic molecules formed by combination of H, the alkali metal Li, Na or K, and the halogen F, Cl or Br atoms: experimental and calculated ionic characters, q_{IC}; atomic charges calculated by natural atomic orbital (NAO) analysis, and by topological analysis of the electron densities (AIM) [2]. All charges in atomic units. Note that the chemical formulae have been written in such a way that the atom carrying net positive charge is listed first.

	q_{IC}(exp)	q_{IC}(calc)	q_{NAO}	q_{AIM}		q_{IC}(exp)	q_{IC}(calc)	q_{NAO}	q_{AIM}
Li–F	0.84	0.83	0.96	0.92	K–Na	0.16	0.16	0.19	0.25
Li–Cl	0.73	0.72	0.93	0.92	Cl–F	0.11	0.13	0.35	0.38
Li–Br	0.69	0.68	0.91	0.91	Br–F	0.15	0.18	0.42	0.46
Na–F	0.88	0.87	0.97	0.91	Br–Cl	0.05	0.05	0.11	0.13
Na–Cl	0.79	0.78	0.93	0.88	Li–H	0.77	0.75	0.81	0.90
Na–Br	0.76	0.73	0.92	0.87	Na–H	–	0.67	0.73	0.72
K–F	0.82	0.81	0.96	0.88	K–H	–	0.71	0.81	0.75
K–Cl	0.80	0.79	0.95	0.88	H–F	0.42	0.42	0.55	0.72
K–Br	0.78	0.77	0.94	0.87	H–Cl	0.18	0.19	0.26	0.21
Li–Na	0.03	0.00	0.01	0.13	H–Br	0.12	0.13	0.19	0.07
K–Li	0.22	0.17	0.22	0.31					

associated with the atom. Addition of its nuclear charge gives the "natural atomic charge," q_{NAO} [3].

The so-called Atoms In Molecules (AIM) approach developed by Bader and coworkers [3] is radically different. Topological analysis of the electron density is used to draw boundary surfaces between the atoms, and the negative charge on each atom is found by integration of the total electron density within its boundaries. Addition of the nuclear charge gives the net charge, q_{AIM}.

Consider a homonuclear diatomic molecule like F_2. See Fig. 8.6. If one starts at the midpoint of the F–F bond and moves to the left or right along the z-axis, the electron density will increase until a nucleus has been reached. If one starts at the midpoint and moves away from it along any trajectory perpendicular to the z-axis, the electron density will decrease. The midpoint of the F–F bond thus represents a saddle point on the three-dimensional electron density surface: the three derivatives $\partial\rho/\partial z$, $\partial\rho/\partial x$ and $\partial\rho/\partial y$ are all equal to zero, the second derivative $\partial^2\rho/\partial z^2$ is positive while second derivatives $\partial^2\rho/\partial x^2$ and $\partial^2\rho/\partial y^2$ both are negative. Such a saddle point on the electron density surface is referred to as a bond critical point (BCP). A line beginning at the BCP and following a path of steepest increase (ascent) of electron density will end up at one of the nuclei. If we make an infinitesimal displacement away from the BCP and the z-axis and then follow the direction of the steepest decrease (descent) of electron density, we obtain a straight line perpendicular to the z-axis. If we rotate the line about the z-axis it will sweep out the boundary surface between the two atoms according to the AIM model: a plane perpendicular to the F–F bond and intersecting it at the BCP. This result is, of course, trivial. If a boundary surface is to be drawn between the atoms in a homonuclear diatomic molecule, nobody would do it in a different way.

We now consider a heteronuclear diatomic molecule as, for instance, LiH. The first step in the topological analysis is to find the saddle point (the BCP) on the electron density surface, which for symmetry reasons must lie on the bond axis. Making an infinitesimal displacement away from the bond axis and the BCP and following the direction of steepest descent of the electron density we trace out the bold boundary line shown in Fig. 8.6. The isolated Li^+ cation and the isolated H^- anion both contain two electrons. Because of the greater nuclear charge of the Li cation the electron density at the nucleus is higher and both the electron density, $\rho(r)$, and the slope, $-d\rho(r)/dr$, decrease more rapidly with increasing distance from the nucleus. For this reason the BCP is found to be much closer to the cation than to the anion, and the boundary between the atoms folds back around the cation. If we rotate the molecule about the bond axis, the boundary line will sweep out the AIM boundary surface between the atoms. This boundary surface will extend to infinity in the x- and y-directions. The Li atom (or ion) is thus bounded to the right, but not to the left. Integration of the electron density on the left side of the boundary surface and addition of the nuclear charge (+3 a.u.) yields a net atomic charge of $q_{\mathrm{AIM}} = +0.90$ a.u.

Table 8.1 contains the net atomic charges in 19 diatomic molecules obtained by NAO and AIM analysis. The mean deviation between these net charges is 0.06 a.u., the maximum deviation (in HF) is 0.17 a.u. In all the molecules except LiNa, the net negative charge is carried by the atom with the larger electronegativity coefficient. For LiNa the sign of the calculated dipole moment, q_{NAO}, and q_{AIM} all indicate that the negative charge is carried by the more electropositive Na atom. In Fig. 8.7 we plot the net AIM charges as a function of the NAO charges. Linear regression analysis yields a correlation coefficient of 0.98. In addition, both q_{NAO} and q_{AIM} are strongly correlated with the calculated q_{IC}. (The $q_{\mathrm{NAO}}/q_{\mathrm{IC}}$ and the $q_{\mathrm{AIM}}/q_{\mathrm{IC}}$ correlation coefficients are 0.96 and 0.98 respectively.) Considering that the three estimates have been obtained by very different procedures, the agreement between them is remarkably close!

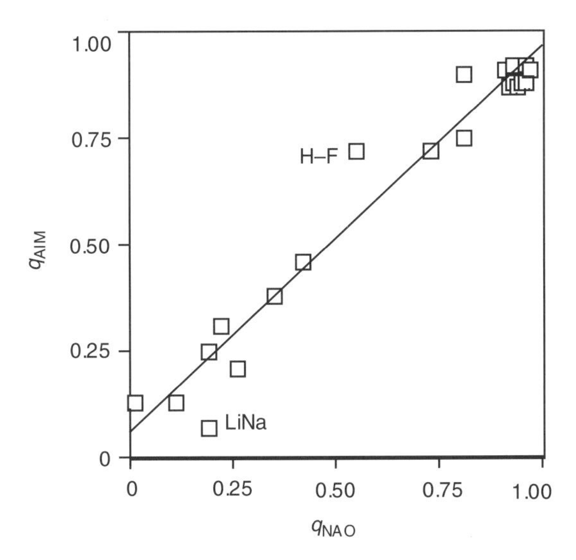

Fig. 8.7. Net atomic charges in 19 heteronuclear diatomic molecules obtained by AIM topological analysis plotted as a function of those obtained by NAO analysis. See Table 8.1.

The relationship between the electric dipole moment and the ionic character of a diatomic molecule on the one hand and the atomic charges obtained by AIM or NBA analysis on the other is not always so straightforward. The carbon monoxide molecule is case in point. The large difference between the electronegativity coefficients of C and O suggests a large dipole moment with the negative pole at the oxygen atom. The experimental electric dipole moment is, however, small and has the negative pole is at the *carbon* atom [5]. This means that if direction of the z-axis is chosen to run from C to O the electric dipole moment is positive, $\mu_{el} = 0.110$ Debye. The ionic character corresponds to a positive charge $q_{IC} = 0.02$ at the oxygen atom.

The quantum chemical calculations described in reference [6] yield a dipole moment of $\mu_{el} = 0.096$ D, also with the positive pole at oxygen. NAO and AIM analysis of the electron density, on the other hand, yield large negative net charges on the O atom, $q_{NAO} = -0.49$ and $q_{AIM} = -1.15$ respectively. The dipole moment, q_{NAO} and q_{AIM} have been obtained by analysis of the same (calculated) electron density. How can the large negative charges on O obtained by AIM or NAO analysis be reconciled with direction and magnitude of the dipole moment?

It should be recalled that the validity of the simple expression for the dipole moment of a diatomic molecule AB

$$\mu_{el,z} = Q_{net,A}(z_A - z_B) \tag{5.5}$$

where $Q_{net,A}$ is the net charge on the positively charged atom, rests on the assumption that the two atoms are spherical. See Section 5.2. Application of this formula to the CO molecule using the net atomic charges obtained by AIM analysis yields a large, negative molecular dipole moment of –6.314 D. A passing glance at the plot of electron densities of CO in Fig. 8.6 is, however, sufficient to show that the atoms created by drawing AIM boundary surfaces are far from spherical. In particular the shape of the C atom is closer to that of a hemisphere than that of a sphere, the center of gravity of the *total* negative charge on the atom is clearly to left of the nucleus. This means that the C atom has a positive dipole moment $\mu_C = Q_-\ (z_- - z_+)$ where Q_- is the *total* negative charge on the C atom and z_+ and z_- are the z-coordinates of the nucleus and of the center of gravity of the electron density respectively. See Section 3.13. It may not be obvious from the figure, but the center of gravity of the total negative charge on the O atom is also to the left of the nucleus. The oxygen atom thus carries a dipole moment with the same sign and about half as large as that of the C atom. The total electric dipole moment of the molecule is given by

$$\mu_{el,z} = Q_{net,O}(z_O - z_C) + \mu_C + \mu_O$$

Insertion of numbers gives

$$\mu_{el} = -6.314\ \text{D} + 4.203\ \text{D} + 2.208\ \text{D} = 0.097\ \text{D}$$

in good agreement with the experimental value [6]. According to the AIM analysis then, the sign of the dipole moment in the CO molecule is *not* determined by the net atomic charges, but by the polarity of the atoms. A similar conclusion is reached by NAO analysis.

The 21 diatomic molecules listed in Table 8.1 are all singly bonded. It is quite possible that the large difference between the ionic character and the net atomic charges determined by NAO and AIM analysis for CO is in large part due to the nature of the multiple bond in this particular molecule (see Section 20.1). One should nevertheless exercise great care when using atomic charges obtained by AIM or NAO analysis to assess bond polarity. The best use of such charges is probably to carry out calculations on a series of closely related molecules and to use the calculated atomic charges to estimate trends, i.e. the direction and magnitude of the change from one molecule to the next.

References

[1] D. J. Shorokhov and A. Haaland, unpublished results. The calculations were carried out with the GAUSSIAN program package with a standard STO 3G basis.
[2] G. Eickerling, T. Strenalyuk and A. Haaland, unpublished results. The electron densities have been calculated for the optimized structures at the B3LYP/6-311++G** level using the GAUSSIAN program package. Topological analysis of the electron densities were carried out using the AIM-PAC program, J. R. Cheseman, T. A. Keith, and R. F. W. Bader, McMaster University, Ontario, 1994.
[3] A. E. Reed, L. A. Curtiss, and F. Weinhold, *Chem Rev.* 88 (1988) 899. See also F. Weinhold and C. R. Landis, *Valency and Bonding*, Cambridge University Press, Cambridge, 2005.
[4] R. F. W. Bader, *Atoms in Molecules: A Quantum Theory*, Oxford University Press, Oxford, 1990.
[5] W. L. Meerts, F. H. de Leeuw, and A. Dymanus, *Chem. Phys.*, 22 (1977) 319.
[6] C. F. Matta and R. J. Gillespie, *J. Chem. Ed.*, 79 (2002) 1141.

Chapter 9

Homonuclear diatomic species of second-period elements from Li_2 to Ne_2

Introduction

In the preceding chapters we have discussed diatomic molecules where the atoms are joined by single bonds. In the following chapters we shall discuss the structure of molecules in which a main group atom is surrounded by two or more singly bonded hydrogen atoms, halogen atoms, or methyl groups. Before doing so we shall, however, make a detour to discuss the homonuclear diatomic species that may be formed from the second-period elements from lithium to neon. The word "species" is used in this context, because in one or two cases the interaction between the two atoms is so weak that it would not normally be described as a chemical bond, nor would the atom pair normally be described as a molecule.

9.1 The dilithium molecule

The bond distance in gaseous Li_2 is $R_e = 267.3$ pm, the dissociation energy is $D_0 = 100$ kJ mol^{-1}. Other spectroscopic parameters are listed in Table 4.1.

The Lewis structure for this molecule would presumably be Li : Li corresponding to a single, covalent bond. The strength of the bond is only a quarter of the strength of the bond in H_2, presumably because of the larger size and more diffuse atomic orbitals on Li.

The first reasonably accurate molecular orbital calculations published nearly 50 years ago [1] yielded the following MO for the two bonding electrons

$$\Psi = 0.35(2s)_A + 0.12(2p_\sigma)_A + 0.35(2s)_B - 0.12(2p_\sigma)_B \tag{9.1}$$

This MO may be rewritten as a linear combination of two $sp_\sigma^{0.12}$ hybrids, one at each atom.

The p to s mixing ratio, 0.12, is so small that as a first approximation we may describe the bonding MO in Li_2 as a combination of the $2s$ orbitals:

$$\Psi = (2s)_A + (2s)_B/\sqrt{D} \tag{9.2}$$

An electron in a $2s$ or a $2p_\sigma$ AO has zero orbital angular momentum about the z-axis. An electron in the molecular orbitals (9.1) or (9.2) will therefore have zero orbital angular momentum about the bond axis. Such an electron is referred to as a σ electron, and the MO is referred to as a σ orbital. Finally the bond that is formed when two electrons occupy a bonding σ orbital is referred to as a σ bond.

The bonding MO formed mainly from the $2s$ AOs in a homonuclear diatomic molecule is often denoted by $2s\sigma_b$;

$$2s\sigma_b = [(2s)_A + (2s)_B]/\sqrt{D} \tag{9.3}$$

while the antibonding combination of the $2s$ orbitals is denoted by $2s\sigma_a$.

For the optimal description of the bonding in LiH we had to employ an $sp_\sigma^{0.66}$ hybrid orbital on the Li atom, in the Li_2 molecule the p to s mixing ratio is reduced to 0.12. The difference may be rationalized by evoking the inherent weakness of the Li–Li bond: The energy required for formation of a hybrid with larger p-character would not be compensated by a greater release of energy during bond formation.

The energy of the $1s^2 2p^1$ configuration of a Li atom is 178 kJ mol^{-1} higher than that of the $1s^1 2s^1$ configuration. In the Be atom the difference between the [He]$2s^1 2p^1$ and [He]$2s^2$ configurations is 263 kJ mol^{-1}. As we move further to the right in the periodic table the energy required to promote an electron from a $2s$ to a $2p$ orbital continues to grow. See Table 3.1 or Fig. 3.1.

When discussing molecules in which a "central atom" forms bonds to several others, as in CH_4, NH_3 or H_2O, we shall find it very useful to construct suitable sp^λ hybrid AOs at the central atom, but we shall find it unnecessary to do the same at "terminal", i.e. mono-coordinate, atoms.

9.2 The beryllium dimer

No Be_2 molecule has been detected in the mass spectra of the vapor above beryllium melts. Any bond between the atoms in Be_2 is obviously too weak for the molecule to survive at temperatures above 1300°C. The bond might be strong enough to prevent dissociation at room temperature, but how to produce and study Be_2 in the gas phase at this temperature where the vapour pressure above the metal is too low to be measured?

A species of composition Be_2 was finally detected in 1984 by vaporizing beryllium metal by intense laser radiation in a stream of He gas cooled to 77 K [2]. The Be atoms produced by the laser beam are initially very "hot," any collision with another metal atom will lead to recoil. Their high kinetic energy will, however, be gradually be dissipated by collisions with the cold He atoms. Two of these "cold" Be atoms may then collide with one another to form a Be_2 molecule before they hit the wall of the reaction chamber and stick there. A spectroscopic investigation has shown that the bond dissociation energy of Be_2, measured from the bottom of the potential well, is $D_e = 9.5$ kJ mol^{-1} and that the internuclear distance is $R_e = 245$ pm. This dissociation energy is so low that most chemists would hesitate to describe the interaction as a chemical bond or the Be_2 "species" as a molecule. The internuclear distance is slightly smaller than the bond distance in Li_2. Since the Be atom is considerably smaller than the Li atom, a single Be–Be bond would be expected to be considerably shorter than a Li–Li bond.

The Be_2 species have no unpaired electrons in their ground state, and the logical Lewis formula is probably Be::Be, corresponding to a double bond. However, the absence – or near absence – of bonding indicates a Lewis structure with two *non-bonding* electron pairs, :BeBe: !

No dihelium molecule has been detected in the gas phase at room temperature: the potential energy curve of a pair of He atoms is in fact strongly repulsive for $R < 140$ pm. (As we shall see below there is an extremely weak attraction between the atoms at larger distances.) The non-existence of the He_2 molecule is commonly explained by assuming that two electrons in the antibonding MO formed by combination of the two $1s$ AOs

$$\Psi_a = [(1s)_A - (1s)_B]/\sqrt{D} \tag{9.4}$$

is more than sufficient to cancel the stabilization due to the two electrons in the bonding orbital Ψ_b.

Calculations show that the ground state electron configuration of Be_2 is $2s\sigma_b^2 2s\sigma_a^2$. The reason why there is a slight stabilization of the Be_2 species is probably that both the $2s\sigma_b^2$ and the $2s\sigma_a^2$ orbitals are slightly stabilized through *sp* hybridization. The Lewis formula **:BeBe:** is thus in good agreement with the molecular orbital description if it is understood that the two pairs of non-bonding electrons represent the net effect of one electron pair in a bonding and another electron pair in an antibonding MO.

9.3 The diboron molecule

B_2 has been detected in high-temperature vapours. The dissociation energy is $D_0 = 290 \pm 10$ kJ mol^{-1}, and the bond distance $R_e = 159.0$ pm. The bond is thus about three times stronger and 40% shorter than in Li_2. While Li_2 and Be_2 are closed shell species, i.e. contain no unpaired electrons, B_2 molecules in the electronic ground state have *two electrons with parallel spins*. This observation suggests that the two electrons occupy two degenerate MOs.

Spectroscopic investigations also show that the two unpaired electrons have orbital angular momentum about the bond axis. The magnitude of this orbital angular momentum is

$$|L_z| = h/2\pi$$

Such electrons are referred to as π electrons.

We begin our discussion of the bonding by considering possible Lewis structures: **B:::B** corresponding to a triple bond; perhaps **.B::B.** corresponding to a double bond and two unpaired non-bonding electrons; or perhaps **:B:B:** corresponding to a single bond. As we shall see, only the double-bond structure is consistent with the presence of two electrons with parallel spins.

In Fig. 9.1 we indicate how the $2p_z$ AOs on two equal second period element atoms may be combined to form one bonding and one antibonding σ orbital, $2p\sigma_b$ and $2p\sigma_a$ respectively:

$$2p\sigma_b = [(2p_z)_A - [(2p_z)_B]/\sqrt{D} \tag{9.5}$$

$$2p\sigma_a = [(2p_z)_A + [(2p_z)_B]/\sqrt{D}$$

The two $2p_x$ orbitals may be also be combined to give a bonding and an antibonding MO:

$$2p\pi_b = [(2p_x)_A + [(2p_x)_B]/\sqrt{D}$$

$$2p\pi_a = [(2p_x)_A - [(2p_x)_B]/\sqrt{D}$$

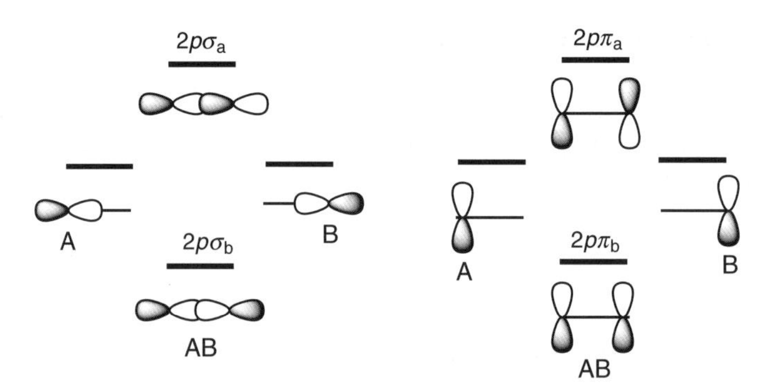

Fig. 9.1. Left: combination of the $2p_z$ atomic orbitals of two equal atoms to form one bonding and one antibonding σ MO. Note that the minus sign in (9.5) modifies the $(2p_z)_B$ function in such a way that the two $2p_z$ orbitals have the same sign in the overlap region. Right: combination of a pair of $2p_\pi$ AOs ($2p_x$ or $2p_y$) to form one bonding and one antibonding π MO. Bold lines indicate energy levels, narrow lines the A–B bond axis.

Just like an electron in one of the $2p_x$ AOs, an electron in an MO formed by linear combination of the two has an orbital angular momentum about the z-axis equal to

$$|L_z| = h/2\pi$$

The two molecular orbitals are therefore labeled $2p\pi_b$ and $2p\pi_a$ respectively. Combination of the $2p_y$ orbitals yields a bonding and an antibonding π MO which are degenerate with those formed from the $2p_x$ AOs.

Note that while the probability density due to an electron in a σ MO is particularly high on the axis connecting the two nuclei, the probability density due to an electron in a π MO is identically equal to zero along the molecular axis.

Four of the six valence electrons in B_2 may be assigned to the $2s\sigma_b$ and $2s\sigma_a$ MOs, but what about the fifth and the sixth? One might be tempted to guess that the energy of the $2p\sigma_b$ orbital is lower than the energy of the $2p\pi_b$ orbital, but the observation that the B_2 molecule has two π electrons with parallel spin, show that the valence shell electron configuration is in fact $2s\sigma_b^2 2s\sigma_a^2 2p\pi_b^2$: according to the rule of maximum spin (Section 2.12) the two electrons have entered different $2p\pi_b$ orbitals with parallel spins.

Since the bonding effect of two electrons in the $2s\sigma_b$ orbital presumably is cancelled by the antibonding effect of two electrons in the $2s\sigma_a$ orbital, the net bonding must be due to the two electrons in the $2p\pi_b$ orbitals. The chemical bond in B_2 may therefore be described as a single π bond. This would be consistent with the Lewis structure **:B:B:**, but the Lewis structure is inadequate insofar as it fails to explain why the two bonding electrons have parallel spins.

Problem 9.1 Show that the electron density due to the two π electrons in B_2 is independent of the angle ϕ, i.e. that the electron density has cylinder symmetry. Hint: Write down the atomic orbitals as functions of the polar coordinates r_A, θ_A and ϕ_A, r_B, θ_B and ϕ_B. Note that if the x and y-axes on the two atoms are parallel, then $\phi_A = \phi_B = \phi$. See Fig. 7.2.

9.4 The dicarbon and dinitrogen molecules

Like B_2, C_2 has been investigated in high-temperature vapors. The dissociation energy is $D_0 = 599$ kJ mol^{-1}, the bond distance $R_e = 124.5$ pm. The bond is thus twice as strong and 20% shorter than in B_2. C_2 contains no unpaired electrons.

The valence electron configuration is $2s\sigma_b^2 2s\sigma_a^2 2p\pi_b^4$ corresponding to a double π bond. The Lewis formula **:C::C:** leads to the prediction of a doubly bonded molecule with no unpaired electrons, but does not tell us that the bonding electrons are all π electrons.

The dissociation energy of the dinitrogen molecule is $D_0 = 942$ kJ mol^{-1}. This is one of the strongest chemical bond known to man. The bond distance is $R_e = 109.8$ pm. The N_2 molecule is repelled by a magnetic field, i.e. it is pulled in the direction of decreasing magnetic field strength. Such molecules are referred to as *diamagnetic*. Diamagnetic molecules have no net spin, i.e. they contain only electron pairs. The valence electron configuration is $2s\sigma_b^2 2s\sigma_a^2 2p\pi_b^4 2p\sigma_b^2$ corresponding to a single σ-bond and a double π-bond.

The Lewis structure **:N:::N:** is thus in agreement with all the molecular properties.

9.5 The dioxygen and diflourine molecules

The dissociation energy of O_2 is $D_0 = 494$ kJ mol^{-1}, the bond distance is 120.8 pm. Compared to N_2 the bond strength has been reduced by nearly 50% and the bond distance increased by 10%. The oxygen molecule is attracted by magnetic fields, i.e. it is pulled in the direction of increasing magnetic field strength. Such molecules are referred to as *paramagnetic*. Paramagnetic molecules contain one or more unpaired electrons. Since O_2 contains an even number of electrons, the number of unpaired electrons must also be even. Closer examination shows that there are *two* unpaired π electrons. The O_2 molecule appears to be the only main group element compound with two unpaired electrons that is stable in the gas phase at ordinary temperatures and pressures.

The valence electron configuration is $2s\sigma_b^2 2s\sigma_a^2 2p\pi_b^4 2p\sigma_b^2 2p\pi_a^2$. The two electrons in the antibonding π MOs presumably cancel the bonding effect of two of the four electrons in the corresponding bonding π MOs. The bond in O_2 is therefore described as a double bond consisting of a σ bond plus a π bond. The Lewis structure **::O::O::** leads to the prediction of a double bond, but, as in the case of B_2, it fails to indicate that the molecule contains two unpaired electrons.

F_2 is diamagnetic. The dissociation energy is $D_0 = 155$ kJ mol^{-1} and the bond distance $R_e = 141.3$ pm. Compared to O_2 the bond strength has been reduced by nearly 70% and the bond distance increased by 17%. The two additional electrons have entered the $2p\pi_a$ orbital, canceling the bonding effect of two more electrons in the bonding π MOs. The chemical bond is due to the two electrons in the $2p\sigma_b$ orbital and it is described as a single σ bond.

Calculations show that the energy of two Ne atoms placed less than 200 pm apart is *higher* than the energy at infinite distance: Two electrons must now be placed in the antibonding $2p\sigma_a$ orbital and their presence more than cancels the bonding effect of the two electrons in the $2p\sigma_b$ orbital: the potential energy curve is in fact repulsive for internuclear distances smaller than 300 pm.

9.6 A molecular orbital energy level diagram for the species from Li_2 to Ne_2

The valence shell electron configurations of all the species from Li_2 to Ne_2 may thus be rationalized on the basis of the schematic energy level diagram shown in Fig. 9.2. This fact does not imply that the diagram depicts the correct order of the energy levels for each species, since exchange of the energies of two filled orbitals would leave the electron configuration unchanged.

Constant probability density contours of molecular orbitals derived from reasonably accurate quantum chemical calculations first became available in the 1960s. Such constant probability density contours of the filled valence shell MOs in F_2 are shown in Fig. 9.3.

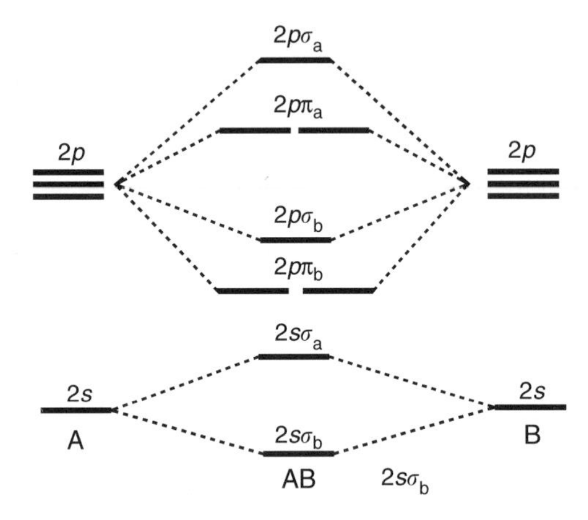

Fig. 9.2. A schematic molecular orbital energy diagram which, together with the *Aufbau* principle leads to the prediction of the correct electron configuration of all the diatomic homonuclear species of second-period elements from Li_2 to Ne_2.

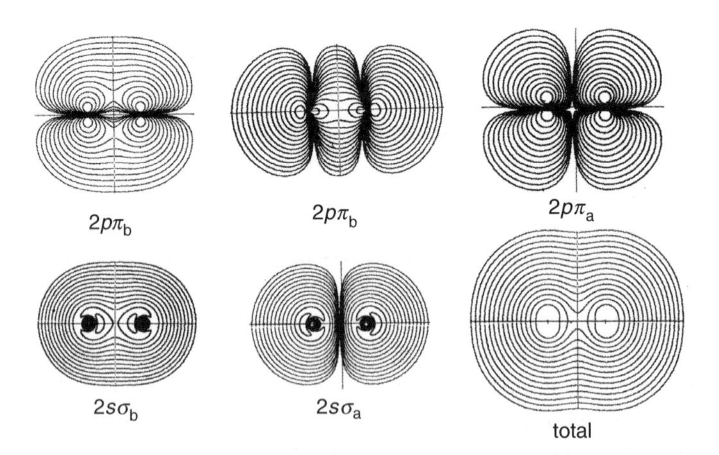

Fig. 9.3. Constant probability density contours of filled valence shell molecular orbitals in F_2. Total valence electron density. Reproduced with permission from A. C. Wahl, *Science*, 151 (1966) 961.

Table 9.1. Dissociation energies and bond distances in gaseous homonuclear diatomic species (charged or uncharged) formed from the second period elements from lithium to neon.[a]

A–A	D_0 (kJ mol^{-1})	R_e (pm)	Comment
Li–Li	100	267.3	
Be–Be	9.5[b,c]	245[c]	
B–B	290 ± 10	159.0	Two unpaired π electrons with parallel spins
C–C	599	124.5	
$(C–C)^-$	818 ± 10	126.8	
$(N–N)^+$	841	111.6	
N–N	942	109.8	
$(N–N)^-$	765	119	
$(O–O)^+$	643	111.6	
O–O	494	120.8	Two unpaired π electrons with parallel spins
$(O–O)^-$	395	134[d]	
F–F	155	141.3	
Ne–Ne	0.3[b,e]	309[e]	

[a] Data reference [3] unless otherwise noted. [b] Dissociation energy from bottom of potential well, D_e. [c] Reference [2]. [d] Reference [4]. [e] Reference [5].

In Table 9.1 we have collected the dissociation energies and bond distances in the neutral species discussed above, along with the corresponding parameters for some related cations and anions.

Problem 9.2 Use the schematic orbital energy diagram in Fig. 9.2 to assign the valence shell electron configurations of the molecular ions in Table 9.1. Try to find general answers to the two questions

(i) What happens to the dissociation energy and bond distance of a neutral diatomic molecule when you remove one of the electrons to form a cation?
(ii) What happens to the dissociation energy and bond distance of a neutral diatomic molecule when the molecule absorbs one electron to form an anion?

9.7 The long-range attraction between neon atoms

While Ne atoms repel one another at short range, there is a very weak attractive force when $R > 310$ pm. If no such attraction existed, neon would never condense to a liquid or freeze to a solid. (Condensation of a gas to form a liquid or a solid is *not* an effect of gravity. Why would water condense on the ceiling if it were?) This interatomic attraction also leads to deviation from ideal gas behavior at low temperatures and high pressures.

Information about the attraction between a pair of noble gas atoms may be derived from measurements of gas properties, from spectroscopic investigations, or from molecular beam experiments. A molecular beam consists of molecules traveling with equal velocity along parallel paths. Information about interactions between the atoms or molecules in two beams may be obtained by analyzing the scattering patterns when two beams are crossed.

Relatively reliable information may also be obtained from quantum chemical calculations provided they include electron correlation effects.

A curve showing the potential energy of two noble gas atoms as a function of internuclear distance R has the same general shape as the potential energy curve for a diatomic molecule. However, we denote the internuclear distance corresponding to the lowest (minimum) energy by R_m rather than R_e, and the well depth by D_m rather than by D_e.

The potential energy of a pair of Ne atoms is at a minimum when the internuclear distance is about $R_m = 309$ pm. At this distance the energy is $D_m = 0.34$ kJ mol^{-1} lower than when the atoms are infinitely far apart. There are two vibrational levels in the well: the dissociation energy from the lowest level is $D_0 = 0.20$ kJ mol^{-1}, the dissociation energy from the second level is about 0.03 kJ mol^{-1} [5]. These dissociation energies may be compared to the average kinetic energy of gas molecules at room temperature, KE $= (3/2)RT = 3.7$ kJ mol^{-1}: the energy involved in a collision with the walls or between two molecules is so large, that an atom pair is extremely unlikely to survive the shock.

The well depths and minimum energy distances R_m of other homonuclear dimers of the noble gas atoms are displayed in Table 9.2. Note that both the well depth and the minimum-energy distance increase as the group is descended.

An attraction between atoms that is too weak to be classified as a chemical bond is often referred to as a "van der Waals attraction." The minimum energy distance R_m is thought to correspond to a situation where the atoms barely touch, and half this distance is referred to as the van der Waals radius of the atom, r_W. These radii are listed in the last column of Table 9.2.

Division of the potential energy $V(R)$ by the well depth yields the dimensionless function, $V(R)/D_m$. This quantity may now be plotted as a function of R/R_m to yield a so-called "reduced" potential energy curve. The reduced potential energy curves of HeHe, NeNe and ArAr are displayed in Fig. 9.4. It is seen that the three curves have very nearly the same shape. When R/R_m is decreased below unity, the energy begins to rise. The crossover point, Σ, where the potential energy is equal to zero, is found at $\Sigma = 0.90R_m$. If R is further reduced the energy increases rapidly. The parameter $\Sigma/2 = 0.90r_W$ may therefore

Table 9.2. Homonuclear pairs of noble gas atoms: well depths, D_m; minimum-energy internuclear distances, R_m, and van der Waals radii, r_W.[a]

A···A	D_m (kJ mol^{-1})	R_m (pm)	r_W (pm)
He···He	0.09	296	148
Ne···Ne	0.35	309	154
Ar···Ar	1.19	376	188
Kr···Kr	1.67	401	201
Xe···Xe	2.35	436	218

[a] Data from reference [6]. Well depths are probably accurate to about 5%, minimum-energy distances probably to about 2%.

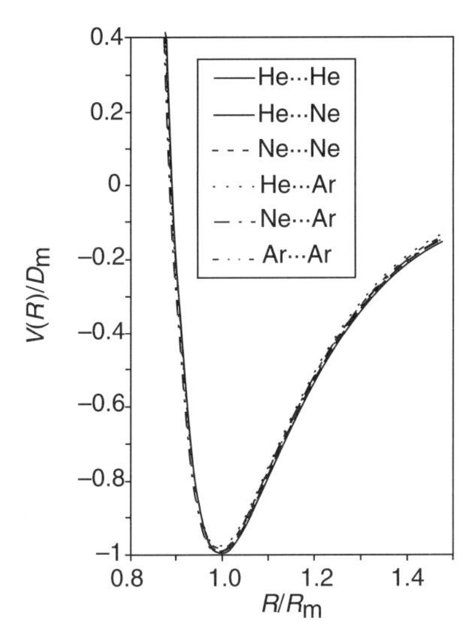

Fig. 9.4. Comparison of the reduced van der Waals potential curves, $V(R)/D_m$, of noble gas atom pairs. Reproduced with permission from T. A. Halgren, *J. Am. Chem. Soc.*, 114 (1992) 7827. Copyright (1997) American Chemical Society.

Table 9.3. Heteronuclear pairs of noble gas atoms: well depths, D_m; minimum-energy internuclear distances, R_m, and the sum of their van der Waals radii, $r_W(A) + r_W(B)$.[a]

A···B	D_m (kJ mol^{-1})	R_m (pm)	$r_W(A) + r_W(B)$ (pm)
He···Ne	0.17	304	302
He···*Ar*	0.25	348	336
He···Kr	0.25	369	349
He···Xe	0.23	397	366
Ne···Ar	0.56	349	342
Ne···Kr	0.60	362	355
Ne···Xe	0.62	386	372
Ar···Kr	1.39	388	389
Ar···Xe	1.57	407	406

[a] Data from references [6] and [7]. Well depths are probably accurate to about 5%, minimum-energy distances probably to about 2%.

be regarded as the radius of a sphere which the noble gas atom fills completely, and from which it tries to exclude all other atoms.

The reduced potential energy curves of the three heteroatomic pairs HeNe, HeAr and NeAr are included in Fig. 9.4. They are seen to have the same shape as the curves for homoatomic pairs.

The well depth and minimum energy distances for atom pairs formed from different noble gases are listed in Table 9.3. The minimum energy distances are seen to be reasonably close to the sum of the van der Waals radii of the two atoms.

A potential energy curve for a pair of methane molecules has been determined by molecular beam experiments under the assumption that the CH_4 molecules may be described as spheres: the well depth was found to be 1.66 kJ mol^{-1} and the minimum energy distance $R_m = 402$ pm.[8] The zero energy crossing point is $\Sigma = 362$ pm. These parameters are very close to those obtained for the noble gas krypton.

Problem 9.3 Compare the parameter $\Sigma/2$ of He with the classical radius of the He atom, equation (2.20).

9.8 The helium dimer

Experiments carried out on He gas cooled to temperatures below 10^{-3} K have demonstrated the presence of bound atom pairs with an average internuclear distance of 6200 ± 1000 pm, nearly one hundred times larger than the bond distance in H_2! Calculations indicate that there is just one vibrational level below the dissociation energy threshold and that D_0 is about 0.01 kJ mol^{-1} [9]. Since R_m is about 300 pm, the huge average internuclear distance suggests that vibrations allow the internuclear distance to increase to more than 10000 pm before the atoms must turn back.

9.9 van der Waals interactions: dipole–dipole interactions; dipole–induced dipole interactions; and dispersion interactions

We have seen (in Problem 5.9) that the Coulomb interaction energy of two electric dipoles lying head to tail at a distance R from one another is equal to

$$E = -2\frac{\mu_1\mu_2}{4\pi\varepsilon_0}\left(\frac{1}{R^3}\right)$$

In the gaseous or liquid phase the molecules will be rotating. The relative orientation of the dipoles will not be fixed, and instants with positive interaction energies will alternate with instants with negative interaction energies. The interaction energy will, however, not average to zero, since orientations with negative interaction energies will be favored. The resulting net, average interaction energy is given by

$$E = -\frac{2}{3\,kT}\frac{\mu_1\mu_2}{(4\pi\varepsilon_0)^2}\left(\frac{1}{R^6}\right) \tag{9.6}$$

Dipole–dipole attractions are therefore active in both liquids and gases.

If a rotating molecule with a permanent dipole moment is brought close to a non-polar molecule, it may induce an instantaneous dipole moment in the latter in such a manner that the net effect averaged over the rotational motion is a lowering of the energy given by

$$E = -\frac{\mu_1^2 \alpha_2}{(4\pi\varepsilon_0)^2}\left(\frac{1}{R^6}\right) \tag{9.7}$$

where μ_1 is the permanent dipole moment of the polar molecule and α_2 the polarizability of the non-polar.

Finally we consider interactions between two non-polar atoms or molecules, for instance two noble gas atoms. Their permanent dipole moments are zero, but for a brief fraction of a second the arrangement of electrons in one of them may be such that their centre of gravity does not coincide with the nucleus. The atom will then have an instantaneous dipole moment, and this instantaneous dipole will induce a dipole on the other atom in such a manner that the instantaneous energy is lowered. As time passes, the movement of the electrons in the two atoms is correlated in such a manner that there is a longer-lasting, net lowering of the energy. Such attractive interactions between instantaneous dipoles are referred to as *dispersion* forces.

The dispersion interaction increases with the product of the polarizabilities of the two molecules or atoms, and deceases with the distance between them:

$$E = -\frac{c\alpha_1\alpha_2}{R^6} \tag{9.8}$$

The energy of interaction between rotating dipoles, between a permanent and an induced dipole, and dispersion interactions are among those lumped together under the label "van der Waals forces." In gases and liquids the stronger of the three will normally be the dispersion forces. Note, however, that all these interaction energies vary as $1/R^6$. It is therefore often assumed that the potential energy due to forces between simple molecules in gases or liquids is given by an expression of the form:

$$V(R) = -\frac{C}{R^6} + \frac{B}{R^{12}}$$

where the first term represent van der Waals attractions (equations 9.6, 9.7 and 9.8) and the second Born repulsion at close range. C and B are empirical constants. Such a potential curve is referred to as a Lennard-Jones 6–12 potential.

9.10 van der Waals radii from crystal structures

We have already noted that the formation of liquids and solids at low temperatures is due to intermolecular attractions. Solid state Ne forms a cubic close-packed lattice: each atom is surrounded by 12 nearest neighbors at a distance of 316 pm, about 2% longer than the R_m distance obtained experimentally by molecular beam studies. The crystal structure of methane at 35 K is also cubic close packed with twelve nearest neighbors at $R(C\cdots C) =$ 416 pm [10] or about 3% longer than the R_m distance of 402 pm. These results indicate that information about the van der Waals radii of atoms may be obtained from the distances

Table 9.4. van der Waals radii of atoms (in pm) derived from crystal structures.[a]

H: 120			
C: 170	N: 155	O: 152	F: 147
Si: 210	P: 180	S: 180	Cl: 175
Ge	As: 185	Se: 190	Br: 185
Sn	Sb	Te: 206	I: 198
Methyl group radius: 200			
Thickness of aromatic ring: 170			

[a] Data from reference [11].

Fig. 9.5. A space-filling model of the methane molecule.

between non-bonded atoms in molecular crystals. A set of van der Waals radii estimated in this manner is listed in Table 9.4.

Problem 9.4 It is sometimes said that the van der Waals radius of an atom is twice the bonding radius. Compare the radii of the halogen atoms in Table 9.4 with the bond distances of the dihalogens. How accurate is this rule of thumb?

The "space-filling" drawing of the methane molecule shown Fig. 9.5 has been made by drawing spheres around the nuclei with radii corresponding to the van der Waals radii of C and H atoms respectively. The minimum extension of the molecule is equal to the radius of the C atom, 170 pm, the maximum extension is 229 pm, equal to the C–H bond distance (109 pm) plus the van der Waals radius of the H atom, (120 pm). It is, therefore, not unreasonable to approximate the molecule by a sphere with radius equal to 200 pm, as indicated by the molecular beam experiments.

Problem 9.5 Suggest a reasonable estimate for the van der Waals radius of the neopentane molecule, $C(CH_3)_4$.

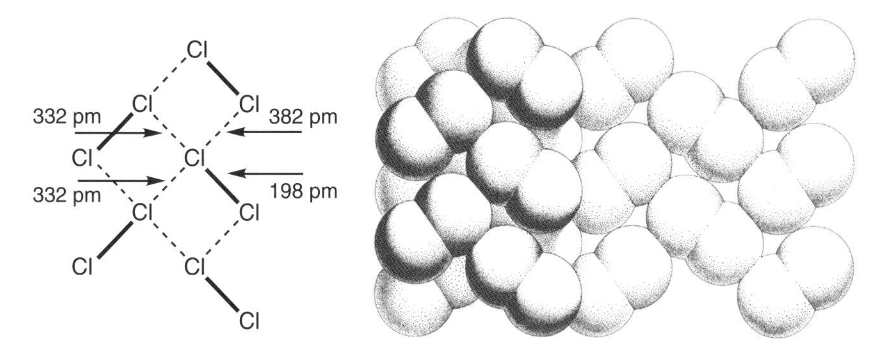

Fig. 9.6. The crystal structure of Cl_2. Reproduced with permission from J. Donnohue, *The Structures of the Elements*, John Wiley & Sons, New York, 1974. Copyright (1972) John Wiley & Sons.

9.11 The crystal structure of chlorine

In Fig. 9.6 we display a section of the crystal structure of Cl_2. The crystal consists of layers of Cl_2 molecules arranged in a herring-bone pattern. Each Cl atom is bonded to another at a distance of 198 pm, not significantly different from the bond distance in gas phase. In addition it appears to touch three atoms in neighboring molecules in the same layer at distances of 332 (twice) and 382 pm. The distances to nearest neighbor Cl atoms in other layers range from 374 to 397 pm. The average of the three non-bonded distances in the layer is 349 pm, i.e. twice as large as the van der Waals radius of Cl (175 pm) derived by examination of nonbonded Cl$\cdots$Cl distances in a large number of organic chlorine compounds.

9.12 The liquid structures of neon and methane

Crystalline neon melts at 24.5 K. In the crystalline phase the atoms vibrate about fixed equilibrium positions, in the liquid phase they are relatively free to move. Their kinetic energy increases with increasing temperature, and as collisions between them become more violent, they are knocked further apart and small cavities are opened between them. The surroundings of any given atom are constantly changing, some nearest neighbors move away and others move into the vacant spaces. A neutron diffraction study of liquid Ne just below its normal boiling point of 26.1 K indicates that at a given instant each neon atom is surrounded by an average number of 9.5 ± 0.5 nearest neighbors at distances ranging from about 270 to about 400 pm [12].

Crystalline methane melts at 91 K. A study by X-ray diffraction study of the liquid at 96 K showed that each CH_4 molecule is surrounded by 11.4 nearest neighbors at an average distance of 424 pm, about 2% longer than the distance in the solid state at 35 K. At 186 K the average kinetic energy per molecule has been doubled, the average number of nearest neighbors has been reduced from 11.4 to 7.2 and the average distance increased from 424 to 435 pm [13].

References

[1] B. J. Ransil, *Rev. Mod. Phys.*, 32 (1960) 239.
[2] V. E. Bondybey, *Chem. Phys. Letters*, 109 (1984) 436.
[3] K. P. Huber and G. Herzberg, *Molecular Spectra and Molecular Structure*, Vol. IV, *Constants of Diatomic Molecules*, Van Nostrand, New York, 1979.
[4] R. J. Celotta, R. A. Bennet, J. L. Hall, M. W. Siegel, and J. Levine, *Phys. Rev., A* 6 (1972) 631.
[5] R. A. Aziz, W. J. Heath, and A. R. Allnatt, *Chem. Phys.* 78 (1983) 295.
[6] T. A. Halgren, *J. Am. Chem. Soc.*, 114 (1992) 7827.
[7] M. Waldmann and A. T. Hagler, *J. Comput. Chem.*, 9 (1993) 1077.
[8] B. P. Reid, M. J. O'Loughlin, and R. K. Sparks, *J. Chem. Phys.*, 83 (1985) 5656.
[9] F. Luo, C. F. Giese, W. R. Gentry, *J. Chem. Phys.*, 104 (1996) 1151.
[10] S. C. Greer and L. Meyer, *Angew. Phys.* 27 (1969) 198.
[11] A. Bondi, *J. Phys. Chem.*, 68 (1964) 441.
[12] M. C. Bellissent-Funel, U. Buontempo, A. Filabozzi, C. Petrillo, and F. P. Ricci, *Phys. Rev. B*, 45 (1992) 4605.
[13] J. I. Petz, *J. Chem. Phys.*, 43 (1965) 2238.

Chapter 10

Structure and bonding in the gaseous dihalides and dialkyl derivatives of the Group 2 and Group 12 metals

Introduction

All the Group 2 metals, beryllium, magnesium, calcium, strontium and barium, have the electron configuration [Ng]ns^2 where Ng denotes one of the noble gases in Group 18, while the Group 12 metals, zinc, cadmium and mercury, have [Ng]$(n-1)d^{10}ns^2$ electron configurations. Only the two *ns* electrons are described as valence electrons: the energy of the $(n-1)d$ electrons is too low for them to participate in the formation of chemical bonds. The addition of the 10 *d* electrons has, however, two important effects. The concomitant increase of the nuclear charge leads to a significant reduction of the atomic radius and an equally significant increase of the ionization energies and the electronegativity coefficient. In the modern version of the periodic table, Be and Mg are placed above Ca, Sr and Ba, but in the first (1869) version of Mendeleev's periodic table, which was based on chemical similarities and atomic weights, they were placed above Zn and Cd.

10.1 Are the gaseous dihalides of the Group 2 and 12 halides linear?

All the Group 2 and 12 metals form halides of composition MF_2, MCl_2, MBr_2 and MI_2. At room temperature all these compounds are solids, but the gaseous molecules can be studied by high temperature techniques. As expected the atoms are joined in the sequence XMX, and the two M–X bond distances are assumed and found to be equal. But are the molecules linear or angular (bent)?

In Fig. 10.1 we show two curves that indicate how the potential energy of the molecule might vary with the valence angle ∠XMX. Curve A has a minimum at ∠XMX = 180°, in such a case the *equilibrium* structure of the molecule is linear. The molecule will, however, undergo bending vibrations, and unless we happen to observe the molecule at the particular instant when it passes through the equilibrium point, the molecule is in fact bent. The frequencies of such bending vibrations normally fall in the range 10^{12}–10^{13} vibrations per second, which means that the time required for a vibrational cycle lies between 0.1 and 1 picosecond. (1 ps = 10^{-12} s.)

Whenever the molecule is linear, the electric dipole moment will be zero. Whenever it is bent, it will have an instantaneous dipole moment. During one vibrational cycle the electric dipole moment will alternate between positive and negative values, but if we observe the molecule over a longer time than it takes to complete a vibrational cycle, that is for longer than 100 ps, the dipole moment will average out to zero: the molecule has no *permanent* dipole moment.

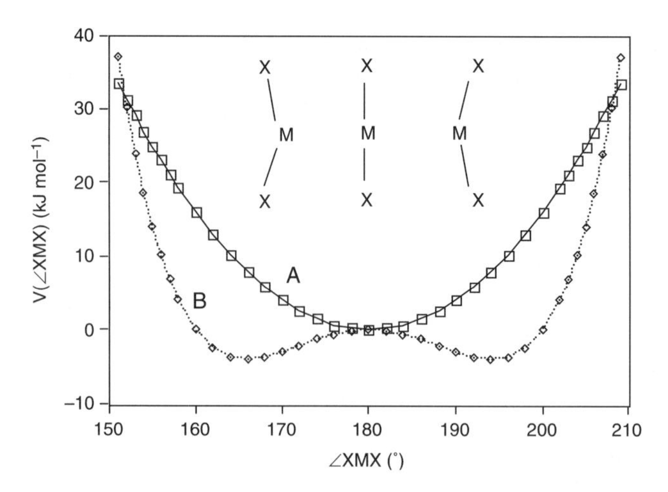

Fig. 10.1. The potential energy of a metal dihalide molecule XMX as a function of the valence angle ∠XMX. Curve A: the equilibrium structure is linear. Curve B: the equilibrium structure is bent.

Curve B in Fig. 10.1 has a local *maximum* at 180°, and two equivalent minima in the plane of the paper corresponding to two equivalent angular configurations. In such a case the equilibrium structure is bent. Bending vibrations will not change the shape: By observing the molecule at different instants, we will obtain various values for the valence angle ∠XMX, but only very rarely will we observe a molecule which has enough energy to open the valence angle to 180°, and passes through that point just when we happen to observe it.

If the molecule is angular in the equilibrium configuration, the electric dipole moment will not average to zero even if we observe for a time that is longer than a vibrational cycle. Such a molecule has a permanent dipole moment.

If a molecule with a dipole moment is placed in an inhomogeneous electric field, it will be subject to a force pulling it towards or away from those regions where the field is most intense. The direction and magnitude of the force is determined by the magnitude of the dipole moment and its direction relative to the electric field gradient. In an *electric deflection* experiment we send a beam consisting of molecules traveling along straight, parallel lines through an inhomogeneous, so-called quadrupolar electric field. In a typical experiment the molecular velocities are of the order of 10 to 100 meters per second and the length of the field of the order of 10 cm. This means that the time required for a molecule to pass through the field is of the order of 1 millisecond, much longer than required for a rotational cycle. For a molecule with a *linear* equilibrium structure the bending vibration will ensure that a moment of attraction due to the instantaneous dipole moment will immediately be followed by a moment of equally large repulsion. The effect of the electric field therefore averages out to zero, and the molecules emerge from the field with their direction of motion unaltered.

If molecules with *permanent* electric dipole moments are passed through the same field, the rotation of the molecules will cause orientations leading to attraction into the field to alternate with orientations leading to repulsion. For most of the molecules in the beam the effects of attraction and repulsion will cancel out and they will emerge from the field with unchanged direction. However, for some of the molecules in particular rotational states, the moments of attraction and repulsion will *not* cancel. These molecules will be deflected and travel along curved trajectories. If the deflection is large enough to be detected, this proves that at least some of the molecules in the beam have permanent dipole moments and, in the case of metal dihalides, that they cannot have linear equilibrium structures. The converse statement is, however, not true: we cannot conclude that a molecule is linear even if we observe no deflection, the permanent dipole moment may be too small for the deflection to be detected.

Another way to establish the shape of the molecule is to carry out an investigation by gas electron diffraction. During such an experiment, the molecules in a molecular beam are bombarded by high-velocity ($v \approx 10^7$ m s^{-1}) electrons. The electrons are scattered and analysis of the scattering pattern allows us to determine the M–X bond distances as well as the nonbonded X···X distance. This, in turn, allows us to calculate the $\angle$XMX valence angle. Since the length of the molecule is less than 1000 pm, it takes the electron less than 10^{-16} seconds to pass through, much smaller than the vibrational period. This means that even if the equilibrium structure is linear, the electron will register the instantaneous, angular structure. During a gas electron diffraction experiment we record the results of a large number of electron–molecule collisions, and processing of the data yields the average of all the instantaneous M–X and X···X distances. Because of the bending vibrations, the average X···X distance will be less than twice the M–X bond distance and the average angle calculated from it always less than 180° even if the equilibrium structure is linear. If the experiment is carried out at a high temperature, say above 1000 kelvin, and if the molecule does not require much energy to bend, the average angle found by gas electron diffraction may be smaller than 150° even if the equilibrium angle is 180°.

Thus the different time scales of the molecular deflection and electron diffraction processes have a profound effect on the outcome of the investigation of a molecule with linear equilibrium structure: the gas electron diffraction experiment with a time scale much shorter than the vibrational period, registers the average of the instantaneous, angular structures, while the molecular deflection experiment with a time scale much longer than the vibrational period, registers the linear equilibrium structure.

In principle it is possible to use additional information about the molecule to correct the average angle obtained by gas electron diffraction, i.e. to calculate the difference between the equilibrium and average angle. This is, however, not easy, and the additional information is not always available.

Another point that needs to be made concerns the uncertainty that will always adhere to a measured quantity like the $\angle$XMX valence angle. Let us assume that we have determined the valence angle in an alkaline earth metal halide by gas electron diffraction and corrected for the difference between the average and equilibrium angles. The data then indicate that the equilibrium valence angle is 177°, but we realize that the number carries an uncertainty of $\pm$5°. This means that we have succeeded in proving that the valence angle is larger than 172°, but *not* that it is exactly 180° as required for a strictly linear molecule. The best way to describe the result in words is probably to say that the equilibrium structure of the molecule is either linear or nearly linear. If the experimental data indicate that the equilibrium valence angle is 140° with an estimated error limit of $\pm$8°, the situation is clear: we have proven that the molecule is bent. Examination of the molecular deflection and gas electron diffraction experiments thus leads to the same conclusion: *Asymmetry (non-linearity) can be proven experimentally, but symmetry (linearity) cannot.*

Table 10.1. The equilibrium geometries of the monomeric Group 2 metal dihalides in the gas phase. L = linear, A = angular.

	F	Cl	Br	I
Be	L	L	L	L
Mg	L	L	L	L
Ca	A	L	L	L
Sr	A	A	L	L
Ba	A	A	A	A

This conclusion can be generalized to molecules with more than two atoms. Thus neither the trigonal planar symmetry of AB_3 molecules (all BAB valence angles equal to 120°) or the tetrahedral symmetry of AB_4 molecules (all BAB angles equal to 109.47°) can be proven, while deviation from such symmetry can. What happens in practice in such cases is that we describe the molecule as symmetric if our models make us believe that they are so, and no experiment indicates the opposite.

Problem 10.1 Find an approximate expression for the permanent dipole moment of MX_2 in terms of the net atomic charge on the metal atom, the M–X bond distance and the $\angle$XMX valence angle.

10.2 Molecular shapes

On the basis of electric deflection and gas electron diffraction experiments as well as studies by other techniques, there is now general agreement that the beryllium- and magnesium-dihalides all have linear equilibrium structures. So do $CaCl_2$, $CaBr_2$ and CaI_2 as well as $SrBr_2$ and SrI_2. The equilibrium structures of all the barium dihalides are, however, angular, and so are the equilibrium structures of CaF_2, SrF_2 and $SrCl_2$ as indicated in Table 10.1. There is a clear pattern: large, electropositive metals and small, electronegative halogens favor bending.

The equilibrium structures of the dihalides of the Group 12 metals zinc, cadmium and mercury all appear to be linear or nearly so: in no case has an angular structure been proven. The difference between the dihalides of Ca and Zn, between Sr and Cd and between Ba and Hg suggests an explanation of the bent equilibrium structures in terms of the vacant valence shell *d* orbitals of the Group 2 metals.

10.3 Bond distances in metal dichlorides

The M–Cl bond distances determined by gas electron diffraction are listed in Table 10.2. The bond distance determined by this method does not correspond to the equilibrium bond distance, but to the average bond distance during vibrations. The bond distances in the table are therefore denoted by R rather than by R_e. The difference between the average and equilibrium bond distances may in unfavorable cases be as large as 3 pm.

The bond distances have been determined by least-squares calculations on the electron diffraction data. Such calculations also yield estimated standard deviations (e.s.d.s) for the

Table 10.2. Gaseous, monomeric dichlorides of the Group 2 and 12 metals, $MCl_2(g)$: bond distances, R, valence angles; ∠ClMCl, mean bond energies, MBE; the bond dissociation energies of MCl, BE1; the bond dissociaton energy of MCl_2, BE2; and the ns^2 to ns^1np^1 excitation energies of the metal atoms, ΔE^* (see text). The uncertainties of BE1, BE2 and MBEs are probably about ±10 kJ mol^{-1}.[a]

MCl_2	R (pm)	∠ClMCl (deg)	MBE (kJ mol^{-1})	BE1 (kJ mol^{-1})	BE2 (kJ mol^{-1})	ΔE^* (kJ mol^{-1})
$BeCl_2$	179.8(2)	180	461	383	540	262
$MgCl_2$	217.9(2)	180	389	321	460	262
$CaCl_2$	248.3(3)	180	444	402	486	182 (243[d])
$SrCl_2$	263.0(2)	155(1)	438	407	469	173 (218[d])
$BaCl_2$	276.8(3)[b]	127(3)[b]	458	441	476	151 (110[d])
$ZnCl_2$	207.2(2)	180	317	229[c]	405[c]	389
$CdCl_2$	228.4(2)	180	278	204	352	367
$HgCl_2$	225.2(2)	180	223	102	344	471 (837[d])

[a] Bond distances and valence angles from reference [1] unless otherwise noted; mean bond energies and bond rupture energies have been calculated from data in reference [3] unless otherwise noted; excitation energies from data in reference [5]. [b] Reference [6]. [c] Reference [7]. [d] s^2 to s^1d^1 excitation energies. See text.

bond distances. In the absence of systematic errors, the probability of a bond distance being in error by more than two e.s.d.s is less than 5%, the probability of the error being greater than three e.s.d.s is less than 1%. Most workers in the field expand the e.s.d.s to include estimated systematic errors. In Table 10.2 such expanded estimated standard deviations are listed behind the bond distances in units of the last digit. The probability of the true value of R(Mg–Cl) falling outside the range 217.9 ± 0.6 is thus less than 1%.

In the rest of this book we shall usually quote bond distances to the nearest pm and dispense with estimated standard deviations unless they are greater than 0.5 pm or with error limits, unless they are greater than 1 pm.

Problem 10.2 Compare M–Cl bond distances as Groups 2 and 12 are descended. Compare M–Cl bond distances between Group 2 and Group 12 elements in the same period. How would you explain the observed variations?

10.4 Bond energies of metal dichlorides

The most straightforward way to describe the strength of the M–Cl bonds in a gaseous metal halide molecule MX_k is probably to use the *mean bond energy*, MBE, which is obtained by dividing the standard energy of atomization with the number of M–X bonds k:

$$MX_k(g) \rightarrow M(g) + kX(g) \quad \Delta U^0 = \Delta U^0_{atom}$$

$$MBE = \Delta U^0_{atom}/k$$

Thus the mean Ca–Cl bond energy in $CaCl_2$ is defined as half the standard energy of the reaction

$$CaCl_2(g) \rightarrow Ca(g) + 2Cl(g) \quad \Delta U^0_{atom} = 888 \, \text{kJ mol}^{-1}$$

$$\text{MBE} = \Delta U^0_{atom}/2 = 444 \, \text{kJ mol}^{-1}.$$

Many tables of thermochemical properties list the standard enthalpies of formation of metal atoms, halogen atoms and gaseous metal halides at 298 or 300 K [2,3,4]. Standard atomization energies may then be calculated from

$$\Delta U^0_{atom} = \Delta H^0_f(M(g)) + k\Delta H^0_f(X(g)) - \Delta H^0_f(MX_k(g)) - k\,RT$$

Unless otherwise noted, atomization energies and mean bond energies listed in this and later chapters, all refer to 298 or 300 K.

Mean bond energies for the metal dichlorides are listed in Table 9.2. The values are probably accurate to within $\pm$ 10 kJ mol^{-1}. Contrary to expectation, the MBEs do *not* decrease monotonically as Group 2 is descended, the M–Cl bonds in $BeCl_2$, $CaCl_2$, $SrCl_2$ and $BaCl_2$ are approximately equal in strength while the bond in $MgCl_2$ is considerably weaker. Clearly the bond energies are determined by more factors than just the size of the metal atom or ion. The MBEs of the Group 12 dichlorides, however, decrease down the group as expected.

The standard energy of the reaction

$$Ca(g) + Cl(g) \rightarrow CaCl(g) \tag{10.1}$$

or the standard energy of formation of the first CaCl bond from the gaseous atoms, is $\Delta U^0 = -402$ kJ mol^{-1}. The standard energy of the reverse process

$$CaCl(g) \rightarrow Ca(g) + Cl(g) \tag{10.2}$$

i.e. the dissociation energy of CaCl, is denoted by *D*(Ca–Cl) or BE1(Ca–Cl) = 402 kJ mol^{-1}.

Similarly the standard energy of the reaction

$$CaCl(g) + Cl(g) \rightarrow CaCl_2(g) \tag{10.3}$$

is $\Delta U^0 = -486$ kJ mol^{-1}. The standard energy of the reverse reaction

$$CaCl_2(g) \rightarrow CaCl(g) + Cl(g) \tag{10.4}$$

is referred to as the bond dissociation energy of $CaCl_2$ and denoted by *D*(ClCa–Cl) or BE2(Ca–Cl) = 486 kJ mol^{-1}.

Experimental values for MCl bond rupture energies are listed in Table 9.2. *The energies show that formation of the first M–Cl bond always is accompanied by a smaller release of energy to the surroundings than formation of the second. Starting with* MCl_2 *it requires more energy to break the first bond than the second.*

This means that the reactions

$$2MCl(g) \rightarrow MCl_2(g) + M(g)$$

are very exothermic and that in each case the equilibrium lies to the right. This is in agreement with our general experience that molecules with an odd number of electrons tend to be thermodynamically less stable than molecules with an even number of electrons.

The fact that the dihalides form stable molecules, that the monohalides are much less stable than the dihalides, and that tri- and higher halides are unknown, is, of course, in perfect agreement with the Lewis electron pair model. Each of the two atoms forming the bond contributes one electron to a bond pair that moves in the space between and around the two kernels: X : M : X.

Problem 10.3 Assume that a metal dihalide molecule is ionic, i.e. that it consists of a M^{+2} cation and two X^- anions. Assume that the molecule is linear and denote the M–X bond distance by R. Use the spherical ion model to show that the potential energy curve may be written as

$$V(R) = \frac{-Me^2}{4\pi\varepsilon_0 R} + C\frac{B}{R^n} + \text{IE1(M)} + \text{IE2(M)} - 2\,\text{EA(X)}$$

where $M = 3.5$ and $C = 2$. Disregard the vibrational energy and show that the *MBE* is given by

$$\text{MBE} = \frac{1}{2}\left\{M\left(\frac{e^2}{4\pi\varepsilon_0 R_e}\right)\left(1 - \frac{1}{\text{n}}\right) - \text{IE1(M)} - \text{IE2(M)} + 2\,\text{EA(X)}\right\} \qquad (10.5)$$

10.5 The spherical ion model and the mean bond energies of the Group 2 and 12 metal dichlorides

We have seen that calculations based on the spherical ion model reproduce the experimental bond energies of the alkali metal halide monomers with an average deviation of about 9%. Similar calculations on the dichlorides of the Group 2 and and 12 metals using equation (10.5) yield the mean bond energies listed in Table 10.3. The most startling result is probably that the mean bond energy in $HgCl_2$ is calculated to be *negative*. This implies that the dissociation process would be exothermic. The calculations thus show that the energy required to produce the Hg^{+2} ion is too large to be compensated by the electron affinities of the Cl atoms and the Coulomb interaction energies of the three ions. Since $HgCl_2$ is in fact a stable molecule in the gas phase, it is clear that the molecule must be covalently bonded. The calculated MBEs of $ZnCl_2$ and $CdCl_2$ are less than 30% of the experimental value. Bonding in $ZnCl_2$ and $CdCl_2$ may be polar, but is obviously far from ionic.

Table 10.3. Comparison of mean bond energies of the gaseous Group 2 and 12 metal dichlorides calculated from equation (10.5) with their experimental counterparts. All energies in kJ mol^{-1}.

MCl_2	MBE (calc)	MBE (exp)	MCl_2	MBE (calc)	MBE (exp)
$BeCl_2$	247	461			
$MgCl_2$	259	389			
$CaCl_2$	363	444	$ZnCl_2$	85	317
$SrCl_2$	374	438	$CdCl_2$	57	278
$BaCl_2$	405	458	$HgCl_2$	−88	223

The calculated MBEs of the Group 2 dichlorides range from 54% of the experimental in $BeCl_2$ and 67% of the experimental in $MgCl_2$, to 82% in $CaCl_2$, 85% in $SrCl_2$ and 89% in $BaCl_2$. Beryllium dichhloride is clearly not an ionic molecule, barium dichloride may well be. The best general description of the gaseous Group 2 dichlorides may be to regard the bonding as polar covalent with the polarity increasing towards the ionic limit as the group is descended.

10.6 The polarizable ion model and the shape of the heavier Group 2 metal halides

As long as we use the spherical ion model, the Coulomb interaction energy will obviously be at a minimum, and the molecule most stable, if the distance between the anions is as large as possible, i.e. if the molecule is linear. Would this still be so if we allow the ions to become polarized?

Polarization of the anion will increase the electron density in the region between M and X. This will not, however, not lead to destabilization of a linear equilibrium structure.

Polarization of the cation in such a way that negative charge is displaced in a direction perpendicular to the X···X vector, may, however, lead to a stabilization of a bent molecule since the negatively charged halogen atoms or ions will be closer to the positive than to the negative end of the dipole on the cation. See Fig. 10.2. The Coulomb interaction between the charges of the anions and the dipole of the cation is given by:

$$E = -2e\mu, \cos(\angle \mathrm{XMX})/(4\pi\varepsilon_0 R^2) \qquad (10.6)$$

where R is the distance from the charge to the dipole, i.e. from the nucleus of the anion to the nucleus of the cation.

If the cation is sufficiently polarizable (i.e. if α_+ is large enough), and the anions are sufficiently close to the cation to produce a strong electric field (i.e. if the anion is small enough) this stabilization may compensate both for the energy required for the polarization of the cation and for the increased repulsion between the anions in an angular equilibrium structure. The polarizabilities of the M^{+2} ions do indeed increase from Be to Ba, and radii of the anions decrease from I^- to F^- [8]. It appears, therefore, that the bent structures of CaF_2, SrF_2, $SrCl_2$ and the four Ba dihalides may be due, at least in part, to polarization effects.

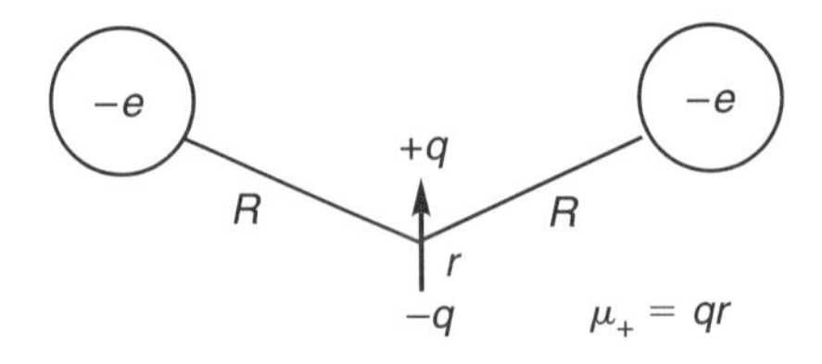

Fig. 10.2. The Coulomb interactions between the charges on the anions and the induced dipole moment of the cation.

10.7 The Valence Shell Electron Pair Repulsion (VSEPR) model

The VSEPR model is probably the most successful and the most widely used model for predicting the shapes of simple non-ionic molecules. It builds directly on the Lewis formula of the molecule, but has been influenced by quantum mechanics in so far as the electrons are allowed to move. A succinct description of the model has been given in a recent textbook by Gillespie and Hargittai [9]:

According to the VSEPR model of molecular geometry, the arrangement of the covalent bonds around an atom depends on the total number of electron pairs in the valence shell of the atom, including those that are non-bonding or lone pairs. If there are n atoms X attached to the central atom by single bonds, and there are m nonbonding or lone pairs, then there are a total of $n+m$ electron pairs in the valence shell of A. The shape of the molecule AX_nE_m, where E represents a non-bonding or lone pair, depends on the arrangement adopted by the $n+m$ electron pairs in the valence shell of A. The basic assumption of the VSEPR model is that the arrangement of a given number of electron pairs in the valence shell of A is that which maximizes the distance between them.

If the number of electron pairs in the valence shell of A is two, the most stable arrangement is linear, if the number of electron pairs is three, the most stable arrangement is trigonal planar, if the number of electron pairs is four, the most stable arrangement is tetrahedral. The most stable arrangement of five electron pairs is trigonal bipyramidal, but a square pyramidal arrangement is assumed to be only slightly less stable. Finally the most stable arrangement for six electron pairs is octahedral.

Qualitative prediction of deviations from ideal bond angles can be made by taking into account the differences of the sizes and shapes of the electron pair domains in a valence shell. The electron pair domains in a valence shell are not all equivalent for the three important reasons:

1. Non-bonding or lone pairs have larger domains than bonding pairs.
2. Bonding domains in the valence shell of the central atom decrease in size with increasing electronegativity of the ligand X [–].
3. Double-bond and triple-bond domains that consist of two or three electron pairs, respectively, are larger than single-bond domains.

The most stable arrangement of electron pairs which are not all equivalent is deformed in such a way that distances between the electron pairs with the larger domains are increased at the expense of the distances between electron pairs with smaller domains.

10.8 The VSEPR model and the structures of the Group 2 and 12 metal dihalides

According to the VSEPR model, the two valence electron pairs surrounding the metal atom, repel each other and should occupy regions at opposite sides of M, leading to ∠XMX angles of 180°. As we have seen, all the Group 12 and most of the Group 2 metal dihalides are linear, but some of them are angular. Thus we find that the VSEPR model fails with respect to some members of the first family of compounds to which we apply it. This is probably just as well, the VSEPR model is remarkably successful for the prediction of molecular shapes, but no model can be 100% effective. A model is constructed by simplifying reality, and sooner or later those aspects of reality that have been excluded from the model, will assert themselves.

10.9 The hybridization model and two-center molecular orbitals

According to Lewis' description a bond is formed between two atoms when they share an electron pair. In a diatomic molecule the pair of electrons may be described by a molecular orbital which in turn may be approximated by a linear combination of a – possibly hybridized – atomic orbital from each atom. This two-center, two-electron or 2*c*, 2*e* bonding model may be carried over to a polyatomic molecule by forming a set of hybridized atomic orbitals on the central atom in such a way that each hybrid orbital on the central atom is orthogonal to the others and overlaps with one and only one ligand orbital. Thus bonding in the MX_2 molecules discussed in this chapter may be described by hybridizing the valence shell *s*-orbitals on M with vacant atomic orbitals in the valence shell. In the case of Be, Mg and the Group 12 metals the only vacant orbitals in the valence shell are the *p* orbitals. Hybridization of the *s* orbital with one of the *p* orbitals, say p_Z, under the requirement that the hybrid orbitals formed should be orthogonal and have the same shape, yields two degenerate sp^1 hybrids which point in opposite directions. See Fig. 10.3.

Each of the two sp^1 hybrid orbitals on the metal atom may be combined with the p_Z orbital on one of the halogen atoms to form a two-center orbital which accommodates two electrons. See Fig. 10.3. Since bonds are stronger when the overlap between central atom and ligand atomic orbitals is large, the hybridization model, like the VSEPR model, provides an explanation for the linearity of the Be, Mg and Group 12 dihalides.

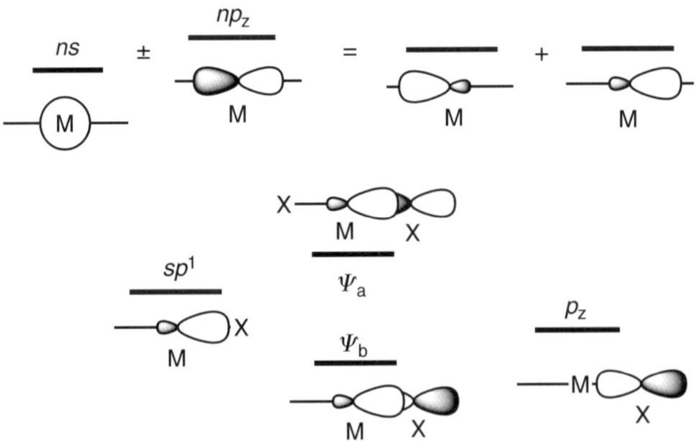

Fig. 10.3. Above: hybridization of the *ns* and np_Z atomic orbitals on the metal atom to yield two sp^1 hybrids pointing in opposite directions. Bold lines indicate energy levels, narrow lines the XMX molecular framework. Below: combination of one such sp^1 hybrid orbital on the metal with the valence shell p_Z orbital on a halogen atom to form one bonding and one antibonding MO. The bonding MO is occupied by two electrons.

When considering Ca, Sr and Ba, the situation is less clear-cut. Should the valence shell *s* orbital on the metal atom be hybridized with the valence shell *d* or *p* orbitals? We have seen that hybridization requires energy, and that the required energy increases with increasing energy of the formerly vacant orbital that is mixed in. Which vacant orbitals have the lower energy, the *d* or the *p* orbitals? Since the element following Ca in the periodic system, scandium, has the valence shell electron configuration s^2d^1, one might conclude that the *d* orbitals have lower energy than the *p*. Nevertheless, it requires more energy to promote a neutral Ca atom to an s^1d^1 configuration than to an s^1p^1 configuration. See Table 10.2.

In the case of Ba, however, the situation is the opposite: the energy required for excitation to a valence shell s^1d^1 configuration, 110 kJ mol^{-1}, is much smaller than the energy required for excitation to an s^1p^1 configuration, 151 kJ mol^{-1}.

Hybridization of the valence shell *s* and d_{xz} orbitals to form two equivalent sd^1 hybrid orbitals is illustrated in Fig. 10.4: the combination $s + d_{xz}$ has particularly large values in the first and third quadrants, the combination $s - d_{xz}$ has particularly large values in the second and fourth. When each of the two sd^1 hybrids on the metal atom is combined with a valence shell *p* orbital on a halogen atom, the resulting valence angle would be 90°, or perhaps a little bit larger due to repulsion between the halogen atoms. See Fig. 10.4.

The *sd* hybridization model does not enable us to predict the magnitude of the valence angles in the barium dihalides, but it does allow us to rationalize the fact that the molecules are non-linear. For all the other metals under consideration the energy of the *p*-orbitals appear to be lower than the energy of the *d* orbitals. The energy difference is, however, small for M = Ca or Sr. Under such conditions it is perhaps not unreasonable to assume that the valence shell *s* orbital is mixed with both *p* and *d* orbitals to form hybrid orbitals of composition sp^λ d^κ. Depending on the relative magnitudes of λ and κ, such hybrids could be made to point in directions favoring any valence angle between 90 and 180°.

The hybridization model also offers a rationalization of the fact that the formation of the first M–X bond results in a smaller energy release than the second: formation of the first bond is accompanied by rehybridization which requires energy. In Fig. 10.5 we plot

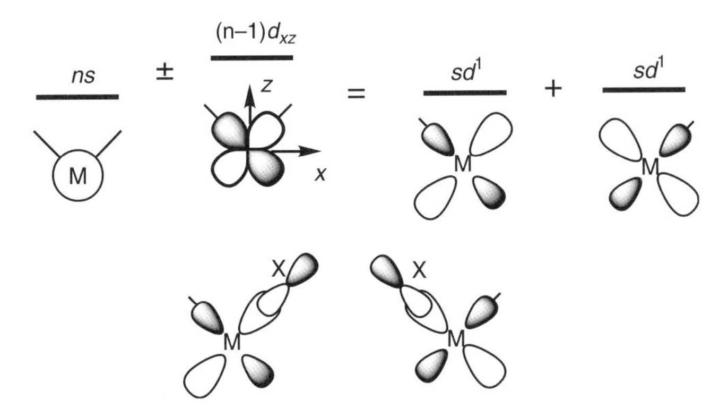

Fig. 10.4. Above: hybridization of the *ns* and $(n - 1)\,d_{xz}$ atomic orbitals on the metal atom to form two orthogonal and degenerate sd^1 hybrid AOs. The *z*-axis bisects the XMX angle, the plane of the paper coincides with the *xz* plane. Below: formation of two two-center bonding molecular orbitals by combination of each sd^1 hybrid on M with a valence shell *p* AO on one of the halogen atoms.

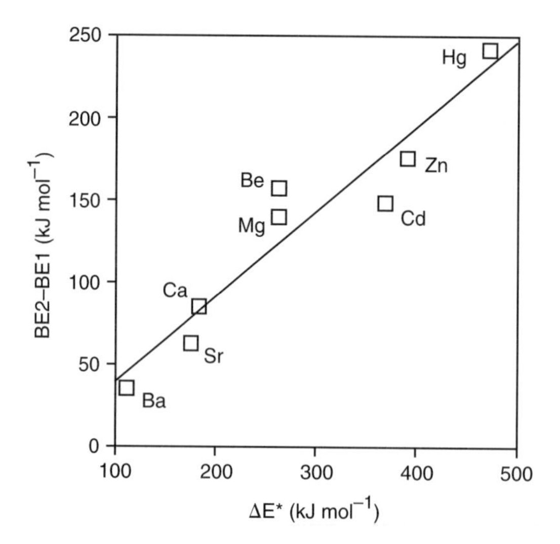

Fig. 10.5. The difference between second and first bond dissociation energies of MCl_2; BE2 – BE1 as a function of the valence shell $s^2 \to s^1p^1$ excitation energies of the metal atom for M = Be, Mg, Ca, Sr, Zn, Cd and Hg and the $s^2 \to s^1\,d^1$ excitation energy for M = Ba. Linear correlation coefficient $\rho = 0.95$.

the difference between the bond dissociation energies of MCl and MCl_2 as a function of the energy ΔE^* required to excite the M atom from the s^2 ground state to the lowest s^1p^1 or s^1d^1 state. For each metal the difference between the bond rupture energies is equal to about half the excitation energy, suggesting that about 3/4 of the excitation energy must be invested when the first bond is formed.

10.10 Delocalized molecular orbitals

Accurate molecular orbital calculations on the metal dihalides are carried out without preconceived notions about hybridization. We first consider the case when the molecule is linear and only the s and p_z valence shell orbitals on the central atom contribute towards formation of molecular orbitals. Calculations show that each of the two valence AOs on the metal atom combines with the p_z atomic orbitals on both ligands to form one bonding and one antibonding orbital. If we place the metal atom at the origin of our coordinate system, halogen atom number 1 at the negative z-axis and halogen atom number 2 at the positive, and fix the signs of the p_z orbitals on the halogen atoms in such a manner that both of the turn their positive lobe towards the metal atom, the two bonding orbitals may be written as

$$\Psi_{b1} = c_1(s)_M + c_2(p_z)_1 + c_2(p_z)_2 \tag{10.7}$$

and

$$\Psi_{b2} = a_1(p_z)_M - a_2(p_z)_1 + a_2(p_z)_2 \tag{10.8}$$

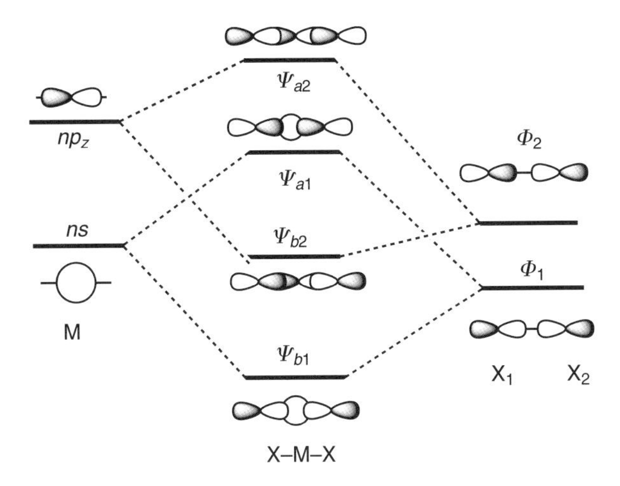

Fig. 10.6. Formation of 3*c* molecular orbitals in MX_2 from the valence shell *s* and p_z orbitals on the metal atom and the ligand symmetry orbitals, Φ_1 and Φ_2. Bold lines indicate energy levels, narrow lines the XMX molecular framework, and stippled lines orbital interactions.

where $(s)_M$ and $(p_z)_M$ are the valence shell *s* and p_z orbitals on the metal atom, and $(p_z)_1$ and $(p_z)_2$ valence shell p_z orbitals on the halogen atoms. For a schematic representation of the two orbitals see Fig. 10.6.

Molecular orbitals which have been formed by linear combination of AOs from three different atoms are referred to as three-center or 3*c* orbitals. Note that in both these molecular orbitals the coefficients in front of two the ligand AOs are equal in magnitude (but not necessarily in sign).

Each of the two bonding MOs contains two electrons, hence four electrons occupy bonding orbitals, just as in the description in terms of 2*c*, 2*e* bonding in the preceding section. What is new is that each of the four electrons is distributed over the entire molecule.

One may imagine the combination of AOs to form three-center bonding and antibonding MOs to take place in two steps: First the ligand AOs combine as they would in an X_2 molecule with an internuclear distance equal to the distance between the halogen atoms in MX_2:

$$\Phi_1 = [(p_z)_1 + (p_z)_2]/\sqrt{D} \quad \text{and} \quad \Phi_2 = [-(p_z)_1 + (p_z)_2]/\sqrt{D}$$

These orbitals may be described as ligand symmetry orbitals. Since the halogen atoms are far apart, the symmetry orbitals will have nearly the same energy. In the second step the ligand symmetry orbitals are combined with the metal AOs to form two bonding and two antibonding 3*c* MOs. This process is presented pictorially in Fig. 10.6.

If the coefficients c_1 and a_1 in (10.6a) and (10.6b) are equal, that is if the metal *s* and p_z orbitals contribute equally to the 3*c* MOs, it can be shown that the description in terms of two 3*c* bonding MOs is identical to the description obtained by *sp* hybridization followed by formation of two 2*c* MOs. This condition is generally not fulfilled. In such cases hybridization introduces an arbitrary restraint on the molecular orbitals, and if employed in calculations, it would reduce the accuracy of the results.

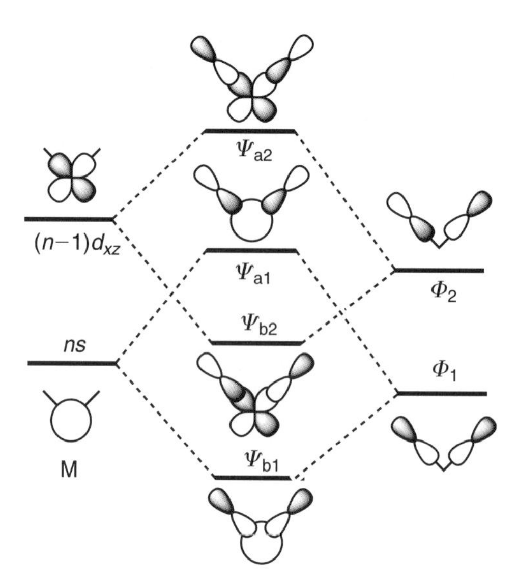

Fig. 10.7. The combination of s and d_{xz} atomic orbitals on the metal atom with ligand symmetry orbitals to form three-center molecular orbitals.

We now consider the case when the molecule is bent and only s and d orbitals on the central atom contributes to formation of occupied molecular orbitals. The combination of ligand symmetry orbitals with metal atom s and d orbitals is visualized in Fig. 10.7.

Analysis of molecular orbitals obtained by accurate calculations on the dihalides of the heavier Group 2 metals Ca, Sr, and Ba indicate that valence shell d orbitals are involved in the bonding and that the atomic kernels (corresponding to M^{+2} ions) are polarized in the molecules that are angular [10]. It would seem that both the polarizable ion and the hybridization model have captured a part of reality.

10.11 Metal dialkyls and dihydrides

The gas phase structures of the dimethylderivatives $M(CH_3)_2$, M = Be, Zn, Cd or Hg and of bis(*neo*-pentyl)magnesium have been determined by gas electron diffraction. All have valence angles of 180° at the metal atom. The dialkyl derivatives of the Group 12 metals are volatile liquids and are believed to be monomeric in the crystalline phase. Dimethylberyllium and bis(*neo*-pentyl)-magnesium are solid at room temperature. The crystal structure of dimethylberyllium will be discussed in Chapter 12.

All the metals in Groups 2 and 12 form solid, non-volatile dihydrides. The dihydrides of Be, Mg and the Group 12 metals have, however, been synthesized in the gas phase by electric discharge through mixtures of H_2 and gaseous metal atoms, and the structures determined by IR emission spectroscopy [11]. All these dihydrides are found to be linear. The structures of gaseous CaH_2, SrH_2 or BaH_2 are unknown.

The M–C and M–H bond distances in the gaseous MR_2 or MH_2 monomers are listed in Table 10.4. This table also includes the M-H bond distances in the gaseous radicals·MH. The M–Cl bond distances of the gaseous dichlorides of the Group 2 and 12 metals and the

Table 10.4. Hydrogen and methyl derivatives of the Group 2 and 12 metals: M–C bond distances in gaseous, monomeric dialkyls MR_2, and M–H bond distances in gaseous, monomeric dihydrides MH_2 or diatomic radicals $\cdot$MH. All distances in pm.[a]

M	*R*(M–C)	*R*(M–H)[b]	*R*(M–H)[c]	M	*R*(M–C)	*R*(M–H)[b]	*R*(M–H)[c]
Be	170	133	134				
Mg	213	170	173				
Ca			200	Zn	193	154	160
Sr			215	Cd	211	168	178
Ba			223	Hg	209	165	177

[a] M–C bond distances from reference [12], M–H bond distances in metal dihydrides from reference [11]; M–H bond distances in monohydrides from reference [13]. [b] In MH_2. [c] In $\cdot$MH.

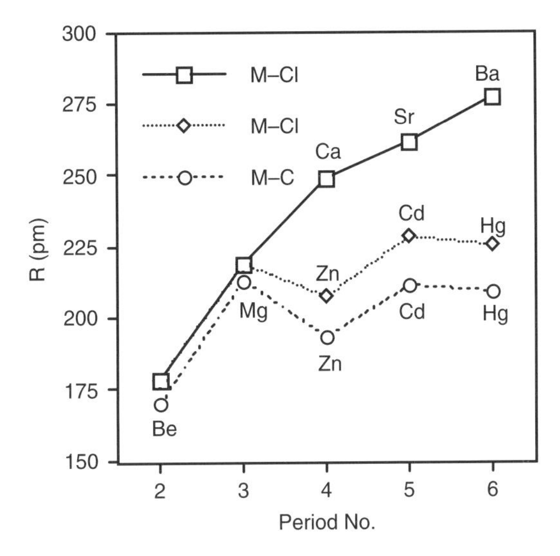

Fig. 10.8. M–Cl bond distances (in pm) in the gaseous dichlorides of the Group 2 and 12 metals. M–C bond distances in gaseous $M(CH_3)_2$, M = Be, Zn, Cd and Hg and $Mg(CH_2CMe_3)_2$.

M–C bond distances in the dialkyl derivatives of Be, Mg, Zn, Cd and Hg are displayed in Fig. 10.8.

10.12 Group variation of bond distances

Both the M–Cl and the $\cdot$MH bond distances increase monotonically as Group 2 is descended. As one moves along the fourth period from Ca to Zn, the nuclear charge increases by +10 while 10 electrons are filled into the 3*d* orbitals. We have already seen that this leads to a substantial decrease of the size of the metal atom (*d*-block contraction). Comparison of M–Cl bond distances in the dichlorides or M–H bond distances in the monohydrides indicates that the bonding radius of Zn is not only about 40 pm smaller than that of Ca, but also 12 pm smaller than that of the third-period element above Ca, viz. Mg.

On going from Sr to Cd in the fifth-period, the nuclear charge again increases by 10 units while 10 electrons are filled into the 4*d* orbitals. At the same time the bonding radius of the metal appears to decrease by about 36 pm, and Cd atom is smaller than both the Sr and Ca atoms.

On going from Ba to Hg in the sixth period, the nuclear charge increases by +24 units while 24 electrons are filled into the 4*f* and 5*d* orbitals. At the same time the M–Cl and M–H bond distances decrease by 52 and 46 pm respectively. As a result, Hg is not only smaller than Ba and Sr, but also two or three pm smaller than Cd. Mercury is in fact unique among main group elements in being smaller than the element above it in the periodic table. The reason is probably the combined effect of *f*-block and *d*-block contractions plus relativistic effects. When the Schrödinger equation of an atom is modified to include relativistic effects such as the increase of the mass of the electron with increasing velocity, this leads to a contraction of atomic orbitals. This relativistic contraction shows a general increase with increasing nuclear charge, but the contraction is not monotonic, it appears to be particularly large for gold and the neighboring elements platinum, mercury and thallium [14].

The mean M–C bond energies in $M(CH_3)_2$, M = Zn, Cd, or Hg, defined as half the standard energies of the reactions

$$M(\mathrm{CH_3})_2(\mathrm{g}) = \mathrm{M(g)} + 2\mathrm{CH_3(g)}$$

are 184, 146, and 128 kJ mol^{-1} respectively. Thus, Hg is abnormal insofar as it forms shorter bonds, but normal insofar that it forms weaker bonds than the element above (Cd).

References

[1] For a review of the structures of metal halides in the gas phase see M. Hargittai, *Chem. Rev.*, 2000, 100, 2233.

[2] D. D. Wagman, W. H. Evans, V. B. Parker, R. H. Schumm, I. Halow, S. B. Bailey, K. L. Churney, and R. L. Nuttall, *The NBS tables of Chemical Thermodynamic Properties*, *J. Phys. Chem. Ref. Data*, 11 (1982) Supplement No. 2.

[3] I Barin, *Thermochemical Data of Pure Substances*, Parts I and II, VCH Publishers, Weinheim, 1989.

[4] *NIST-JANAF Thermochemical Tables*, Fourth Edition, M. W. Chase, Jr., Ed., *J. Phys. Chem. Ref. Data*, Monograph No. 9, 1998.

[5] C. E. More, *Atomic Energy Levels*, National Bureau of Standards, Circular 467, Washington, 1949.

[6] V. P. Spiridonov, A. B. Altmann, A. G. Gershikov, and G. V. Romanov, Abstracts, Second Conference on the Determination of Molecular Structure by Microwave Spectroscopy and Electron Diffraction, Tübingen, 1980.

[7] D. L. Hildenbrand, K. H. Lau, and J. W. Roos, *J. Chem. Phys.*, 111 (1999) 1337.

[8] M. Guido and G. Gigli, *J. Chem. Phys.*, 65 (1976) 1397.

[9] R. J. Gillespie and I Hargittai, *The VSEPR Model for Molecular Geometry*, Allyn and Bacon, Boston, 1991. See also R. J. Gillespie and R. S. Nyholm, *Quart. Rev. Chem. Soc.*, 11 (1957) 339, and R. J. Gillespie and E. A. Robinson, *Angew. Chem. Int. Ed. Engl.*, 35 (1996) 495.

[10] M. Knaupp, P. v. R. Schleyer, H. Stoll, and H. Preuss, *J. Am. Chem. Soc.*, 113 (1991) 6012.

[11] A. Shayesteh, S. Yu, and P.F. Bernath, *Chem. Eur. J.* 2005, 11, 4709.

[12] A Haaland, Organometallic Compounds of the Main Group Elements, in *Stereochemical Applications of Gas-Phase Electron Diffraction*, Part B, *Structural Information for Selected Classes of Compounds*, Eds. I. Hargittai and M. Hargittai, VCH Publishers, Weinheim, 1988.

[13] K. P. Huber and G. Herzberg, *Molecular Spectra and Molecular Structure*, Vol. IV, *Constants of Diatomic Molecules*, Van Nostrand, New York, 1979.

[14] For discussions of relativistic effects see P. Pyykkö and J.-P. Desclaux, *Accounts Chem. Res.*, 12 (1979) 267; K. S. Pitzer, *Accounts Chem. Res.*, 12 (1979) 271; or L. J. Norrby, *J. Chem. Ed.*, 68 (1991) 110.

Chapter 11

Structure and bonding in the gaseous monochlorides, trichlorides and trimethyl derivatives of the Group 13 elements

Introduction

The first two elements in Group 13, boron and aluminum, have the electron configurations [Ng]ns^2np^1 while the heavier elements, gallium, indium and thallium, have electron configurations [Ng]$(n-1)d^{10}ns^2np^1$. The energy of the d electrons is too low for them to participate in the formation of chemical bonds, but as we have seen, the increase of the nuclear charge that accompanies the addition of the 10 d electrons leads to a reduction of the atomic radii and an increase of their ionization energies and electronegativities. The properties of the heavier elements in the Group, Al, Ga, In and Tl, are such that they are classified as metals, while boron is classified as a non-metal or metalloid. We shall nevertheless use the symbol M to denote any member of the group.

11.1 Mono- and tri-chlorides of the Group 13 elements

The first four elements in the group, B, Al, Ga, and In, form trichlorides that are stable at normal temperatures and pressures. Boron trichloride is a volatile liquid at room temperature, and monomeric in the gas phase. $AlCl_3$, $GaCl_3$ and $InCl_3$ are solid at room temperature, and investigations by X-ray chrystallography show that each metal atom is surrounded by six chlorine atoms. Evaporation at moderate temperature yields mixtures of monomeric (MCl_3) and dimeric (M_2Cl_6) species. The amount of monomer increases with increasing temperature and decreasing pressure. In this chapter we shall discuss the structures of the monomers, the structures of the dimers will be discussed in Chapter 16.

At high temperatures the gaseous trichlorides suffer partial decomposition to form monochlorides and Cl_2:

$$MCl_3(g) = MCl(g) + Cl_2(g) \tag{11.1}$$

The standard reaction enthalpies are positive, the standard reaction entropies negative: the amount of monochloride increases with increasing temperature and decreasing pressure. For a given temperature and pressure the equilibrium shifts to the right as the group is descended from boron to indium.

Thallium trichloride can be made, but decomposes at temperatures above 40°C to yield the solid monochloride and chlorine gas:

$$TlCl_3(s) = TlCl(s) + Cl_2(g)$$

The mass spectra of the gas formed by evaporation of solid $TlCl_3$ shows that it contains significant amounts of the monochloride and chlorine in addition to the trichloride.

Table 11.1. Gaseous, monomeric mono- and tri-chlorides of the Group 13 elements: M–Cl bond distances, R (in pm), and bond energies, D_0 or MBE (in kJ mol^{-1}). The Tl–Cl bond distance and the mean bond energy of $TlCl_3$ are unknown, presumably because of the low stability of the gaseous trichloride relative to the monochloride. Gaseous monomeric trimethyl derivatives: M–C bond distances and mean bond energies.[a]

	MCl(g)		MCl_3(g)		$M(CH_3)_3$(g)	
M	R(M–Cl) (pm)	D_0(M–Cl) (kJ mol^{-1})	R(M–Cl) (pm)	MBE(M–Cl) (kJ mol^{-1})	R(M–C) (pm)	MBE(M–C) (kJ mol^{-1})
B	172	508[b]	174	440	158	371
Al	213	500	207	424	196	281
Ga	220	461	211	353	197	254
In	240	433	229	324	216	199[c]
Tl	249	367			221	

[a] Bond distances in the monochlorides from reference [1]; bond distances in the trichlorides from reference [2]; M–Cl bond energies calculated from data in reference [3] unless otherwise indicated; M–C bond distances from reference [4] and M–C bond energies from reference [5] unless otherwise indicated. [b] Reference [6]. [c] Reference [7].

We shall refer to a Group 13 atom that forms three single bonds as predicted by the Lewis electron pair model as Lewis-valent and to a Group 13 atom that forms just one single bond as *subvalent*. M–Cl bond distances and bond dissociation energies of the gaseous monochlorides are listed in Table 11.1.

11.2 The molecular structures of the monomeric trichlorides

As expected, the Group 13 atom M is surrounded by three ligating Cl atoms. The three M–Cl bond distances are believed to be identical, and the three valence angles $\angle$ClMCl to be equal to 120°: no experimental observation indicates otherwise. The magnitude of the valence angles implies that the molecules are planar.

If we place the molecule in a Cartesian coordinate system with the origin at the metal atom and the z-axis perpendicular to the molecular plane, rotation about the z-axis through 120, 240 or 360° superposes a picture of the molecule on itself. The z-axis is therefore referred to as a *threefold symmetry axis*. In addition the molecule is characterized by three twofold symmetry axis: rotation about an axis through one of the three M–Cl bonds through 180 or 360° again superposes a picture of the molecule on itself. Finally the plane of the molecule constitutes a symmetry plane. Such molecules are described as *trigonal planar* or as having D_{3h} symmetry.

M–Cl bond distances and mean bond energies of the trichlorides are listed in Table 11.1.

11.3 Trimethylderivatives of the Group 13 elements

All the Group 13 elements form methyl derivatives of composition $M(CH_3)_3$. The trimethyl derivatives of B, Ga, In and Tl are monomeric in the gas phase. Gaseous trimethylaluminum, however, forms a mixture of monomeric and dimeric species in amounts depending on temperature and pressure. In this chapter we shall discuss the structure of the monomers, the structure of dimers will be discussed in Chapter 12. The MC_3 frameworks of the monomeric trimethyl compounds are found to be trigonal planar. Bond distances and mean bond energies are listed in Table 11.1.

11.4 Group variation of M–C and M–Cl bond distances and bond energies

The M–C bond distances in the trimethyl derivatives and the M–Cl bond distances in both trichlorides and monochlorides are displayed in Fig. 11.1. The M–C bond distances are seen to increase from B to Al, but to remain essentially constant from Al to Ga. The unexpected shortness of the Ga–C bond reflects the *d*-block contraction. The break at Ga is followed by a normal increase from Ga to In and a smaller increase from In to Tl. The unexpected shortness of the Tl–C bond reflects the *f*-block contraction as well as relativistic effects.

The variation of M–Cl distances in the trichlorides is very similar to the variation of M–C bond distances. To the extent that M–Cl bonds are shortened by polarity effects, the shortening appears to be nearly constant down the group.

Both the mean M–C bond energies in the trimethyls and the mean M–Cl bond energies in the trichlorides decrease monotonically as the group is descended. It is noteworthy that even though the Al–C and Ga–C bond distances differ by only 1%, the Ga–C bond is 10% weaker. Similarly, even though the Al–Cl and Ga–Cl bond differ by only 2%, the Ga–Cl bond is 17% weaker.

Comparison of bond distances and bond energies for subvalent and Lewis-valent chlorides show that the B–Cl bond in the trichloride BCl_3 is slightly longer and considerably weaker than in the monochloride BCl. Al–Cl, Ga–Cl and In–Cl bonds in the trichlorides are,

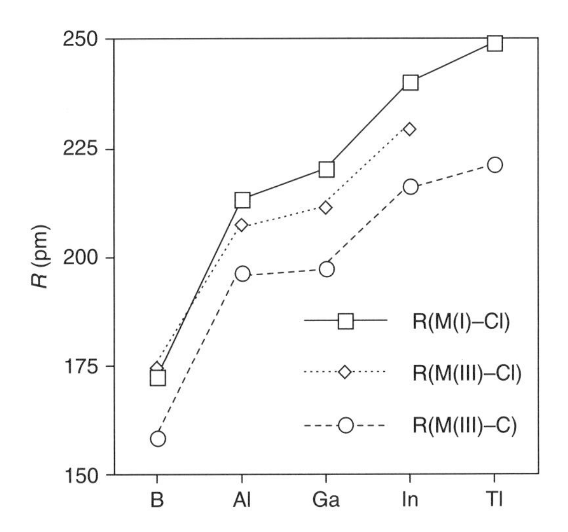

Fig. 11.1. Group variation of bond distances in the subvalent Group 13 element chlorides, MCl(g), in the Lewis-valent trichlorides, MCl_3(g) and in the trimethyl derivatives $M(CH_3)_3$.

however, both *shorter* and *weaker* than in the monochloride. This observation is surprising, since a shorter bond distance normally indicates a stronger bond.

11.5 The spherical ion model and the bond energies of the Group 13 element chlorides

The dissociation energy of BCl calculated from a spherical ion model, equation (5.15), is about 45% smaller than the experimental. Similar calculations on AlCl, GaCl, InCl and TlCl yield dissociation energies about 30% smaller than the experimental. For the 20 alkali metal halides the largest deviations between the experimental dissociation energy and that calculated for a spherical ion model was 15%. The electric dipole moments of gaseous monochlorides InCl and TlCl, 3.79 and 4.52 Debyes respectively, correspond to ionic characters smaller than 0.50. We conclude that bonding in the gaseous monochlorides should be described as polar covalent rather than ionic.

If we assume that the gaseous trichlorides consist of spherical M^{+3} cations and Cl^- anions and proceed as in Problem 10.3, we find that the mean M–Cl bond energy is equal to

$$\text{MBE} = \frac{1}{3}\left\{M\left(e^2/4\pi\varepsilon_0 R_e\right)(1-(1/n)) - \text{IE1(M)} - \text{IE2(M)} - \text{IE3(M)} + 3\text{EA(Cl)}\right\}$$

where $M = 2.423$ and IE1, IE2 and IE3 are the first three ionization energies of M. For BCl_3, $GaCl_3$ and $InCl_3$ the mean bond energies thus obtained are *negative*: the Coulomb interaction energy is too small to compensate for the energy required to produce the M^{+3} cation. The mean bond energy obtained for $AlCl_3$ is positive but 80% smaller than the experimental value. It would clearly be absurd to describe trichlorides of the of the Group 13 elements as ionic.

Problem 11.1 Is there any reason to believe that the gaseous monoflourides of the Group 13 elements may be ionic even if the monochlorides are not? The M–F bond distances in AlF, GaF, InF and TlF are 165.4, 177.4, 199.8, and 208.4 pm respectively and the dissociation energies 674, 583, 515, and 445 kJ mol^{-1}. Use the spherical ion model, equation (5.15), to estimate the dissociation energies and compare with the experimental values. The electric dipole moments of AlF, GaF, InF and TlF are 1.53, 2.45, 3.40 and 4.19 Debyes respectively. Calculate the ionic characters. Would you describe any of the gaseous monoflourides as ionic?

11.6 The Valence Shell Electron Pair Repulsion model

The Group 13 atom in a trichloride or a trimethyl derivative is surrounded by three bonding electron pairs. According to the VSEPR model the three valence shell electron pairs surrounding the metal atom repel one another and occupy domains near the corners of an equilateral triangle with the metal atom at the center. The observed trigonal planar structures are thus in accord with the VSEPR model, but the model provides no explanation for the observation that the bonds in the trichlorides are shorter and weaker than in the monochlorides.

11.7 sp^2 hybridization and localized molecular orbitals

We now turn to the molecular orbital model. If the three M–Cl or M–C bonds in the trichlorides or trimethyls are to be described as a $2c, 2e$ bonds, we need to hybridize the valence shell s and p orbitals at the central atom in such a manner that we obtain three equivalent and orthogonal hybrid AOs pointing towards the three ligands, i.e. forming angles of 120° with respect to one another.

We have seen that combination of valence shell s and p_z AOs yield hybrid AOs with maximum electron density on the positive or negative z-axis: combination with an s orbital does not change the direction of the p orbital, it only transfers electron density from one side to the nucleus to the other. If we wish to form sp^λ hybrid orbitals on the central atom in a trichloride that point in the directions of the three ligands, we must first construct pure p orbitals pointing in the right directions.

According to equation (1.24) the three valence shell p orbitals may be written on the form

$$(p_q) = (R(r)/r)\sqrt{3/4\pi}\, q$$

where $q = x$, y, or z.

We first introduce a new coordinate system for points in the x,y plane. The new coordinates are defined as

$$q' = x\cos\alpha + y\sin\alpha \quad \text{and} \quad q'' = -x\sin\alpha + y\cos\alpha$$

See Fig. 11.2A.

Then we form a new p orbital by linear combination of the p_x and p_y orbitals:

$$\begin{aligned}(p') &= \cos\alpha(p_x) + \sin\alpha(p_y) \\ &= \cos\alpha[R(r)/r]\sqrt{3/4\pi}\, x + \sin\alpha[R(r)/r]\sqrt{3/4\pi}\, y \\ &= [R(r)/r]\sqrt{3/4\pi}(x\cos\alpha + y\sin\alpha) \\ &= [R(r)/r]\sqrt{3/4\pi}\, q'\end{aligned} \tag{11.2}$$

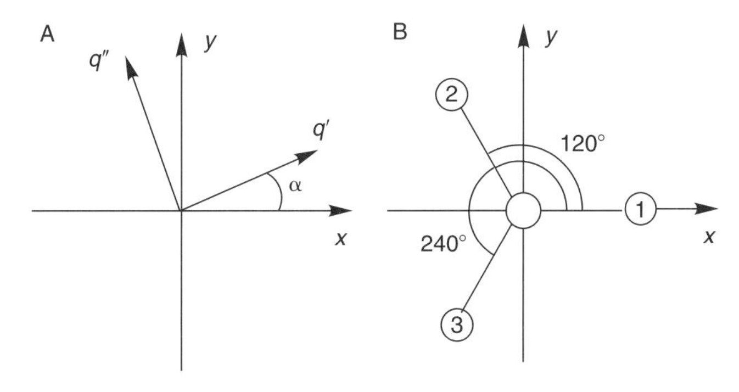

Fig. 11.2. A: the two coordinate systems (x, y) and (q', q''). B: Coordinate system for construction of sp^2 hybrid orbitals on the metal atom pointing towards the ligating atoms in a trigonal planar molecule.

We see that the new orbital has the same shape as the original p_x and p_y orbitals, but is pointing along the q' axis.

Problem 11.2 Form a new p orbital (p'') pointing in the direction of the q'' axis by linear combination of the p_x and p_y orbitals. Show that since the p_x and p_y orbitals are orthogonal, the p' and p'' orbitals are also orthogonal. Show that both p' and p'' are normalized.

Using equation (11.2) we can construct pure p type orbitals pointing in the directions of the three ligands in a trigonal planar MX_3 molecule:

$$p_1 = (p_x)$$

$$p_2 = \cos(120^\circ)(p_x) + \sin(120^\circ)(p_y) = (-1/2)(p_x) + \sqrt{3/2}\,(p_y)$$

$$p_3 = \cos(240^\circ)(p_x) + \sin(240^\circ)(p_y) = (-1/2)(p_x) + -\sqrt{3/2}\,(p_y)$$

(These orbitals are normalized, but *not* orthogonal.)

We now form three equivalent hybrid orbitals with the correct directions:

$$h_1 = \frac{(s) + a(p_1)}{\sqrt{1+a^2}}$$

$$h_2 = \frac{(s) + a(p_2)}{\sqrt{1+a^2}}$$

$$h_3 = \frac{(s) + a(p_3)}{\sqrt{1+a^2}}$$

Finally we fix the value of the constant a by insisting that the hybrids are orthogonal:

$$\int h_1(x,y,z)h_2(x,y,z)\,\mathrm{d}\tau = 0$$

$$\left(\frac{1}{1+a^2}\right)\int [(s) + a(p_x)]\left[(s) + a\left(-\frac{1}{2}\right)(p_x) + a\sqrt{\frac{3}{2}}(p_y)\right]\mathrm{d}\tau = 0$$

Multiplication by $(1+a^2)$ gives

$$\int [(s) + a(p_x)]\left[(s) + a - \frac{1}{2}(p_x) + a\left(\sqrt{\frac{3}{2}}\right)(p_y)\right]\mathrm{d}\tau$$

$$= \int \left[(s)^2 - a\frac{1}{2}(s)(p_x) + a\sqrt{\frac{3}{2}}(s)(p_y) + a(s)(p_x) - a^2\frac{1}{2}(p_x)^2 + a^2\sqrt{\frac{3}{2}}(p_x)(p_y)\right]\mathrm{d}\tau$$

$$= 1 + 0 + 0 + 0 - \frac{1}{2}a^2 + 0 = 0,$$

or

$$a^2 = 2 \text{ and } a = \sqrt{2}$$

As usual we indicate the $p : s$ mixing ratio by writing $\lambda = a^2 = 2$ as a superscript; and denote the three hybrids by sp^2. Each of these hybrids may now be combined with a valence shell p-orbital on the appropriate Cl atom (or with an sp^3 hybrid on the C atom in the methyl group) to form a two-center bonding orbital containing two electrons.

Problem 11.3 Show that the electron density calculated for an atom with one electron in each of the three sp^2 hybrid orbitals, is equal to the electron density calculated for an atom with a valence shell $s^1p_x^1p_y^1$ electron configuration.

Here we have used the hybridization model *after* the molecular structure had been determined experimentally. Could the hybridization model – like the VSEPR model – be used to predict the structure? The answer is yes: we could have begun by asking if it is possible to combine the s, p_x and p_y orbitals to form three new hybrids which are mutually orthogonal and have the same shape. The derivation is more involved, but it can be shown that the only possible candidates are the sp^2 hydrides that we have just described, and which form angles of 120° relative to one another. There is, however, not much reason to carry out the derivation, since the VSEPR model leads us directly and quickly to the same conclusion.

We have seen that some Group 2 dihalides are found to be angular even though the VSEPR model predicts that they should be linear, and that the angular structures could be rationalized in terms of the hybridization model. Perhaps the hybridization model would suggest that some Group 13 trihalides are pyramidal? The answer is no: the angular structures of some dihalides of the heavier Group 2 metals were rationalized by evoking hybridization of the valence shell s orbitals with *vacant* valence shell d orbitals: since the valence shell d orbitals in the Group 13 elements Ga, In and Tl are filled by 10 electrons, such hybridization is not possible.

The ground state valence electron configuration of the Group 13 elements is ns^2np^1. Formation of the first M–Cl bond may presumably take place without hybridization: the M–Cl bonding orbital may be formed by combination of an unhybridized p orbital on M with a p orbital on the Cl atom. Such a description is in agreement with the observation that the M–Cl bond in the trichlorides of Al, Ga and In are shorter than in the monochlorides: we have seen that sp-hybridization increases the electron density between the two bonded atoms.

Why, then, are the bonds in the trichlorides weaker than in the monochlorides?

Under the atomic orbital approximation the equality of the electron densities of the valence shell electron configurations $s^1p_x^1p_y^1$ and $h_1^1h_2^1h_3^1$ implies that the energies of the two configurations are equal. The lowest atomic state arising from the $s^1p_x^1p_y^1$ configuration is ^4P where the three electrons have parallel spins. The energies required to excite the Group 13 atoms from the ground state to the ^4P valence state are listed in Table 11.2.

We may describe the formation of a gaseous trichloride from the gaseous atoms as proceeding in two steps. In the first step the Group 13 element is excited from the ground state s^2p^1 to the$s^1p_x^1p_y^1$ valence state under absorption of the energy ΔE^*(M), and in the second step the excited atom reacts with the three Cl atoms and releases an energy equal to $\Delta E^*(M)$ + 3MBE. See Fig. 11.3. The mean M–Cl bond energy when the trichloride is formed from (or dissociated to give) M atoms in the ^4P state is thus given by

$$\text{MBE}^* = \text{MBE} + \frac{1}{3}\Delta E^*(\text{M}) \tag{11.3}$$

Table 11.2. Mean bond energies in MCl_3(g). Ground state to ^4P excitation energies for the Group 13 atoms, ΔE^*(M), and mean bond energies of MCl_3(g) when dissociating to the ^4P state, MBE*. Standard energies of the decomposition reaction (11.1) calculated from experimental enthalpies of formation at 298 K or estimated from equation 11.4. All energies in kJ mol^{-1}.

M	MBE	$\Delta E^*(M)^a$	MBE*	ΔU°_{dec}(exp)	ΔU°_{dec}(est)
B	440	345	555	542	553
Al	424	347	540	531	533
Ga	353	454	504	359	339
In	324	418	463	301	312
Tl	–	540	–	–	42

[a] Reference [8].

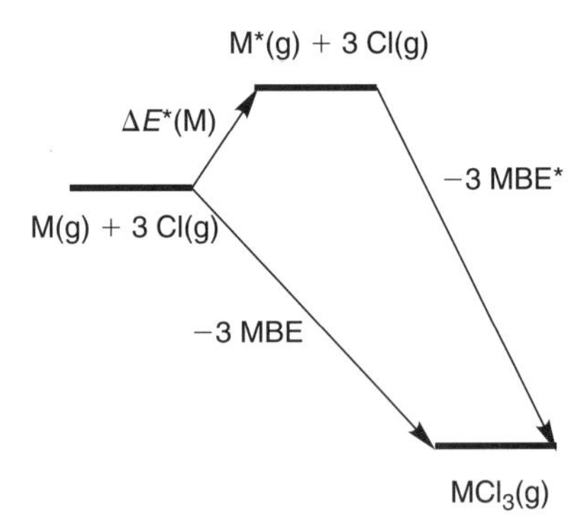

Fig. 11.3. The formation of a gaseous MCl_3 molecule from the constituent atoms.

The mean bond energies of BCl_3, $AlCl_3$, $GaCl_3$ and $InCl_3$ estimated in this manner are listed in Table 11.2. They are all about 8% larger than the dissociation energy of the corresponding monochloride. We conclude, therefore, that the M–Cl bonds in the trichlorides are weaker than the bonds in the monochlorides because a significant part of the energy released by bond formation has been expended in promoting the metal atom to the valence state.

11.8 The relative stabilities of the Lewis-valent and subvalent chlorides

Why do the gaseous trichlorides become increasingly unstable relative to the monochlorides as the Group is descended? The standard energies of the decomposition reactions

$$MCl_3(g) = MCl(g) + Cl_2(g) \quad \Delta U^0 = \Delta U^0_{dec} \tag{11.1}$$

calculated from experimental enthalpies of formation, are listed in Table 11.2. The decomposition energies may also be estimated as the difference between the bond energies of the reactant and products:

$$\Delta U^0_{dec}(\text{est}) = 3\,\text{MBE} - D_0(\text{M–Cl}) - D_0(\text{Cl–Cl})$$

or

$$\Delta U^0_{\text{dec}}(\text{est}) = 3\text{MBE}^* - \Delta E^* - D_0(\text{M–Cl}) - D_0(\text{Cl–Cl}) \tag{11.4}$$

The decomposition energy and hence the thermodynamic stability of the trichloride is seen to *increase* with the energy required to break the three bonds in the trichloride in such a manner that the metal atom is left in the excited ^4P state, and to *decrease* with the amount of energy required to promote the metal atom to the valence state, $\Delta E^*(\text{M})$.

We now simplify equation (11.4) by assuming that the mean bond energy required for dissociation to the excited ^4P state is 8% larger than the dissociation energy of the monochloride, i.e. that $\text{MBE}^* = 1.08D_0(\text{M–Cl})$

$$\Delta U^0_{\text{dec}}(\text{est}) = [2.24D_0(\text{M–Cl}) - D_0(\text{Cl–Cl})] - \Delta E^*(\text{M}) = \Delta D_0 - \Delta E^*(\text{M}) \tag{11.5}$$

The estimated decomposition energies thus obtained are listed in Table 11.2.

There is good agreement between the estimated and experimental decomposition energies for M = B, Al, Ga and In: the difference between them is barely visible when they are plotted on the scale of Fig. 11.4. Comparison with the curves showing the variation of the terms ΔD_0 and $\Delta E^*(\text{M})$ show that the falling energy of decomposition, i.e. the falling thermodynamic stability of the trichloride, is due both to a monotonic decrease of the inherent strength of the M–Cl bonds, and to a less regular increase in the energy required to promote metal atom to the valence state.

The thermodynamic data required to calculate the decomposition energy of $TlCl_3$ are not available. The estimated decomposition energy is only 42 kJ mol^{-1}. The reduction by about 270 kJ mol^{-1} compared to $InCl_3$ is seen to be due in equal measure to an increase of the hybridization energy and a decrease in the strength of the M–Cl bonds.

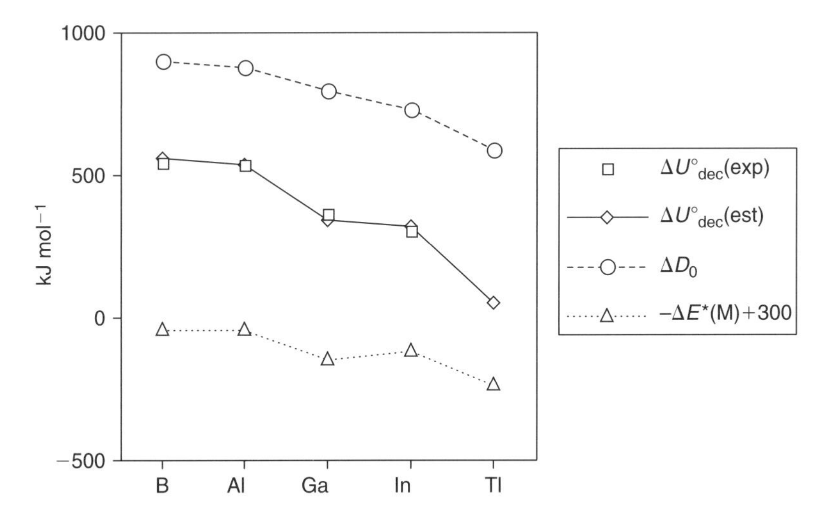

Fig. 11.4. The experimental and estimated decomposition energies of the gaseous trichlorides of Group 13 elements. For definition of ΔD_0 see equation (11.5). Note that the negative of the ground state to ^4P excitation energies, $-\Delta E^*(\text{M})$, has been plotted with an offset of 300 kJ mol^{-1}.

References

[1] K. P. Huber and G. Herzberg, *Molecular Spectra and Molecular Structure*, Vol. IV, *Constants of Diatomic Molecules*, Van Nostrand, New York, 1979.

[2] A. Haaland, A. Hammel, K.-G. Martinsen, J. Tremmel, and H. V. Volden, *J.C.S. Dalton*, (1992), 2209 and references therein.

[3] I Barin, *Thermochemical Data of Pure Substances*, Parts I and II, VCH Publishers, Weinheim, 1989.

[4] Haaland, Organometallic Compounds of the Main Group Elements, in *Stereochemical Applications of Gas-Phase Electron Diffraction*, Part B, *Structural Information for Selected Classes of Compounds*, Eds. I. Hargittai and M. Hargittai, VCH Publishers, Weinheim, 1988.

[5] G. Pilcher and H. A. Skinner, Thermochemistry of Organometallic Compounds, in F. R. Hartley and S. Patai, *The Chemistry of the Metal–Carbon Bond*, Wiley, New York, 1982.

[6] C. W. Bauschlicher, *J. Phys. Chem.*, 103 (1999) 6429.

[7] D. L. Hildenbrand, *J. Chem. Phys.*, 105 (1996) 10507.

[8] ^2P (ground state) to ^4P (valence state) excitation energies of the Group 13 elements were calculated from data in C. E. More, *Atomic Energy Levels*, National Bureau of Standards, Circular 467, Washington, 1949.

Chapter 12

Electron deficient molecules: three-center, two-electron bonds

Introduction

In an ionic molecule like a gaseous, monomeric alkali metal halide one electron has been transferred from the electropositive to the electronegative atom to form an ion pair which is subsequently held together by Coulomb attraction. Single bonds in the molecules that we have referred to as covalent or polar covalent are, according to Lewis, described as due to "one pair of electrons held in common by two atoms," according to the more advanced MO model the two electrons occupy a two-center bonding molecular orbital formed by linear combination of one atomic orbital from each atom. There are, however, known molecules where the $2c, 2e$ model fails completely: if the number of bonds is larger than half the number of valence electron pairs in the molecule, there are simply not enough valence electrons to go round. Such molecules are commonly described as "electron deficient."

12.1 The trihydrogen cation

Like the dihydrogen cation, H_2^+, the trihydrogen cation, H_3^+, is known only in the gas phase. Both ions have, however, been found in interstellar space. The structure of H_3^+ is that of an equilateral triangle with the three H–H distances equal to 87.5 pm as compared to 74 pm in H_2 or 106 pm in H_2^+. These distances are so short as to imply the existence of three H–H bonds, though the total number of electrons in the molecule is just two! The dissociation energy, defined as the energy of the reaction

$$H_3^+(g) \rightarrow H_2(g) + H^+(g) \text{ is } D_0 = 456 \text{ kJ mol}^{-1}.$$

(Dissociation to yield $H_2^+(g) + H(g)$ requires more energy).

Problem 12.1 Show that the following reaction is strongly exothermic:

$$H_3^+(g) + e^- \rightarrow H_2(g) + H(g)$$

12.2 A molecular orbital description of H_3^+

Consider the molecular orbital

$$\Psi(x, y, z) = \frac{[c_1\Phi_1(x, y, z) + c_2\Phi_2(x, y, z) + c_3\Phi_3(x, y, z)]}{\sqrt{D}} \tag{12.1}$$

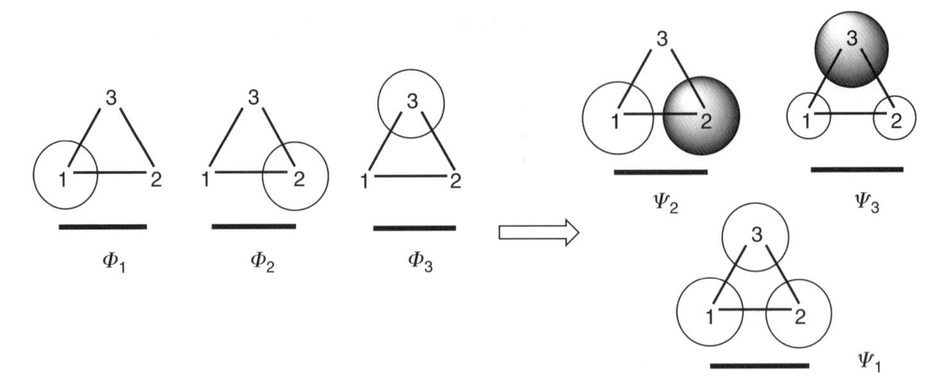

Fig. 12.1. The formation of three-center molecular orbitals in H_3^+.

where Φ_1, Φ_2 and Φ_3 are the 1*s* orbitalson hydrogen atoms number 1, 2 and 3 respectively. See Fig. 12.1. We shall use simplified Hückel calculations to obtain an estimate of the orbital energy and the orbital coefficients c_1, c_2, and c_3.

The one-electron Hamiltonian operator $\hat{h}$ contains one term representing the kinetic energy of the electron, one term representing the energy of repulsion to the other electron and three terms representing the energies of attraction to the three protons. The orbital energy of the electron is given by

$$\varepsilon = \int \Psi(x, y, z)\hat{h}\Psi(x, y, z)\,\mathrm{d}\tau$$

Introducing the integrals

$$H_{i,j} = \int \Phi_i(x, y, z)\hat{h}\Phi_j(x, y, z)\,\mathrm{d}\tau$$

and

$$S_{i,j} = \int \Phi_i(x, y, z)\Phi_j(x, y, z)\,\mathrm{d}\tau$$

we obtain

$$\varepsilon = \frac{N(c_1, c_2, c_3)}{D(c_1, c_2, c_3)} \tag{12.2}$$

where

$$N(c_1, c_2, c_3) = c_1^2 H_{1,1} + c_2^2 H_{2,2} + c_3^2 H_{3,3} + 2c_1c_2H_{1,2} + 2c_1c_3H_{1,3} + 2c_2c_3H_{2,3} \tag{12.2a}$$

and

$$D(c_1, c_2, c_3) = c_1^2 + c_2^2 + c_3^2 + 2c_1c_2S_{1,2} + 2c_1c_3S_{1,3} + 2c_2c_3S_{2,3} \tag{12.2b}$$

Calculating the derivatives of ε with respect to each of the coefficients c_1, c_2, and c_3, setting the three derivatives equal to zero in order to find the lowest possible value of ε, and proceeding as outlined in Section 8.1, we obtain the three secular equations:

$$\begin{aligned}(H_{1,1}-\varepsilon)c_1+(H_{1,2}-S_{1,2}\,\varepsilon)c_2+(H_{1,3}-S_{1,3}\varepsilon)c_3&=0\\(H_{2,1}-S_{2,1}\varepsilon)c_1+(H_{2,2}-\varepsilon)c_2+(H_{2,3}-S_{2,3}\varepsilon)c_3&=0\\(H_{3,1}-S_{3,1}\varepsilon)c_1+(H_{3,2}-S_{3,2}\varepsilon)c_2+(H_{3,3}-\varepsilon)c_3&=0\end{aligned} \tag{12.3}$$

Since the three atoms are identical, the three integrals $H_{1,1}$, $H_{2,2}$ and $H_{3,3}$ are equal. Since the three atoms occupy the corners of an equilateral triangle, all the resonance integrals must also be equal: $H_{1,2} = H_{2,1} = H_{1,3} = H_{3,1} = H_{2,3} = H_{3,2}$. If we replace all overlap integrals S by zero, the secular equations simplify to

$$\begin{aligned}(H_{1,1}-\varepsilon)c_1+H_{1,2}c_2+H_{1,2}c_3&=0\\H_{1,2}c_1+(H_{1,1}-\varepsilon)c_2+H_{1,2}c_3&=0\\H_{1,2}c_1+H_{1,2}c_2+(H_{1,1}-\varepsilon)c_3&=0\end{aligned} \tag{12.4}$$

These homogeneous linear equations will have non-trivial solutions if the secular determinant is equal to zero:

$$\begin{vmatrix}(H_{1,1}-\varepsilon) & H_{1,2} & H_{1,2}\\ H_{1,2} & (H_{1,1}-\varepsilon) & H_{1,2}\\ H_{1,2} & H_{1,2} & (H_{1,1}-\varepsilon)\end{vmatrix}=0 \tag{12.5}$$

Equation (12.5) has three roots:

$$\varepsilon_1 = H_{1,1} + 2H_{1,2} \quad \varepsilon_2 = H_{1,1} - H_{1,2} \quad \varepsilon_3 = H_{1,1} - H_{1,2} \tag{12.6}$$

Since the resonance integrals $H_{1,2}$ are negative, the first energy is that of a bonding orbital, the second and third those of two degenerate, antibonding orbitals.

Insertion of ε_1 and $c_1 = 1$ into the secular equations yield the coefficients $c_2 = c_3 = 1$. The normalization constant $D = c_1^2 + c_2^2 + c_3^2 = 3$, and

$$\Psi_1 = (\Phi_1 + \Phi_2 + \Phi_3)/\sqrt{3} \tag{12.7}$$

The orbital energies ε_2 and ε_3 yield the orbitals

$$\Psi_2 = (\Phi_1 - \Phi_2)/\sqrt{2} \tag{12.8a}$$

and

$$\Psi_3 = (\Phi_1 + \Phi_2 - 2\Phi_3)/\sqrt{6} \tag{12.8b}$$

See Fig. 12.1.

The two valence electrons in H_3^+ occupy the bonding molecular orbital Ψ_1. The resulting total electron density is found to be particularly high at the three nuclei and in the region between each nucleus and the center of the equilateral triangle defined by the positions of the nuclei. The bond uniting the three atoms is therefore described as a three-center, two-electron or $3c, 2e$ bond.

Problem 12.2 Consider a linear H_3^+ ion. Go back to the general secular equations (12.3), neglect all overlap integrals and assume that atoms 1 and 3 are so far apart that $H_{1,3}$ is equal to zero. Show that the energy of the lowest molecular orbital is $H_{1,1} + \sqrt{2}H_{1,2}$, and determine the orbital coefficients. Is this orbital energy consistent with the observation that the equilibrium structure of the molecule is triangular rather than linear?

Problem 12.3 The Nobel Prize in chemistry for 1994 was awarded to George A. Olah for his work on hydrocarbon cations. One of these ions is "methonium," or CH_5^+, whose structure may be described as that of H_3^+ with one H atom replaced by a methyl group.

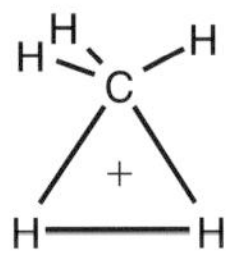

H H H C + H H

How would you describe the bonding molecular orbitals in CH_5^+?

12.3 Jahn–Teller distortion

The H_3^+ ion is a very symmetric molecule; like a monomeric Group 13 metal trichloride it is characterized by a symmetry plane, a threefold symmetry axis (perpendicular to the ring plane and through the center of the ring), three twofold symmetry axes (each through the ring center and one H atom). Though it may not be obvious, it is important to realize that the degeneracy of the two antibonding molecular orbitals is a consequence of the molecular symmetry: non-linear molecules can only have degenerate molecular orbitals if they possess a threefold (or even higher) symmetry axis.

The orbital energy diagram in Fig. 12.1 may be used to investigate what would happen if we add another electron to the H_3^+ ion to form the neutral H_3 molecule. The electron might enter either of the two antibonding MOs Ψ_2 or Ψ_3. As long as the molecule remains trigonal, i.e. as long as it has a threefold symmetry axis perpendicular to the plane defined by the three nuclei, Ψ_2 or Ψ_3 will remain degenerate and the H_3 molecule will have an orbitally degenerate ground state. (The expression "orbitally degenerate" is used to make it clear that the degeneracy is *not* due to an unpaired electron that may have two different, α or β, spin functions.) But will the molecule remain trigonal after the electron has been added?

Imagine that the electron is added to Ψ_2. This would lead to antibonding, that is to repulsion between hydrogen atoms 1 and 2, and we would expect the H(1)–H(2) bond distance to increase. An increase of this distance would in turn reduce the antibonding interactions and the Ψ_2 orbital energy would fall.

If we add the electron to Ψ_3 we get bonding between H(1) and H(2), but antibonding between both these atoms and H(3). As a consequence we would expect the bond distance H(1)–H(2) to decrease and the distances H(1)–H(3) and H(2)–H(3) to increase. These distortions would in turn increase the bonding and decrease the antibonding interactions, and the Ψ_3 orbital energy would fall. The conclusion is that it does not matter to which of the two degenerate orbitals the electron is added, in either case the molecule will distort, the molecular symmetry will be broken and the degeneracy of the Ψ_2 and Ψ_3 orbitals disappear.

The H_3 molecule is an example of a general theorem: *No nonlinear molecule can have an orbitally degenerate ground state*. If calculations indicate that a non-linear molecule would have an orbitally degenerate ground state under a given symmetry, the molecule will in fact distort in such a way that the symmetry is broken. The general statement is referred to as the Jahn–Teller theorem after the scientists who first proved it.

How large would the distortion be? The answer depends on the forces that would tend to oppose the change, i.e. on number of bonding electrons and the nature of the molecular orbitals they occupy. In the special case of the H_3 molecule, the two bonding electrons in Ψ_1 are not sufficient to prevent the distortion to continue until we are left with a H_2 molecule and a H atom.

12.4 The first electron deficient molecule, diborane

The structure of the B_2H_6 molecule remained a mystery for more than two decades after it had first been prepared shortly after 1920. Not only was the structure of the molecule unknown, its very existence presented an enigma: each boron atom has three valence electrons, each hydrogen atom one. Each boron atom presumably forms three electron pair bonds to three hydrogen atoms, but that leaves no electrons for formation of the B–B bond! Or are the two BH_3 fragments connected through one or two H atoms? In that case those hydrogen atoms must be able to form more than one bond! For a summary of the discussion during the two following decades see reference [1].

The molecular structure of diborane as it was finally established in the 1940s is shown in Fig. 12.2. The central part of the molecule consists of a planar four-membered ring consisting of two B and two H atoms. The two H atoms are referred to as *bridging* (br). The remaining four H atoms are bonded to one B atom only and are referred to as *terminal* (t). The four terminal H atoms define a plane perpendicular to the central four-membered ring. A bridging atom is denoted by the Greek letter μ ("mu") and the formula of diborane is written as $H_4B_2(\mu-H)_2$. Note that we use the terms "bridging" and "terminal" to describe the *geometry* of the molecule, not the nature of the bonding.

The diborane molecule is characterized by three twofold symmetry axes at right angles to each other: if we place a Cartesian axis system with the origin at the midpoint of the B–B vector with the x-axis perpendicular to the four-membered ring, the y-axis running through

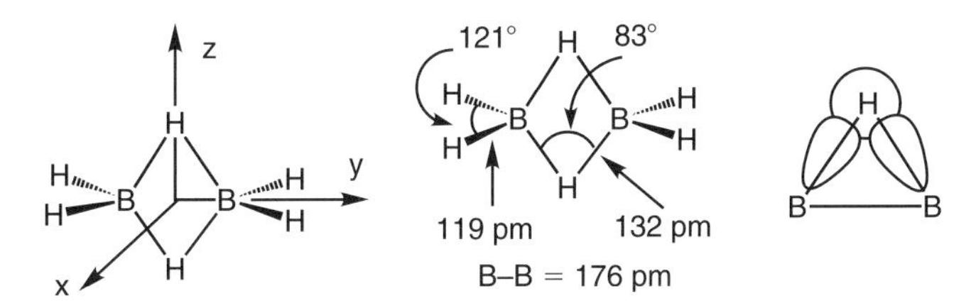

Fig. 12.2. Left and center: the molecular structure of diborane. Right: a three-center bonding molecular orbital in diborane.

the two boron atoms and the z-axis running through the two bridging H atoms, rotation about either of the three coordinate axes by 180 or 360° superposes the picture of the molecule on itself. Similarly mirroring the molecule through either of the three symmetry planes (x, y), (y, z) or (x, z) leaves the picture of the molecule unchanged.

Since the two bridging H atoms are interchanged through a 180° rotation about the x- or y-axis, they are referred to as symmetry equivalent. The four terminal H atoms are interchanged through a 180° rotation about the x-, y- or z-axis and are also symmetry equivalent. Since bridging and terminal H atoms cannot be exchanged through any of the symmetry operations, they are said to be symmetry inequivalent.

The valence angles $\angle H_tBH_t$ are equal to 121°, the valence angles $\angle H_bBH_b$ are equal to 97°, and the angles at the bridging H atoms only 83°. The terminal B–H bond distances are as expected for normal $2c$,$2e$ bonds, but the bridging B–H distances are some 13 pm longer. The B–B distance across the ring is 176 pm as compared to 172 pm in F_2B–BF_2: if we are drawing bonds between the B atoms and the bridging H atoms, we should also draw a bond between the two B atoms. Indeed the assumption that there is bonding between the B atoms, provides a reasonable explanation for the acute $\angle BH_bB$ angles.

We now turn to a description of the bonding. As we shall see in the next chapter, it is possible to hybridize the $2s$ and the three $2p$ orbitals on each B atom in such a way that two equivalent hybrids span an angle of 120° and are pointing towards the two terminal H atoms and another pair of equivalent hybrids span an angle of about 100° and are pointing towards the bridging H atoms. We assume that each hybrid orbital pointing towards a terminal H combines with the H $1s$ orbital to form a $2c$ MO accommodating two electrons. The $1s$ orbital of each bridging H may combine with one hybrid orbital from each B atom to form a bonding three-center orbital which accommodates two electrons. See Fig. 12.2. This means that the central B_2H_2 ring is held together by two $3c, 2e$ bonds.

Problem 12.4 The standard enthalpies of formation of B(g), H(g), BH_3(g), and B_2H_6(g) at 298 K are 560, 218, 92 and 36 kJ mol^{-1} respectively. Calculate the mean bond energy in BH_3(g). Assume that the energy of atomization of diborane may be written as the sum of four terminal B–H bond energies (equal to the MBE in BH_3) and four bridging B–H bond energies and calculate the mean bridging B–H bond energy.

Problem 12.5 See Fig. 12.3. Compare the Al–Al the Al–C_t and Al–H_b bond distances in tetramethyldialane and the corresponding bond distances in digallane with the bond distances calculated

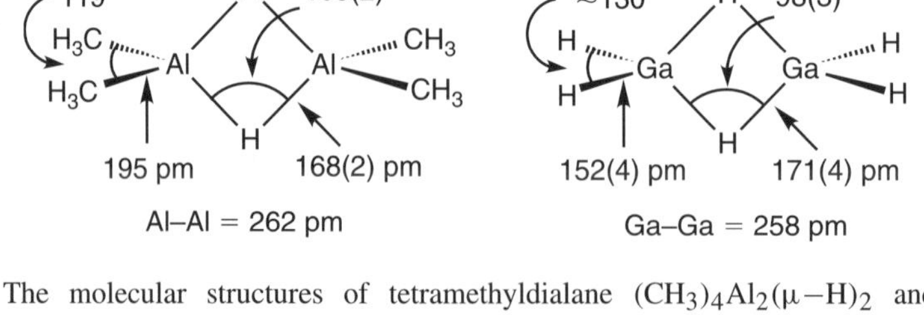

Fig. 12.3. The molecular structures of tetramethyldialane $(CH_3)_4Al_2(\mu{-}H)_2$ and digallane $H_4Ga_2(\mu{-}H)_2$.

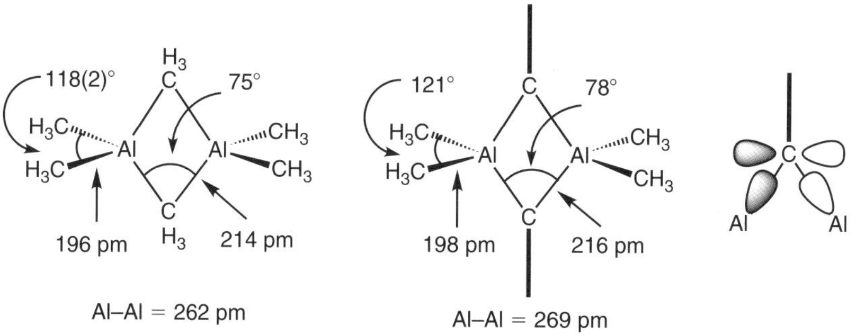

Fig. 12.4. Left: the molecular structures of $(CH_3)_4Al_2(\mu-CH_3)_2$ and $(CH_3)_4Al_2(\mu-Ph)_2$, Ph = phenyl, (the phenyl rings are perpendicular to the plane of the paper). Right: delocalization of the π-orbitals of the phenyl ring into an Al–C_b bonding, but Al–Al antibonding orbital.

from the modified Schomaker–Stevensen rule (equation 6.2). How would you describe the bonding in these molecules?

12.5 Hexamethyldialane: bridging methyl groups

$(CH_3)_3Al$ is dimeric in the solid phase, in hydrocarbon solution and in the vapor phase at low temperatures and high pressures. The structure of the dimer is diborane-like with two bridging and four terminal methyl groups. See Fig. 12.4. The μ–CH_3 groups are trigonal pyramidal and the C–H bonds are pointing away from the center of the four-membered ring. The Al–Al distance is equal to the Al–Al distance in $(CH_3)_4Al_2(\mu-H)_2$. Since bridging Al–C distances are longer than bridging Al–H distances, the ∠AlCAl angles at the bridging C atoms are reduced to 75°. The bonding may be described in the same manner as for diborane, except that the bridging C atom uses an sp^3 hybrid AO pointing towards the midpoint of the Al–Al vector for formation of the 3c MO.

The dissociation energy of $(CH_3)_4Al_2(\mu-CH_3)_2$, defined as the energy of the reaction

$$(CH_3)_4Al_2(\mu-CH_3)_2 \rightarrow 2Al(CH_3)_3$$

has been measured both in the gas phase and in hydrocarbon solutions. The dissociation energies are 83 and 79 kJ mol^{-1} respectively.

Triethylaluminum and other trialkylaluminum compounds also form alkyl-bridged dimers, but the dissociation energies of the dimers decrease with increasing bulk of the alkyl groups: the dissociation energy of dimeric triethylaluminum is 69 kJ mol^{-1}, and the dissociation energy of dimeric tri-iso-butylaluminum 32 kJ mol^{-1}. The latter is 60% dissociated as a pure liquid at 10°C. Tri-tert-butylaluminium appears to be 100% monomeric under all conditions, there is probably not room for four so voluminous alkyl groups around an Al atom.

12.6 Comparison of the strength of hydrogen, methyl and phenyl bridges

The structure of $(CH_3)_4Al_2(\mu-H)_2$ shows that hydrogen atoms form stronger bridge bonds between aluminum atoms than methyl groups do. Trimethylborane, $(CH_3)_3B$, and trimethylgallane, $(CH_3)_3Ga$, are monomeric under all conditions, but tetramethyldiborane, $(CH_3)_4B_2(\mu\text{–}H)_2$, and tetramethyldigallane $(CH_3)_4Ga_2(\mu\text{–}H)_2$ are both stable molecules: It appears that hydrogen atoms generally form stronger bridges between Group 13 atoms than methyl groups do.

The structure of $(CH_3)_4Al_2(\mu\text{–}Ph)_2$, see Fig. 12.4, was initially something of a surprise: since the strength of alkyl bridges decrease with increasing bulk of the alkyl groups, it was expected that the bridging positions in would be occupied by the methyl groups rather than by the phenyl groups. As indicated in the figure, the planes of the phenyl rings are perpendicular to the plane of the $(AlC_b)_2$ ring. The phenyl bridge is believed to be stabilized by interaction between the π-electrons of the ring and one of the empty $3c$ orbitals of the AlC_bAl fragment. See Fig. 12.4.

12.7 Nuclear magnetic resonance spectroscopy and bridge-terminal exchange of methyl groups in $(CH_3)_4Al_2(\mu\text{–}CH_3)_2$

The proton – just like the electron – has an inherent angular momentum or spin, and the magnitude of this angular momentum about the z-axis is given by $S_z = \pm(1/2)h/2\pi$. Just as for the electron we denote the corresponding spin functions by α and β. The presence of the spin endows the proton with a magnetic moment, or to put it more simply, the particle behaves as small rod-shaped magnet. In the absence of a magnetic field the α and β states of an isolated proton are degenerate, but the degeneracy is removed if we place the proton in a magnetic field. If we use the direction of the magnetic field to define the z-direction, then the energy of a β proton will be higher than the electron that of an α proton:

$$\Delta E = E_\beta - E_\alpha = g_I \mu_N B$$

where B is the strength of the magnetic field, g_I the so-called g-factor of the proton, and μ_N ("the nuclear magneton") is a collection of natural constants.

If a collection of such protons are irradiated with electromagnetic waves of frequency satisfying the condition

$$h\nu = \Delta E$$

a part of the radiation will be absorbed while some of the protons change their spin state from α to β.

The proton magnetic resonance (PMR or ^1H NMR) spectrum of a compound is recorded by placing a solution in an intense magnetic field and measuring the absorption of radiofrequency radiation. The spectrum of ethanol obtained in such an experiment consists of three lines, one for the three protons in the methyl group, one for the two protons in the methylene (CH_2) group and one for the hydroxide group proton. The reason why there are three absorption lines rather than one, is that each proton is surrounded by an electron cloud, and that these electron clouds reduce the magnitude of the magnetic field at the nuclei. This reduction is referred to as the shielding. The greater the shielding, the weaker the magnetic field and lower the frequency of the absorbed radiation. The three methyl-group protons have the same electron environment and are thus equally shielded, so are the two methylene protons. The relative intensities of the three absorption lines are 3:2:1; the intensities are directly proportional to the number of symmetry equivalent protons.

Bridging and terminal methyl groups in $(CH_3)_4Al_2(\mu-CH_3)_2$ are symmetry inequivalent. The electron densities around the H atom nuclei in bridging and terminal methyl groups differ, and as a consequence they are expected to give two separate peaks in a proton magnetic resonance spectrum. This is indeed found to be the case in an NMR spectrum recorded at −55 °C. See Fig. 12.5. One peak is found at a frequency $\Delta\nu = 127\,\text{Hz}(1\,\text{Hz} = 1\,\text{s}^{-1})$ higher than the absorption frequency of the methyl groups in the toluene solvent, the other at $\Delta\nu = 162\,\text{Hz}$. (The magnitude of the absorption frequencies obviously depend on the strength of the magnetic field, but this need not concern us here.) The second peak is twice as intense as the first and is therefore assigned to the protons in the terminal methyl groups.

When the temperature of the sample is allowed to rise, the two absorption peaks become broader and at −35 °C they have merged into one. When the temperature is further increased, this very broad peak becomes more narrow, and at −15 °C we are left with a peak with about the same halfwidth as the two original peaks at −55 °C. The frequency of this new peak is equal to the weighted average of the two low temperature peak frequencies, and the

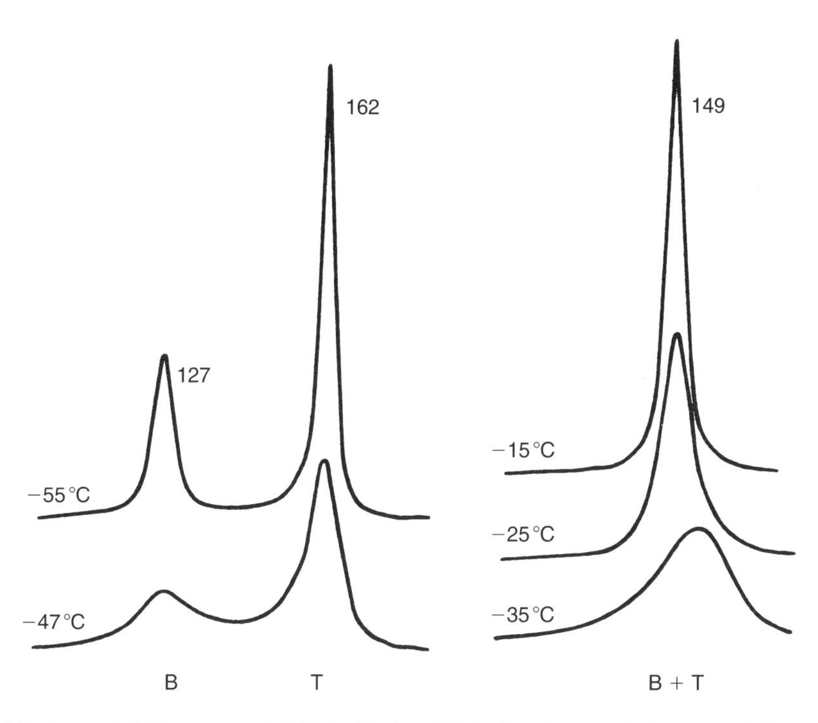

Fig. 12.5. Proton NMR spectra of $(CH_3)_4Al_2(\mu-CH_3)_2$ in toluene solution recorded at various temperatures. B: protons in bridging methyl groups; T: protons in terminal methyl groups; B + T: protons in methyl groups in rapid exchange between bridging and terminal positions. Reproduced from N. S. Ham and T. Mole, in *Progress in Nuclear Magnetic Resonance Spectroscopy*, Vol. 4, J. W. Elmsley, J. Feeney and L. H. Sutcliffe, Eds., Pergamon Press, Oxford, 1969, p 91–192.

intensity is equal to the sum low temperature peak intensities: all protons in the molecule absorb at the same frequency. What is going on?

The reason for the observed temperature variation of the spectrum is that the methyl groups are not staying in the same position: bridging and terminal groups are changing places. The width of the absorption lines at −55 °C is determined by the Heisenberg uncertainty principle which states that the energy of a given quantum state is limited by its "lifetime." If the average lifetime of a methyl group in a given position (bridging or terminal) is equal to τ, the uncertainty of the energy difference ΔE will be

$$\delta \Delta E = \frac{h}{2\pi\tau}$$

and the uncertainty of the absorption frequency

$$\delta \nu = \frac{\delta \Delta E}{h} = \frac{1}{2\pi\tau}$$

The halfwidths of the two peaks at −55 °C is about 3 s^{-1} which corresponds to an average lifetime of

$$\tau = \frac{1}{2\pi\,\delta\nu} = 0.05 \text{ s.}$$

As the rate of exchange between bridging and terminal positions increases with increasing temperature, the average lifetime decreases and the peaks become broader. When the halfwidth is about equal to the distance between the peaks, or 35 s^{-1}, the peaks coalesce. This corresponds to a lifetime of about 0.005 s or 5 milliseconds. Judging from the spectra this occurs somewhere between −47 and −35 °C.

When the temperature is allowed to rise further, the lifetime of the protons in bridging or terminal positions becomes so short that the experiment registers the average for the two environments rather than the difference between them. At −15 °C the average life time in each environment is smaller than a millisecond, the time average has become very accurate and the peak is consequently quite sharp.

The exchange mechanism is believed to be dissociative: the dimer breaks up into monomers that immediately recombine, most often with new methyl groups in the bridging positions. See Fig. 12.6.

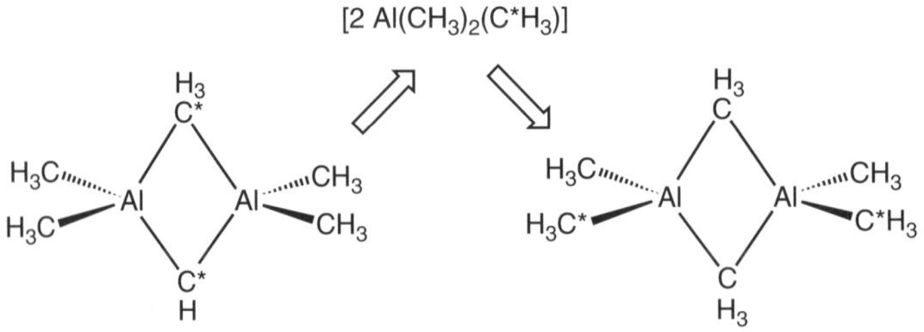

[2 $Al(CH_3)_2(C^*H_3)$]

Fig. 12.6. The mechanism of exchange of bridging and terminal methyl groups in $(CH_3)_4Al_2$ $(\mu{-}CH_3)_2$.

Trimethylgallium is monomeric in hydrocarbon solutions, and the ^{1}H NMR spectrum of a solution of $(CH_3)_3Ga$ and $(CH_3)_6Al_2$ at −50 °C contains three peaks that may be assigned to bridging and terminal methyl groups in $(CH_3)_6Al_2$ and to methyl groups in monomeric $(CH_3)_3Ga$. When the temperature is increased, the spectrum changes, and at 30 °C it contains *one* peak, indicating rapid exchange of methyl groups between bridging and terminal positions in hexamethyldialuminum *and* between $(CH_3)_6Al_2$ and $(CH_3)_3Ga$! The latter exchange presumably occurs in three steps: First dissociation of $(CH_3)_6Al_2$; then combination of monomric $(CH_3)_3Al$ and $(CH_3)_3Ga$ to form $(CH_3)_2Al(\mu\text{–}CH_3)_2Ga(CH_3)_2$; then dissociation of this complex in such a manner that the bridging methyl groups have been exchanged between Al and Ga. At 30 ° C this sequence of events must take place several times per millisecond!

12.8 Clusters

The subvalent chloride BCl has only been observed as a monomer in high temperature vapours, at moderate temperatures it forms tetramers both the gaseous and crystalline phases. The molecular structures of B_4Cl_4 and the analogous alkyl compound $B_4(CMe_3)_4$ are shown in Fig. 12.7. The central part of these molecules consists of a tetrahedral B_4 fragment with edges a few pm shorter than the B–B distance in diborane. Such polyhedral structures are often referred to as clusters. If we assume that the terminal B–Cl or B–C bonds should be described as $2c, 2e$ bonds, the B_4 tetrahedra must be held together by a total of four electron pairs. Since the six edges of the tetrahedra are short enough to imply bonding between the boron atoms, these compounds are electron deficient. (There is, however, no reason to refer to them as subvalent.)

The bonding may be described in terms of three-center, two electron bonds: assume each B atom to be sp^3 hybridized, and that one hybrid points towards the terminal Cl atom or

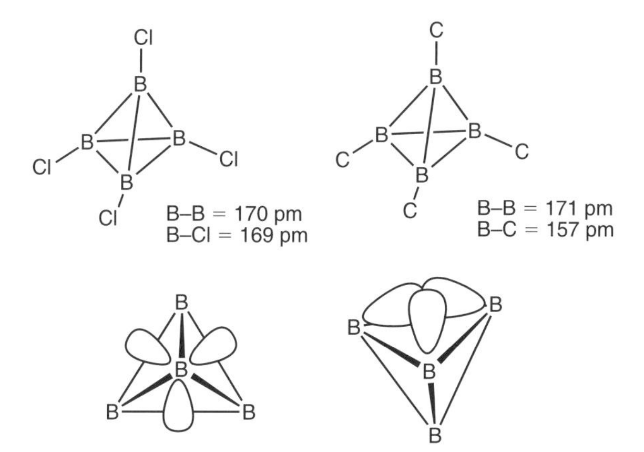

Fig. 12.7. Above: the molecular structures of B_4Cl_4 and $B_4(CMe_3)_4$ (methyl groups excluded). Below: bonding in B_4Cl_4. Left: the three sp^3 hybrids on each B that are directed towards points above the centers of the three neighboring B_3 faces. Right: formation of a three-center bonding orbital over a triangular B_3 face.

alkyl ligand while the three others are pointing towards points above the centers of three triangular faces of the tetrahedron. See Fig. 12.7. We can now form one two-center bonding MO to each terminal Cl or alkyl group and one three-center bonding orbital per triangular face by combining one hybrid orbital from each of the three B atoms at the corners. Each such 3*c* orbital accommodates two electrons.

12.9 Cage compounds: closoboranes and carboranes

Doubly charged anions of composition $B_nH_n^{-2}$ are known for all values of *n* from 6 to 12. Their structures are shown in Fig. 12.8. The boron atoms form *polyhedra with triangular faces only*, they are therefore referred to as closoboranes (closo = cage). Each B atom is also bonded to a terminal H atom (not shown in the figure) with B–H bonds pointing away from the center of the cage. Some of the structures are very symmetric: the boron atoms in the hexaclosoborane (-2) ion $B_6H_6^{-2}$ form an octahedron, the boron atoms in heptaclosoborane(-2) ion a pentagonal bipyramid. Note that the octacloso-borane(-2) ion does *not* form a cube, perhaps because triangular faces are more favorable than squares for the formation of strong bonds. The observed structure is that of a dodecahedron where four B atoms have coordination number five, and four B atoms coordination number six.

The structure of the decaclosoborane(-2) $B_{10}H_{10}^{-2}$ may be described in terms of a tetragonal antiprism, i.e. a cube where the top face has been rotated through 45° relative to the bottom face. The two square faces of the antiprism are capped by BH fragments number

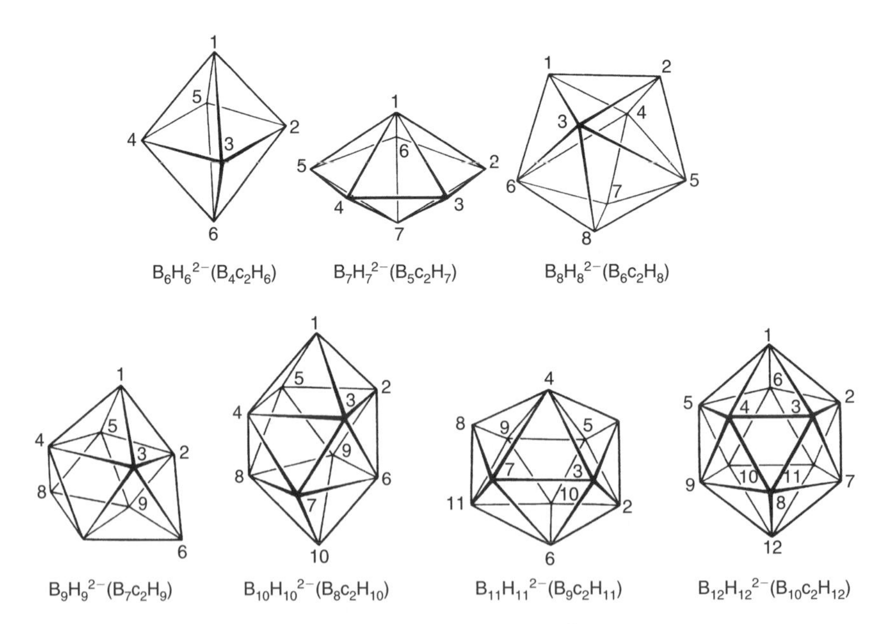

Fig. 12.8. The structures of the closoborane dianions, $B_nH_n^{-2}$ and of the carboranes $c_2B_{n-2}H_n$, $n = 6$ to 12. Reproduced with permission from F. A. Cotton and G. Wilkinson, *Advanced Inorganic Chemistry*, 3rd. Ed., Interscience, New York, 1972. Copyright (1972) John Wiley & Sons.

1 and 10. See Fig. 12.8. In $B_9H_9^{-2}$ one of the capping BH groups has been removed. The structure of $B_{12}H_{12}^{-2}$ is again very symmetric, the twelve B atoms lie at the corner of an icosahedron, one of the Platonic solids where all faces are equilateral triangles.

If we assume the external B–H bonds to be $2c, 2e$ bonds, $n + 1$ electron pairs are left for bonding in the cage. This number is smaller than both the number of edges and the number of faces in each polyhedron. Thus the cage in $B_{12}H_{12}^{-2}$ is held together by 13 electron pairs, while the number of B–B edges is 30 and the number of triangular faces 20! It is clear that the bonding molecular orbitals must be delocalized over more than three atoms. Nevertheless the number of valence electrons appears to be just right: all attempts to prepare the neutral closo-carboranes B_nH_n have failed. Dicarboranes of composition $C_2B_{n-2}H_n$ are, however, known for all n from 6 to 12. Since C has one more valence electron than B, each of these molecules is isoelectronic (i.e. have the same total number of valence electrons) with the borane dianion with the same number of atoms n in the cluster, and adopts the same closo structure.

12.10 The dimethylberyllium polymer and the methyllithium tetramer

Electron deficient bonding is not limited to Group 13 elements like B or Al. Thus dimethylberyllium $(CH_3)_2Be$, forms a solid at room temperature that consists of infinite chains as indicated in Fig. 12.9.

Problem 12.6 Compare the Be–C distance in the polymer with that of the gaseous monomer, and the Be–Be distance with that of gaseous Be_2. Assume each Be atom to be approximately sp^3 hybridized and suggest a description of the bonding in terms of $3c, 2e$ bonds.

The structure of tetrameric $LiCH_3$ is shown in Fig. 12.10: the Li and C atoms occupy alternating corners of a distorted cube. The positions of the four Li nuclei define a perfect tetrahdron with Li–Li distances of 259 pm as compared to 263 pm in Li_2(g). The positions of the four C atoms define an equally perfect, but significantly larger tetrahedron with C–C distances of 362 pm, i.e. more than twice as large as a normal C–C single bond distance. Each C atom occupies a position above the midpoint of a triangular Li_3 face. The $\angle$LiCLi angles are 70°, the $\angle$CLiC angles 107°. The Li–C distances are 226 pm as compared to 196 pm in gaseous, monomeric $LiCH_3$. The staggered orientation of the methyl groups is such as

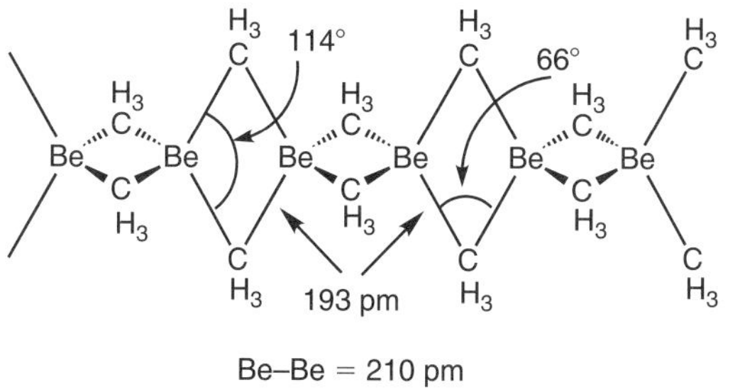

Be–Be = 210 pm

Fig. 12.9. A section of the polymeric structure of crystalline dimethylberyllium.

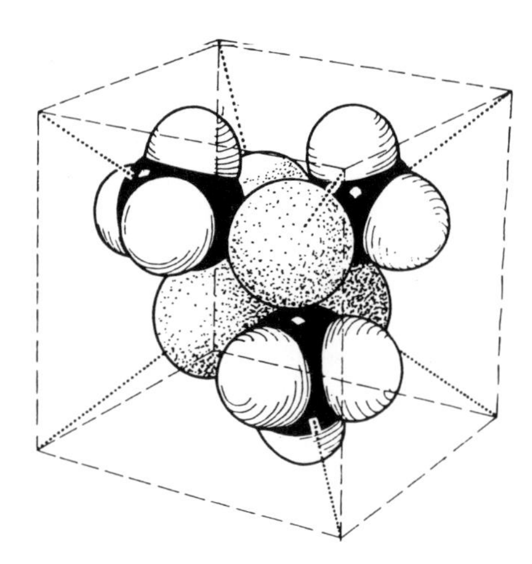

Fig. 12.10. The structure of tetrameric methyllithium, $Li_4(\mu^3\text{–}CH_3)_4$. Reproduced with permission from E. Weiss and E. A. C. Lucken, *J. Organometal. Chem.* 2 (1964) 197. Copyright (1964) Elsevier.

to maximize H··Li distances. The vapor pressure of methyllithium is very low, probably because each Li atom interacts with a methyl group in a neighboring tetramer at a distance of Li–C = 236 pm. The thermodynamic information required for calculation of the energy of dissociation according to

$$Li_4(\mu^3-CH_3)_4(g) = 4\ LiCH_3(g)$$

is unfortunately not available.

The tetramer has been the subject of several studies by quantum chemical calculations [2]. There is universal agreement that the bonding is very polar, i.e. that the litium atoms carry a large positive and the carbon atoms a large negative charge. The controversial question has been whether the bonding is so close to the ionic limit that it may be adequately described solely in terms of electrostatic forces between atoms. As is often the case when such questions arise, the answer will depend on which property of the molecule you wish to explain or predict and how accurate you want your prediction to be.

We begin with a discussion in terms of a molecular orbital model. If we assume that each C atom forms a 2c,2e bond to each H atom, the cluster is held together by four electron pairs, one pair for each triangular Li_3 face.

As in B_4Cl_4, we may assume each Li atom to direct one sp^3 hybrid AO toward a point above the centre of each triangular face of the Li_4 tetrahedron. One such hybrid from each corner meet at a point above the midpoint and combine with an sp^3 hybrid AO of the C atom to form a four-center bonding orbital which contains two electrons. See Fig. 12.10. Since C is significantly more electronegative than Li, there is no doubt that in this 4*c* LCAO molecular orbital the coefficient of the sp^3 orbital of the carbon atom will be significantly larger than those of the Li orbitals. In the ionic limit the coefficient will aproach unity while the coefficients of the three Li AOs will approach zero.

There is no doubt that the observed structure of $Li_4(\mu^3\text{–}CH_3)_4$ is consistent with an completely ionic model: it is indeed very similar to that of gaseous Na_4Cl_4 described in

Problem 5.7. The fact that the ∠LiCLi angles are about 40° smaller than the ∠CLiC angles, would then have to be interpreted as a result of Born repulsion between the CH_3^- anions.

References

[1] P. Laszlo, *Angew. Chem Int. Ed. Engl.*, 39 (2000) 2071.
[2] E. Matito, J. Poater, F. M. Bickelhaupt, and M. Sola, *J. Phys. Chem. B*, 110 (2006) 7189 and references therein.

Chapter 13

Structure and bonding in simple compounds of the Group 14 elements

Introduction

The first two elements in Group 14, carbon and silicon, have the electron configurations [Ng]ns^2np^2 while the heavier elements, germanium, tin, and lead, have the electron configurations [Ng]$(n-1)d^{10}ns^2np^2$. The properties of Sn and Pb are such that they are classified as metals, while Si and Ge are classified as non-metals or metalloids. Carbon is definitely a non-metal. We shall use the symbol E (for Element) to denote any member of the group.

All the Group 14 elements form compounds of composition EH_4, $E(CH_3)_4$ and ECl_4. With the exception of PbH_4 and $PbCl_4$ these compounds are relatively stable and have been extensively studied. All of them are monomeric in the gas phase and in solution. Each molecule contains a central Group 14 atom, E, surrounded by four H atoms, Cl atoms, or methyl groups as ligands. The carbon atoms of the methyl groups that are bonded directly to the central atom, are described as ligating atoms. The coordination geometry of E is defined as the shape of the polyhedron with the ligating atoms at the corners. The coordination geometries of the compounds EH_4, $E(CH_3)_4$ and ECl_4 are all perfectly tetrahedral.

The oxidation numbers of the C atoms in CH_4 and CCl_4 are -4 and $+4$ respectively. The oxidation number of the central carbon atom in $C(CH_3)_4$ is zero. Minimum energy rupture of a bond in CH_4 or CCl_4 or a C–C bond in $C(CH_3)_4$ yields neutral, radical species. These bonds are therefore classified as homolytic bonds. We shall refer to the number of (single) homolytic bonds that an atom forms, as the valency of the atom. The central atoms in the compounds EH_4, $E(CH_3)_4$ and ECl_4 are all tetravalent.

All the tetrachlorides form liquids at room temperature. The C, Si, Ge and Sn tetrachlorides are stable, but the lead analogue decomposes explosively at temperatures above 20°C to yield chlorine and a solid dichloride:

$$PbCl_4(l) \rightarrow PbCl_2(s) + Cl_2(g)$$

Solid dichlorides of tin and germanium are also known. Evaporation yields monomeric Pb, Sn, or Ge dichlorides in the gas phase. Monomeric CCl_2 or $SiCl_2$ may be generated and studied in high temperature vapors. We shall refer to Group 14 atoms that form four homolytic bonds as Lewis-valent and to atoms that form two homolytic bonds as subvalent.

The enthalpy of the reaction:

$$ECl_4(g) \rightarrow ECl_2(g) + Cl_2(g)$$

(at 298 K) is 313 kJ mol^{-1} for E = C, 494 kJ mol^{-1} for Si, 329 kJ mol^{-1} for Ge, 274 kJ mol^{-1} for Sn, and 155 mol^{-1} for Pb: the stability of the Lewis-valent chloride relative to the subvalent decreases as the group is descended from Si to Pb. We have already made the same observation for the chlorides of the Group 13 elements from B to Tl.

13.1 The structures of the subvalent element chlorides

The dichlorides of of the Group 14 elements are all *angular* in the gas phase. The valence angles, bond distances, and mean bond energies are listed in Table 13.1.

As expected, the bond distances increase as the group is descended from C to Pb, while mean bond energies decrease from Si to Pb. The first element in the group, C, displays irregular behavior in so far as the mean bond energy in CCl_2 is smaller than in $SiCl_2$. The greater Si–Cl bond energy is probably due to the lower electronegativity of Si and the concomitant increased polarity of the Si–Cl bond. Also the valence angle in CCl_2 is very close to the tetrahedral value of 109.5° while the valence angles of all the other dichlorides are close to 100°. The wider valence angle in CCl_2 may be due to repulsion between the Cl atoms.

13.2 Bonding models for the subvalent element chlorides

The angular structures of the gaseous dichlorides follow directly from the VSEPR model: the central atom is surrounded by two bonding electron pairs and one nonbonding. The three electron pairs will occupy regions of space which are approximately trigonal planar with respect to the central atom. Since the non-bonding electron pair is – on the average – closer to the central atom than the bond pairs, it requires more space, and the angle ∠ClECl is reduced from 120° to 109° in CCl_2 and to about 100° when E = Si, Ge, Sn or Pb.

In the ground state electron configuration of the Group 14 atoms the two *p* electrons in the valence shell occupy different orbitals and have parallel spins. This atomic state has been labeled ^3P. As a first approximation we assume that the two valence shell *s* electrons remain unaffected by formation of the two E–Cl bonds. Combination of each of the two

Table 13.1. Gaseous dichlorides of the Group 14 elements: E–Cl bond distances, *R*; mean bond energies, MBE; and valence angles, ∠ClECl. Gaseous tetrachlorides of the Group 14 elements: E–Cl bond distances and mean bond energies.

	ECl_2(g)			ECl_4(g)	
E	*R*(E–Cl) (pm)	MBE(E–Cl) (kJ mol^{-1})	∠ClECl (°)	*R*(E–Cl) (pm)	MBE(E–Cl) (kJ mol^{-1})
C	172[a]	373	109[b]	177	323
Si	207	431	101	202	397
Ge	217	394	100	211	337
Sn	235	371	99	228	312
Pb	245	306	98	237	252[b]

[a] Reference [1]. [b] Reference [2].

pure *p* orbitals on the central atom with a suitable *p* orbital on one of the ligating Cl atoms, yields two 2c MOs which accommodate two electrons each. Such a description suggests that the valence angle in the dichlorides should be 90°, or perhaps somewhat larger due to repulsion between the Cl atoms.

13.3 The structures of the Lewis-valent element chlorides

The coordination geometry of each tetrachloride appears to be tetrahedral with all valence angles equal to 109.47°. Bond distances and mean bond energies are listed in Table 13.1. The bond distances are also displayed in Fig. 13.1. Comparison with Fig. 11.1 shows that the variation of E–Cl bond distances as the group is descended, is strikingly similar to that observed for the Group 13 elements, the only difference being that the increase of E–Cl bond distances between the third and fourth period elements (Si and Ge) is larger than the increase between the corresponding elements in Group 13 (Al and Ga). As expected, the mean bond energies decrease from Si to Pb. Again the first element in the group shows irregular behavior: the MBE in CCl_4 is smaller than in $SiCl_4$.

We now turn our attention to a comparison of subvalent and Lewis-valent chlorides. Comparison of bond distances and MBEs shows that the bonds in the tetrachlorides, ECl_4, E = Si, Ge, Sn or Pb are shorter, but weaker than the bonds in the dichlorides. We have already noted a similar difference between the Lewis-valent and the subvalent chlorides of the Group 13 elements.

The C–Cl bond CCl_4 is unusual in being longer than in in CCl_2. Since there are six Cl ··· Cl contacts in CCl_4 as compared to just one in the dichloride, the C–Cl bonds in the tetrachloride may be elongated by Cl ··· Cl repulsions. We shall return to this question at the end of the chapter.

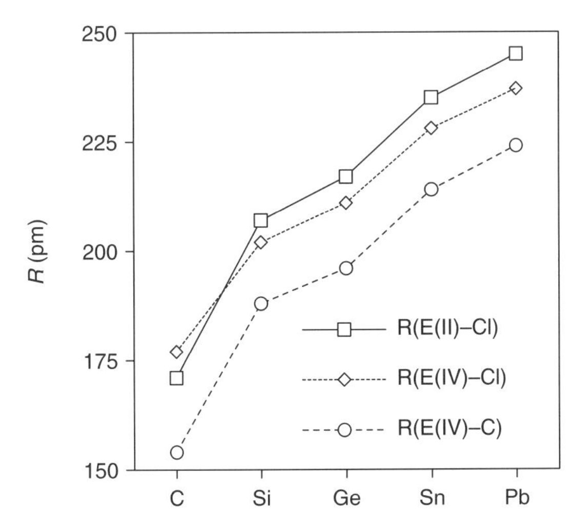

Fig. 13.1. Periodic variation of bond distances in subvalent chlorides, ECl_2, Lewis-valent chlorides, ECl_4, and tetramethyl derivatives, $E(CH_3)_4$, of the Group 14 elements.

Table 13.2. Gaseous tetrahydrogen and tetramethyl derivatives of the Group 14 elements: bond distances and mean bond energies.

	EH_4		$E(CH_3)_4$	
E	R(E–H) (pm)	MBE(E–H) (kJ mol^{-1})	R(E–C) (pm)	MBE(E–C) (kJ mol^{-1})
C	109	413	154	365
Si	148	319	188	318
Ge	153	286	196	256
Sn	171	250	214	226
Pb	–	–	224	161

13.4 Hydrogen and methyl derivatives of the Lewis-valent elements

The bond distances and mean bond energies of the compounds EH_4, E = C to Sn and in $E(CH_3)_4$, E = C to Pb are listed in Table 13.2. The variation of E–C bond distances as the group is descended is shown in Fig. 13.1.

The structures of two gaseous subvalent alkyls, $E[CH(SiMe_3)_2]_2$, E = Ge or Sn, Me = CH_3, are also known. The Ge–C and Sn–C bond distances are about 8 pm longer than in the tetramethyls, and the ∠CEC angles are smaller than tetrahedral. These structures are therefore analogous to the structures of the subvalent chlorides.

13.5 Bonding models for tetravalent compounds

The fact that the Group 14 elements form four single bonds is in agreement with the Lewis electron pair model. In these compounds the central atom is surrounded by four bonding electron pairs and the structure is tetrahedral in agreement with the VSEPR model. The observation that the E–Cl bonds in the tetrachlorides are shorter than in the dichlorides is, however, difficult to rationalize in terms of the VSEPR model.

In order to describe bonding in terms of $2c$, $2e$ bonds, we need to hybridize the AOs on the central atom in such a manner that we obtain four orthogonal hybrid orbitals pointing towards the four ligands. The simplest way to do this is to define a Cartesian coordinate system on the central atom in such a manner that the four ligating atoms occupy alternating corners on a cube with the Group 14 atom at the centre and edges parallel to the coordinate axes. See Fig. 13.2.

The first step is to form pure p type orbitals pointing towards the ligands:

$$(p_1) = [+(p_x) + (p_y) + (p_z)]/\sqrt{D}$$

$$(p_2) = [-(p_x) - (p_y) + (p_z)]/\sqrt{D}$$

$$(p_3) = [-(p_x) + (p_y) - (p_z)]/\sqrt{D}$$

$$(p_4) = [+(p_x) - (p_y) - (p_z)]/\sqrt{D}$$

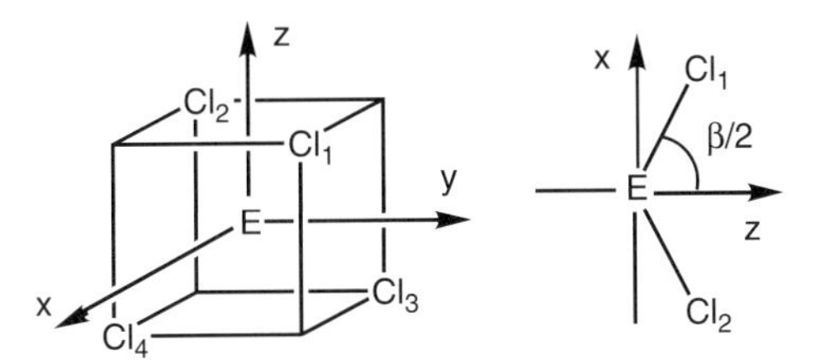

Fig. 13.2. Left: coordinate systems for the construction of sp^3 hybrid atomic orbitals in ECl_4. Right: coordinate system for construction of sp^λ hybrid orbital pointing towards the Cl atoms in ECl_2.

These orbitals are normalized if $D = 3$.

We now form the hybrid orbitals:

$$h_1 = \frac{(s) + a(p_1)}{\sqrt{1+a^2}}$$
$$h_2 = \frac{(s) + a(p_2)}{\sqrt{1+a^2}}$$
$$h_3 = \frac{(s) + a(p_3)}{\sqrt{1+a^2}}$$
$$h_4 = \frac{(s) + a(p_4)}{\sqrt{1+a^2}}$$

These orbitals will have the same shape if all coefficients (a) are equal.

Finally we fix the value of a by insisting that the hybrids are orthogonal:

$$\int h_1(x, y, z)h_2(x, y, z)d\tau = 0 \tag{13.1}$$

Problem 13.1 Show that equation (13.1) yields $a^2 = 3$. The hybrids are therefore denoted by sp^3.

Each of these hybrids may now be combined with the appropriate atomic orbitals on a ligating H, Cl or C atom to form a 2c MO which accommodates two electrons.

It can be shown that the sp^3 hybrids are the only four hybrids that can be formed from the s and the three p orbitals under the condition that the hybrids are orthogonal and have the same shape. The hybridization model might therefore be used to predict the tetrahedral shape of these molecules, but in a much less direct manner than the VSEPR model.

We have already suggested that $2c$,$2e$ bonds formed with with sp^λ hybrid orbitals are shorter than those formed from pure p orbitals. The electron density calculated for an atom with one electron in each of the four sp^3 hybrid orbitals is equal to the electron density calculated for an atom in the $s^1p_x^1\,p_y^1\,p_z^1$ configuration. This result, in turn, implies that the energies of the two configurations are equal. The lowest state arising from the $s^1p_x^1\,p_y^1\,p_z^1$ configuration is ^5S. In the C atom this state is 403 kJ mol^{-1} above the ground state, in Sn 474 kJ mol^{-1}. These hybridization energies are sufficient to explain why the bonds in the tetrachlorides are weaker than in the dichlorides.

13.6 Construction of two orthogonal and equivalent hybrid orbitals spanning an arbitrary valence angle β

While the ∠ClECl valence angles in $SiCl_2$ and the dichlorides of the heavier Group 14 elements are about 100°, the valence angle in CCl_2 is about 109°, and – as we shall see in the following section – the ∠HCH valence angle in the methylene radical CH_2 is 134°. In this section we shall show how one may construct two orthogonal and equivalent hybrid orbitals on the central atom in a triatomic molecule like the dichloride ECl_2 that span any valence angle between 90 and 180°. We begin by defining a Cartesian coordinate system with the origin at the central atom E, the z-axis bisecting the ClECl valence angle, the x-axis parallel to the Cl–Cl vector, and the y-axis perpendicular to molecular plane. See Fig. 13.2.

Problem 13.2 Denote the angle ∠ClECl by β, and show that orthogonal sp^λ hybrids with the same shape and pointing towards the ligating B atoms have the form

$$h = \frac{(s) + a(p)}{\sqrt{1 + a^2}} \tag{13.2}$$

with

$$a^2 = \frac{-1}{\cos\beta} \tag{13.3a}$$

and the s character and p characters:

$$s \text{ char} = \frac{-\cos\beta}{1 - \cos\beta} \tag{13.3b}$$

$$p \text{ char} = \frac{1}{-\cos\beta} \tag{13.3c}$$

Insertion of $\beta = 100°$, corresponding to the valence angles in $SiCl_2$ and the dichlorides of the heavier elements in the group, yields the p to s mixing ratio $\lambda = a^2 = 5.78$, the s character $= 0.15$ and the p character $= 0.85$. Insertion of $\beta = 134°$, corresponding to the valence angle in CH_2, yields the p to s mixing ratio $\lambda = a^2 = 1.44$, the s character $= 0.41$ and the p character $= 0.59$.

Rearrangement of equation (13.3b) gives the angle spanned by the two equivalent hybrids as a function of their s character:

$$\cos\beta = \frac{s - \text{char}}{s - \text{char} - 1} \tag{13.4}$$

In Fig. 13.3 we use equation (13.4) to plot the angle between two equivalent sp^λ hybrids as a function of their s character: the angles thus obtained vary from 90 to 180°. *The angle between two orthogonal sp^λ hybrids increases with increasing s character.*

The non-bonding electron pair on the central atom in Group 14 element dichlorides might be assumed to occupy a hybrid atomic orbital formed by combination of the s, p_z and p_x orbitals:

$$h_3 = \frac{(s) + b(p_z) + c(p_x)}{\sqrt{1 + b^2 + c^2}} \tag{13.5}$$

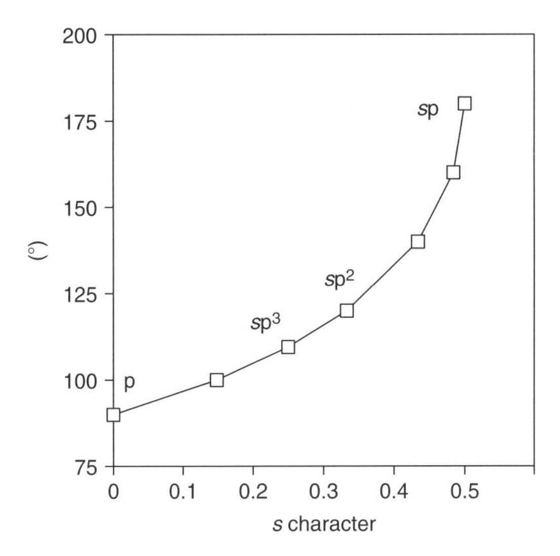

Fig. 13.3. The angle between two equivalent, orthogonal sp^{λ} hybrid orbitals as a function of their s characters.

This hybrid is orthogonal to the two hybrids pointing towards the chlorine atoms only if c $= 0$ and $b = -0.65$ corresponding to an s character of 0.70.

Problem 13.3 Use Fig. 8.4 to make a sketch of the electron density contours of the hybrid orbitals pointing towards the ligands or containing the electron lone pair in $SiCl_2$. Is the direction of the hybrid containing the lone pair consistent with the VSEPR model? Note that the hybrid orbitals in the Fig. 8.4 are labeled with the *p characters*.

13.7 Radical species formed by H atom extraction from methane

Progressive removal of hydrogen atoms from methane yields the radicals CH_3 (methyl), CH_2 (methylene), and CH (methylidyne). These very unstable species may be studied in electric discharges. The C–H bond distances and bond dissociation energies are listed in Table 13.3

The methyl radical contains one unpaired electron. The coordination geometry is trigonal *planar*, and the unpaired electron occupies the $2p_z$ orbital on C. (Remember that the direction of the z-axis is determined by the highest symmetry axis of the molecule.) This structure is *not* in agreement with the VSEPR model since the repulsion between the C–H bonding electron pairs and the unpaired electron in the p_z orbital has no apparent effect on the structure. The unpaired electron is therefore described as stereochemically inactive.

The methylene radical offers another surprise: the ground state is a triplet, i.e. it contains two unpaired electrons with parallel spins! The electronic structure is therefore different from that of CCl_2 which, as we have seen, contains a non-bonding electron pair in an sp_z hybrid AO on the C atom. In the ground state of CH_2 one non-bonding electron remains in

Table 13.3. C–H bond distances, HCH valence angles and C–H bond dissociation energies of CH_n, $n = 1, 2, 3$, or 4.

	R(CH) (pm)	∠ HCH (°)	D(C–H) (kJ mol^{-1})
CH_4	108.7	109.5	(H_3C–H) 436
CH_3	107.7	120	(H_2C–H) 456
CH_2	107.5	134	(HC–H) 423
CH	112.0	–	(C–H) 338

the sp_z hybrid orbital while the other resides in the pure p_y orbital which is perpendicular to the molecular plane. The ∠HCH is valence angle is 134° as compared to 109° in CCl_2.

The first excited electronic state of methylene is a singlet like the ground state of the dichloride CCl_2: the two non-bonding electrons occupy the sp_z AO and have antiparallel spins. The energy of singlet methylene is only 37 kJ mol^{-1} above the triplet ground state. The bond distance is 110.7 pm and the valence angle ∠HCH = 102°. *Note that when we change the electronic structure of a molecule, i.e. the number of electrons in one ore more molecular orbitals, we also change the molecular structure.* The change of the valence angle is easily rationalized within the VSEPR model: *one* electron in the sp_z hybrid AO does not repel the C–H bonding electron pairs as strongly as two electrons do, and the magnitude of the repulsion between an electron in the p_y orbital and the two bonding electron pairs is independent of the angle.

Why does the ground state of CH_2 have the electron configuration $sp_z^1\,p_y^1$ rather than sp_z^2? When discussing the energies of exited He atoms with electron configuration $1s^1 2s^1$ we saw that the energies were not determined by the electron configuration alone. Atoms in the triplet states, i.e. with parallel electron spins, have lower electron–electron repulsion energies and are therefore energetically more stable than atoms in singlet states.

The sp_z and p_y atomic orbitals in CH_2 are, of course not degenerate, and accurate quantum chemical calculations [3] indicate that it requires 96 kJ mol^{-1} excite molecules from the sp_z^2 to the $sp_z^1\,p_y^1$ configuration *if the spins remain antiparallel.* The energy released when their spin are changed from antiparallel to parallel is, however, so large (133 kJ mol^{-1}) that the triplet becomes the ground state.

The methylidyne radical CH has five valence electrons. Two non-bonding electrons may be assigned to the 2s AO on the C atom, and two electrons to a 2c C–H bonding orbital formed by linear combination of the C $2p_z$ orbital with the H 1s orbital. This description is in agreement with the unusually long C–H bond distance. The unpaired electron has an angular momentum around the molecular z-axis, $L_z = h/2\pi$, and must therefore occupy one of the p_π orbitals on the C atom.

13.8 Heteroleptic compounds: the molecular structures of fluorinated methanes and methylsilanes

Up till now we have described compounds of the Group 2, 12, 13, and 14 elements in which all the ligands have been equal. Such compounds are referred to as *homoleptic*. Compounds with different ligands like CH_2F_2 are referred to as *heteroleptic*.

Table 13.4. Equilibrium bond distances and valence angles in the heteroleptic compounds $CH_{4-n}F_n$, $n = 0–4$.

	R_e(C–H) (pm)	R_e(C–F) (pm)	$\angle_e$HCH (°)	$\angle_e$HCF (°)	$\angle_e$FCF (°)
CH_4	108.7(1)	–	109.47	–	–
CH_3F	108.6(2)	138.3(1)	110.2(3)	108.8(3)	–
CH_2F_2	108.4(3)	135.1(1)	112.8(3)	108.6	108.5(1)
CHF_3	109.1(14)	132.8(3)	–	110.3(3)	108.6(3)
CF_4	–	131.5(2)	–	–	109.47

Table 13.4 contains the equilibrium bond distances and valence angles of the five molecules in the series $CH_{4-n}F_n$, with n ranging from 0 to 4, as determined by rotational spectroscopy. Since the variations of bond distances and valence angles through the series are are relatively minor, we have included the estimated standard deviations.

The C–F bond distances in the fluorinated methanes are found to decrease monotonically with increasing number of F atoms attached to the C atom, from 138.3(1) pm in CH_3F to 131.5(1) pm in CF_4. The difference between one member of the series and the next is, however, small. This is particularly true in regard to the last two members of the series, CHF_3 and CF_4. Can we be sure that the difference is real? The best way to answer this question is to calculate the estimated standard deviation of the difference between them. If we have two parameters P_A and P_B determined with estimated standard deviations equal to σ_A and σ_B respectively, the e.s.d. of the difference between them is given by

$$\sigma(P_A - P_B) = \sqrt{\sigma_A^2 + \sigma_B^2}$$

The difference between the two bond distances is 1.3 pm, while its e.s.d. is 0.35 pm. Since the difference is more than three times larger than the e.s.d., the probability that it is in fact zero is less than 1% and may consequently be ignored. Comparison of the other C–F bond distance differences along the series with their e.s.d.s, indicate that they are all real.

The four C–H bond distances in Table 13.5 differ by less than 1 pm. All the differences are smaller than three times their e.s.d. The data in the table thus provide no evidence for variation of C–H bond distances with the number of F atoms.

Table 13.5 contains the bond distances and valence angles of the five molecules in the series, $Si(CH_3)_{4-n}F_n$,$n = 0–4$, as determined by gas electron. These bond distances and valence angles represent average values for the vibrating molecules. Both Si–C and Si–F bond distances decrease markedly with increasing number of F atoms. Si–C bond distances decrease by about 5 pm between between $Si(CH_3)_4$ and $Si(CH_3)F_3$, the Si–F bond distances by 5 pm between $Si(CH_3)_3F$ and SiF_4. Successive shortening has also been observed for Ge–C and Ge–F bond distances in the series $Ge(CH_3)_{4-n}F_n$ and for M–C and M–Cl bonds in the series $M(CH_3)_{4-n}Cl_n$, M = Si, Ge or Sn. There is at present enough evidence to conclude that the phenomenon is general: *when a more electronegative substituent is introduced at the most electropositive of two bonded atoms, the bond distance decreases.*

Since C is more electropositive than F the rule correctly predicts that introduction of more F atoms at the C atom in a fluorinated methane will lead to C–F bond shortening.

Table 13.5. Bond distances and valence angles in the heteroleptic compounds $Si(CH_3)_{4-n}F_n, n = 0–4$.

	R(Si–C) (pm)	R(Si–F) (pm)	∠CSiC (°)	∠CSiF (°)	∠FSiF (°)
$Si(CH_3)_4$	187.6(2)	–	109.47	–	–
$Si(CH_3)_3F$	184.8(1)	160.0(1)	111.5(2)	107.5(2)	–
$Si(CH_3)_2F_2$	183.6(2)	158.6(1)	116.7(6)	108.7	104.6(4)
$Si(CH_3)F_3$	182.8(4)	157.0(1)	–	112.2(5)	106.8(5)
SiF_4	–	155.4(3)	–	–	109.47

H, however, is more electropositive than C, so the rule is silent as to whether introduction of F atoms on the latter should lead to C–H bond shortening.

Since both F and C atoms are more electronegative than Si, the rule predicts that increasing the number of F atoms at Si will lead to reduction of both Si–F and Si–C bond distances in the series of fluorinated methylsilanes.

The coordination geometries in simple heteroleptic derivatives of the tetravalent Group 14 elements remain approximately tetrahedral, but the valence angles may deviate from the ideal value of 109.47°. The ∠FCF angle in CH_2F_2 is 108.5(1)°, the ∠HCH angle is 112.8(3)°. The molecule has a twofold symmetry axis bisecting the ∠HCH and ∠FCF angles, and in addition there are two symmetry planes defined by the C atom and the two H atoms and by the C atom and the two F atoms respectively. The two symmetry planes intersect along the symmetry axis. If we regard the symmetry axis as vertical, then so are the symmetry planes. A molecule with a twofold symmetry axis and two vertical symmetry planes is said to have C_{2v} symmetry. All compounds of type EX_2Y_2 where E is a Group 14 element and X and Y are H, halogen atoms or methyl groups, seem to have this symmetry.

The fluoromethane molecule CH_3F has a threefold symmetry axis along the C–F bond. The vertical direction in a molecule (i.e. the direction of the z-axis) is defined by this symmetry axis. CH_3F also has three vertical symmetry planes. Such a molecule is said to have C_{3v} symmetry. The three ∠HCH valence angles in are slightly larger than tetrahedral, the three ∠HCF angles slightly smaller. See Table 13.5.

The trifluoromethane molecule CHF_3 also has C_{3v} symmetry with the threefold symmetry axis along the C–H bond. The three ∠HCF angles are slightly larger than tetrahedral, the ∠FCF angles slightly smaller. Thus, in each of the three heteroleptic compounds, $CH_{4-n}F_n$ with $n = 1, 2$ or 3, the valence angles are found to decrease in the order ∠HCH > ∠HCF < ∠FCF.

The valence angles in the heteroleptic silanes Si $(CH_3)_{4-n}F_n$ with $n = 1, 2,$ or 3 show larger deviations from the ideal tetrahedral angle of 109.47°, in each member of the series the valence angles decrease in the order ∠CSiC > ∠CSiF > ∠FSiF. Again there is enough evidence from related molecules to allow us to generalize: *The smallest valence angles in a heteroleptic compound are subtended by the most electronegative substituents.* This generalization is agreement with the VSEPR model: the electrons in bonds to more electronegative substituents are pulled further away from and require less space on the surface of the central atom.

Such structures may also be rationalized in terms of the hybridization model. Consider the case of $Si(CH_3)_2F_2$. Since the angle subtended by the F atoms is smaller than tetrahedral, the $2c$ bonding orbitals are presumably formed from hybrid orbitals on the central atom which have less than 0.25 s character. If the hybrids used for forming $2c$ bonding orbitals to C are to be orthogonal to the ones pointing towards the F atoms, they must have more than 0.25 s character. This in turn means that the valence angle subtended by the C atoms must be greater than tetrahedral. This line of reasoning suggests that *the central atom concentrates the s character in the hybrid orbitals pointing towards the more electropositive ligands.* This statement is equivalent to the VSEPR statement that electrons forming bonds to electronegative substituents require less space around the central atom.

Why should the central atom concentrate the s character in the hybrids pointing towards the more electropositive ligands? We have seen that hybridization increases the electron density along the line connecting the bonded atoms and thus leads to the formation of a stronger bond. When forming a $2c$ bonding MO by combination of a hybrid AO on the central atom with an AO on the ligating atom, the coefficient in front of the central atom AO will be larger in the bonding MOs to eletropositive ligands, and the properties of that hybrid will be more important for the bond strength. The lowest energy will therefore be obtained if the central atom invests its s orbital in the MOs where it will do the most good, i.e. in the hybrid orbitals pointing towards the electropositive ligands.

13.9 C–F and C–H bond energies in fluorinated methanes

We have seen that the C–F bonds in CF_4 are 7 pm shorter than in CH_3F. Are they also stronger? One way to answer this question is to compare the C–F bond dissociation energies:

$$CH_3F(g) = CH_3(g) + F(g) \qquad \Delta U^\circ = D(H_3C\text{–}F) = 459\,\text{kJ mol}^{-1}$$

and

$$CF_4(g) = CF_3(g) + F(g) \qquad \Delta U^\circ = D(F_3C\text{–}F) = 540\,\text{kJ mol}^{-1}$$

The shortening of the C–F bond by 5% is thus accompanied by a 17% increase of the bond dissociation energy!

A similar comparison of the C–H bond dissociation energies

$$CH_4(g) = CH_3(g) + H(g) \qquad \Delta U^\circ = D(H_3C\text{–}H) = 438\,\text{kJ mol}^{-1}$$

and

$$CHF_3(g) = CF_3(g) + H(g) \qquad \Delta U^\circ = D(F_3C\text{–}H) = 443\,\text{kJ mol}^{-1}$$

shows that the introduction of three F atoms in methane increases the C–H bond dissociation energy with less than 2%. The difference is just barely larger than the error limits.

The data needed for calculation of the C–F bond dissociation energies of CH_2F_2 or CHF_3 or of the C–H bond dissociation energies of CH_3F or CH_2F_2 are unfortunately not available.

Table 13.6. C–F and C–H bond dissociation energies, thermochemical bond energies and mean bond energies in fluorinated methanes. All energies in kJ mol^{-1}.

	D(C–H)	D(C–F)	MBE	TBE(C–F)
CH_4	438	–	414[a]	–
CH_3F	–	459	–	432
CH_2F_2	–	–	–	462
CHF_3	443	–	–	482
CF_4	–	540	489[b]	–

[a]MBE(C–H). [b]MBE(C–F).

We therefore turn our attention to bond energies that may be calculated from the atomization energies:

$$CH_4(g) = C(g) + 4H(g) \qquad \Delta U^\circ_{atom} = 1654\,\text{kJ mol}^{-1}$$

$$CH_3F(g) = C(g) + 3H(g) + F(g) \qquad \Delta U^\circ_{atom} = 1674\,\text{kJ mol}^{-1}$$

$$CH_2F_2(g) = C(g) + 2H(g) + 2F(g) \qquad \Delta U^\circ_{atom} = 1752\,\text{kJ mol}^{-1}$$

$$CHF_3(g) = C(g) + H(g) + 3F(g) \qquad \Delta U^\circ_{atom} = 1859\,\text{kJ mol}^{-1}$$

$$CF_4(g) = C(g) + 4F(g) \qquad \Delta U^\circ_{atom} = 1956\,\text{kJ mol}^{-1}$$

The atomization energies of CH_4 and CF_4 yield the mean C–H and C–F bond energies listed in Table 13.6.

The atomization energy of CH_3F may be written as the sum of one C–F and three C–H bond energies. Since the length of C–H bonds appear to be little affected by introduction of F atoms, we assume that the C–H bond energies in CH_3F are equal to the mean C–H bond energy in CH_4, and obtain an estimated C–F bond energy of 432 kJ mol^{-1}. The C–F bond energies in CH_2F_2 and CHF_3 may be estimated in a similar manner. See Table 13.6. Mean bond energies in heteroleptic compounds estimated from energies of atomization by transferring one or more bond energies from related molecules, are commonly referred to as "thermochemical bond energies" or TBEs. The C–F bond energies are found increase monotonically with the number of F substituents, from TBE = 432 in CH_3F to 482 kJ mol^{-1} in HCF_3 and to MBE = 489 kJ mol^{-1} in the tetrafluoride. *It is clear that the C–F bonds not only become shorter with increasing number of F substituent, they also become significantly stronger.*

The data needed for calculation of Si–F or Si–C bond energies along the series $Si(CH_3)_{4-n}F_n$ are, unfortunately, not available.

13.10 The polarity of C–F bonds in fluorinated methanes

The length of a C–F single bond may be estimated from the empirical MSS rule (Section 6.5):

$$R(\text{C–F}) = r_C + r_F - 8.3\,\text{pm}|\chi_C - \chi_C|^{1.5}$$

Table 13.7. The net atomic charges (in atomic units) in fluorinated methanes obtained by AIM [4] or NAO [5] analysis of calculated electron densities.

	$q_{AIM}(F)$	$q_{AIM}(H)$	$q_{AIM}(C)$	$q_{NAO}(F)$	$q_{NAO}(H)$	$q_{NAO}(C)$
CH_4	–	−0.03	0.13	–	0.18	−0.71
CH_3F	−0.71	0.01	0.68	−0.43	0.13	0.03
CH_2F_2	−0.71	0.06	1.30	−0.42	0.11	0.62
CHF_3	−0.70	0.12	2.00	−0.40	0.10	1.09
CF_4	−0.69	–	2.78	−0.37	–	1.50

Using the data in Table 6.2 we find that the sum of the bonding radii ($r_C + r_F$) is equal to 155.6 pm, while the third term which represents the contraction due to the bond polarity is −17.6 pm. The estimated C–F bond distance is 138.0 pm in good agreement with the experimental bond distance in CH_3F. Is there any reason to believe that the polarity of the C–F bonds in fluorinated methanes increases with the number of F atoms?

In Table 13.7 we list the atomic charges in the molecules $CH_{4-n}F_n$ obtained by Atoms in Molecules (AIM) or Natural Atomic Orbital (NAO) analyses of calculated charge densities. The atomic charges obtained by the two methods differ significantly with respect to the magnitude of the charge on the F atoms. Both methods agree, however, that the F atoms carry a large negative charge which remains essentially constant through the series; that the net charges on the H atoms are small in comparison; and that the net positive charge on the C atoms consequently increases in an approximately linear fashion with the number of F atoms attached to it.

We found in Chapter 6 that the bond dissociation energies of diatomic molecules increase, and that the bond distances decrease, with increasing bond polarity: the linear correlation coefficient of the excess bond strength and the net atomic charges estimated from their dipole moments (q_{IC}) was $\rho = 0.85$. The net charges of the two atoms in a diatomic molecule must be of opposite sign but equal magnitude. Inspection of Table 13.7 shows that the net charges carried by two bonded atoms in a polyatomic molecule may be of equal or opposite signs and differ in magnitude. We shall assume that the strength of a polar covalent A–B single bond polyatomic molecule increases, and that the bond distance consequently decreases, with increasing negative values of the product of the net atomic charges $q_A \times q_B$. This suggestion is entirely consistent with observed variation of C–F bond lengths and bond energies and the calculated atomic charges of the C and F atoms in the fluorinated methanes.

Before moving on, we pass to note that since the charge carried by the H atoms in each of the compounds $CH_{4-n}F_n$ with $n = 1 - 3$ is positive, Coulomb interaction with the C atoms would be destabilizing rather than stabilizing. However, because the magnitude of the H atom charges in each case is much smaller than the magnitude of the F charge, the destabilization effect is expected to small.

13.11 The polarity of Si–F and Si–C bonds in fluorinated methylsilanes

Atomic charges obtained by AIM or NAO analysis of the calculated charge densities of the molecules $Si(CH_3)_{4-n}F_n$ are listed in Table 13.8. The two methods differ in the magnitudes

Table 13.8. Net atomic charges (in atomic units) in fluorinated alkyl silanes obtained by AIM or NAO analysis of calculated electron densities [6].

	$q_{AIM}(F)$	$q_{AIM}(C)$	$q_{AIM}(Si)$	$q_{NAO}(F)$	$q_{NAO}(C)$	$q_{NAO}(Si)$
$Si(CH_3)_4$	–	−0.64	2.73	–	−0.34	1.41
$Si(CH_3)_3F$	−0.85	−0.67	2.87	−0.72	−0.36	1.69
$Si(CH_3)_2F_2$	−0.84	−0.69	2.98	−0.70	−0.35	1.94
$Si(CH_3)F$	−0.83	−0.70	3.10	−0.66	−0.32	2.18
SiF_4	−0.81	–	3.25	−0.60	–	2.41

of the charges assigned to C or F atoms, but they agree that the F and C atoms in each molecule carry large negative charges; that the negative charge on the F atoms is the larger; and that these charges remain approximately constant through the series. As a consequence the Si atoms carry a net positive charge with magnitude increasing with the number of attached F atoms. The observed variation of bond distances is thus in perfect agreement with the variation of calculated atomic charges.

13.12 The effect of introducing more electronegative atoms at the most electronegative of two bonded atoms

What happens if we introduce a more electronegative substituent on the most electronegative of the two bonded atoms? See Table 13.9.

While F substitution at the more electropositive metal atom in $H_3Si–CH_3$ or $H_3Ge–CH_3$ to form $F_3Si–CH_3$ or $F_3Ge–CH_3$ shortens the Si–C or Ge–C bonds by about 4 pm, F substitution at the more electronegative C atoms to form $H_3Si–CF_3$ or $H_3Ge–CF_3$ *increases* the Si–C or Ge–C bond distances by 5 or 6 pm. Similarly F substitution at the C atoms in $F_3Si–CH_3$ or $F_3Ge–CH_3$ increases the Si–C or Ge–C bond distances by about 7 or 9 pm respectively. Thus substitution of F atoms at the more electronegative of the bonded atom pair *increases* the bond distance. The number of documented cases of the effect electronegative substituents at the most electronegative of the bonded atoms is, however, not large enough to allow us to conclude that the effect is general.

Problem 13.4 Show that the elongation of the Si–C bonds in $H_3Si–CH_3$ or $F_3Si–CH_3$ when three F atoms are introduced at the more electronegative C atom may be rationalized by considering their effect on the charge on the C atom.

The effect of F substitution on the C–C bond distance in ethane is as might be expected from the foregoing: introduction of three F atoms to form $F_3C–CH_3$ makes the C–C bond polar and shortens it, introduction of three more F atoms to form $F_3C–CH_3$ removes the polarity and lengthens it again. See Table 13.9.

Table 13.9. C–C, Si–C and Ge–C bond distances (in pm) in fluorinated ethanes, methylsilanes and methylgermanes: $X_3A - CY_3$, E = C, Si or Ge, X and Y = H or F.

	$H_3E–CH_3$	$F_3E–CH_3$	$H_3E–CF_3$	$F_3E–CF_3$,
C–C	153.3(2)	149.7(3)	–	154.3(3)
Si–C	186.7(3)	182.8(4)	192.5(3)	191.3(6)
Ge–C	194.5(5)	190.4(1)	199.7(6)	197.0(9)

13.13 Inductive versus steric effects

In the preceding sections we have rationalized the variation of bond distances, valence angles and bond energies in the series $CH_{4-n}F_n$ and $Si(CH_3)_{4-n}F_n$ in terms of the electronegativity differences between the central atoms, C or Si, and the ligating atoms, H, C, or F. The observation that the smallest valence angles were those spanned by two F atoms, was explained as a result of the greater tendency of the F atoms to pull the two bonding electrons away from the central atom. The shortening of the C–F bonds in the methane derivatives and of both Si–C and Si–F bonds in the silane derivatives was explained by the greater ability of the F atoms to withdraw negative charge from the central C or Si atoms: this withdrawal increases the net positive charge on the central atoms and thus the polarity of all C–F, Si–C, or Si–C bonds. Since the observed variations of bond distances, bond energies and valence angles are interpreted as a result of changes in the electron distribution of the molecule, they are referred to as "electronic" or more precisely as "inductive" effects.

In Section 13.3 noted that the C–Cl bond distance in CCl_4 is 7 pm longer than in CCl_2 and suggested that the former may be elongated by repulsion between the Cl atoms. Such repulsion would be expected to decrease the C–Cl bond energies and increase C–Cl bond distances in the sequence CH_3Cl, CH_2Cl_2, $CHCl_3$, and CCl_4 as the number of repulsive Cl–Cl contacts increases from zero to 1 to 3 to 6. Similarly Cl–Cl repulsion would be expected to open ClCCl valence angles in CH_2Cl_2 and $CHCl_3$ at the expense of HCH and HCCl angles. Such changes that would depend on the size and proximity of the substituents are referred to as "steric" effects.

The reader may already have noted that the steric effects of the replacement of H atoms in CH_3Cl by an increasing number of Cl atoms, are excactly the opposite of the inductive effects expected from the larger electronegativity of Cl.

The observed ClCCl valence angles in CH_2Cl_2 and $CHCl_3$, 112.3(1)° and 111.3(2)° respectively, indicate that valence angles are determined by steric Cl$\cdots$Cl repulsions rather than the valence shell electron pair repulsions.

In other respects inductive and steric effects appear to cancel out:

(i) The thermochemical bond energies, TBE = 327 kJ mol^{-1} in CH_3Cl; 329 kJ mol^{-1} in CH_2Cl_2; and 326 kJ mol^{-1} in $CHCl_3$ and the mean bond energy of CCl_4, MBE = 323 kJ mol^{-1}, indicate that the C–Cl bond strength remains constant over the series.

(ii) The observed bond distances (177.6(1) pm in CH_3Cl; 176.4(1) in CH_2Cl_2; 175.8(2) in $CHCl_3$; and 176.7(3) pm in CCl_4) fall in the narrow range 176.7 ± 1.0 pm and show that C–Cl bond distances also remain essentially constant.

Due to the larger size of the Si atom, the non-bonded Cl$\cdots$Cl distance in $SiCl_4$ is 330 pm as compared to 288 pm in CCl_4. The increased distance seems to have reduced the Cl$\cdots$Cl repulsions: the ClSiCl valence angles in $Si(CH_3)_2Cl_2$ and $Si(CH_3)Cl_3$ are 107.5(1)° and 108.6(5)° respectively. The van der Waals radius of a *tert*-butyl group (measured from the tertiary C atom) is probably of the order of 350 pm as compared to 200 pm for a methyl group. (See Problem 9.5.) Replacement of the two methyl groups in $Si(CH_3)_2F_2$ by $(CH_3)_3C$ groups increases the CSiC angle by about 10°: we conclude that *steric effects are important when small atoms are surrounded by large atoms or groups.*

References

[1] N. Hansen, H. Mäder, and F. Temps, *Phys. Chem. Chem. Phys.,* 3 (2001) 50.

[2] F. Ya. Kulba, *Zh. Obshch. Khim.,* 24 (1954) 1700.

[3] J. R. Flores and R. J. Gdanitz, *J. Chem Phys.,* 123 (2005) 144316.

[4] The AIM charges were obtained by analysis of electron densities calculated at the HF/6-311++G**//MP2/6-31G* level, K. B. Wiberg and P. R. Rablen, *J. Am. Chem. Soc.,* 115 (1993) 614.

[5] The NAO charges were obtained by analysis of electron densities calculated at the HF/6-311++G**//MP2/6-31G* level, N. Strenalyuk and A. Haaland, unpublished results.

[6] AIM and NAO charges were obtained by analysis of electron densities calculated at the B3LYP/6-311++G** level, N. Strenalyuk and A. Haaland, unpublished results.

Chapter 14

Structure and bonding in some simple hydrocarbons and in ethane and ethene analogues of the heavier Group 14 elements

Introduction

It is generally true that the properties of chemical compounds formed by the second period elements from Li to He differ from the analogous formed by the heavier elements in the same group. Nowhere, however, is the difference so great as between the compounds formed by carbon on the one hand and those of silicon, gemanium, tin, or lead on the other:

(i) Linear alkanes with composition C_kH_{2k+2} with up to 100 C atoms in the chain have been isolated and characterized: there seems to be no upper limit for the stability of such chains. Disilane, Si_2H_6, on the other hand, decomposes slowly at room temperature. The stability of silanes decreases with increasing chain length: no linear or branched silane with more than eight Si atoms appears to have been characterized. The stability of metal–metal bonded H derivatives continues to fall as the group is descended: the highest germane to be characterized is Ge_5H_{12}, the highest stannane is Sn_2H_6. Diplumbane, Pb_2H_6, appears to be unknown.

(ii) While ethene, C_2H_4, and other double-bonded hydrocarbons are well known, analogous compounds of the heavier Group 14 elements have only recently been prepared. These are only stable if protected by very bulky substituents.

(iii) While carbon forms three gaseous oxides, viz. CO, CO_2 and C_3O_2, Si, Ge Sn, and Pb form solid oxides with network structures.

We shall discuss the structures of the three gaseous carbon oxides in Chapter 20. In this chapter we shall begin with a brief description of ethane, ethene, ethyne and benzene, before proceeding to discuss the structure and bonding in the ethene analogues of the heavier Group 14 elements.

14.1 The molecular structures of ethane, ethene, and ethyne

The term "hydrocarbon" covers compounds formed from hydrogen and carbon atoms only. In addition to ethane, there are two more hydrocarbons containing of two carbon atoms, C_2H_4 (ethene) and C_2H_2 (ethyne). The molecular structures of these compounds are shown in Fig. 14.1.

The coordination geometries of the carbon atoms in ethane are distorted tetrahedral, and the molecule has a threefold symmetry axis along the C–C bond. In the equilibrium structure the relative orientation of the two methyl groups is such that distances between C–H bond vectors is at a maximum as indicated in Fig. 14.1. This conformation is referred

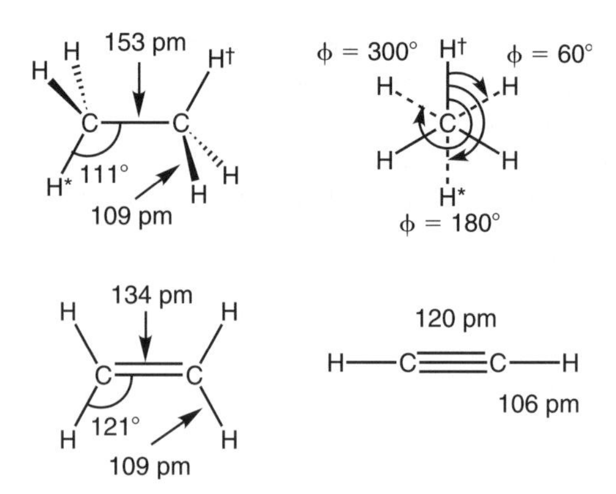

Fig. 14.1. Above: the molecular structure of ethane: side and end views. Below: the molecular structures of ethene and ethyne.

to as *staggered*. The relative orientation of the two methyl groups may be specified by a dihedral angle, i.e. by the angle between a plane containing both carbon atoms and a H atom from one methyl group and another plane containing both carbon atoms and a H atom from the other. Thus the dihedral angle ϕ ($H^{\dagger}CCH^{*}$) is defined as the angle between the two planes containing the atoms $H^{\dagger}CC$ and CCH^{*} respectively. In the equilibrium structure the dihedral angles ϕ(HCCH) are equal to 60°, 180° or 300° (=−60°). See Fig. 14.1.

A careful study of the infrared absorption spectrum of the tri-deuterated molecule CH_3CD_3 has shown that the potential energy of the molecule as a function of a dihedral angle $\phi(H^{\dagger}CCH^{*})$ to a high degree of accuracy is given by

$$V(\phi) = V_3/2[\cos(3\phi) + 1] \tag{14.1}$$

with $V_3 = 12.1$ kJ mol^{-1} [1]. This curve has three equivalent minima $V(\phi) = 0$ for $\phi = 60°$, 180° and 300° (or −60°), and three equivalent maxima $V(\phi) = V_3$ for $\phi = 0$, 120° and 240°. In an end view of a maximum energy structure the C–H bonds in the nearest methyl group covers those in the other. This orientation is described as *eclipsed*. The molecule is described as having a threefold barrier that restricts internal rotation of one methyl group relative to the other.

At room temperature the energy associated with internal rotation is equal to $RT = 2.4$ kJ mol^{-1} and most of the molecules in the gas will be restricted to small twisting oscillations about one of the three minima. But a certain fraction of the molecules will have sufficient energy for the methyl groups to pass over the barrier and continue the internal rotation without change of direction. The number of such molecules will, of course, increase with increasing temperature.

Why is the energy of the eclipsed form of ethane higher than that of the staggered? In order to find an answer to this question, one has to partition the total energy of the molecule into physically meaningful components, and then identify the component (or components) that is higher for the eclipsed form as the origin of the barrier. The difficulty with this approach is, however, that there is more than one way to partition the energy, and that the answer depends on the manner in which the partition has been carried out. Some investigators have interpreted the barrier as the result of increased repulsion between C-H bonds at the two ends of the molecule,[2] but other interpretations are possible, see reference [3].

The barrier to internal rotation in Si_2H_6 is 4.9 kJ mol^{-1}, i.e. less than half the barrier in ethane [4]. The smaller barrier in disilane is, of course, consistent with a rationalization in terms of repulsion between C–H or Si–H bonds, but does not prove it.

Ethene is a *planar* molecule. The coordination geometry of each carbon atom is approximately trigonal planar. The C–C bond distance is about 20 pm shorter than in ethane. The ethyne molecule is linear, the C–C bond distance is about 14 pm shorter than in ethene and about 33 pm shorter than in ethane.

The CC bond strengths in the three compounds may be defined as the bond dissociation energies, i.e. as the standard energies of the reactions

$$C_2H_{2j}(g) \rightarrow 2CH_j(\mathrm{g})$$

The resulting values are,

$$D(\mathrm{C{-}C}) = 367 \text{ kJ mol}^{-1} \text{ for ethane,}$$
$$D(\mathrm{C{=}C}) = 716 \text{ kJ mol}^{-1} \text{ for ethene,}$$
$$D(\mathrm{C{\equiv}C}) = 959 \text{ kJ mol}^{-1} \text{ for ethyne.}$$

These energies are associated with uncertainties of about ± 5 kJ mol^{-1}. Note that the bond dissociation energies vary as 1.0 : 2.0 : 2.6.

Another measure of C–C bond strength is provided the thermochemical bond energies calculated from the energies of atomization in combination with the mean C–H bond energy in methane:

$$\mathrm{TBE(C{-}C)} = 328 \text{ kJ mol}^{-1} \text{ for ethane,}$$
$$\mathrm{TBE(C{=}C)} = 587 \text{ kJ mol}^{-1} \text{ for ethene,}$$
$$\mathrm{TBE(C{\equiv}C)} = 808 \text{kJ mol}^{-1} \text{ for ethyne.}$$

The bond energies obtained in this manner vary as 1.0 : 1.8 : 2.5.

The variation of bond energies and bond distances in the ethane, ethene, ethyne series justifies a description in terms of single, double, and triple CC bonds, respectively.

14.2 Ethane, ethene, and ethyne: model considerations

Lewis formulas would place two electrons between the carbon atoms in ethane, four between the carbon atoms in ethene, and six electrons between the carbon atoms in ethyne. These Lewis structures are, of course, consistent with Lewis' suggestion that two atoms joined by a double bond share four electrons, while two triply bonded atoms share six electrons.

According to the VSEPR model, the four electrons in a double bond, or the six electrons in a triple bond, are regarded as constituting a single electron cloud, but due to the larger electron densities, these clouds are larger than, and require more space on the surface of the central atom than a cloud containing just one electron pair. The trigonal planar coordination geometry of the C atoms in ethene, and the linearity of ethyne are in accord with this model.

The planarity of each C atom in ethene does not necessarily imply that the whole molecule is planar. The planarity of the molecule may be rationalized by the VSEPR model, but it is easier to do so using the molecular orbital model. Bonding in ethene may be described in terms of two-center molecular orbitals if the carbon atoms are assumed to be sp^2 hybridized, with the unhybridized p_z orbitals perpendicular to the molecular plane. Two sp^2 hybrids on each C atom are combined with the $1s$ orbitals on the appended H atoms to form two $2c$ bonding MOs. The third sp^2 hybrid is combined with the corresponding hybrid on the neighbouring C to form another bonding $2c$ MO. Each of these orbitals accommodates two electrons. The description up to this point (which leaves two electrons unaccounted for) corresponds to the formation of single σ bonds between all neighboring atoms in the molecule.

Finally the p_z orbitals on the two C atoms are combined to form one bonding and one antibonding π molecular orbital. The bonding combination orbital accommodates the two remaining valence electrons.

If the two planes containing the CH_2 groups are perpendicular, each C atom carries an unpaired electron in the p_z orbital. Both the overlap integral between the two p_z orbitals and the corresponding resonance integral $H_{1,2}$ are zero by symmetry. The two electrons that occupied the π bonding orbital in the planar form, now occupy two non-bonding atomic orbitals, one at each C atom. See Fig. 14.2. High-level quantum chemical calculations on the perpendicular form yield a bond distance of 150 pm and an energy that is 272 kJ mol^{-1} higher than that of the planar form [5]: The π bond has been broken, while the σ bond remains. Compare the energy required to break the π bond with the difference between the double and single CC bond energies in the preceding section.

Propadiene, $H_2C{=}C{=}CH_2$, is linear at the central C atom, while the planes defined by the terminal methylene groups are perpendicular to one another. See Fig. 14.3. The terminal C atoms are assumed to be sp^2, and the central C atom to be sp^1 hybridized. These hybrids are used to form $2c$ MOs between the central and terminal C atoms and between the terminal C

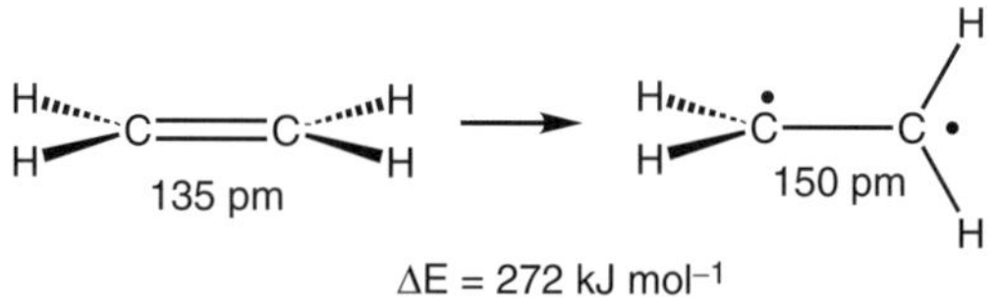

135 pm
150 pm
$\Delta E = 272$ kJ mol^{-1}

Fig. 14.2. CC bond distances and relative energies of planar and perpendicular forms of ethene.

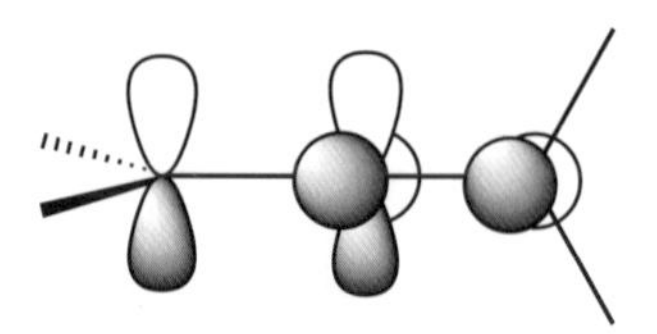

Fig. 14.3. The structure of propadiene and a schematic representation of the two π bonding orbitals.

and H atoms. If each of these σ orbitals is occupied by two electrons, four valence electrons remain unaccounted for. These are assumed to occupy two $2c$ π bonding orbitals between the central and terminal C atoms. Since these π orbitals are formed from different p orbitals on the central atom, it follows that their nodal planes and hence the planes containing the terminal CH_2 groups must be perpendicular. See Fig. 14.3.

A discussion of bonding in ethyne in terms of $2c, 2e$ bonds is possible it the carbon atoms are assumed to be sp^1 hybridized. These hybrids are used to form three bonding $2c$ σ orbitals, one between the two C atoms and one from each C atom to its appended H. These orbitals each accommodate an electron pair, and four electrons are unaccounted for.

If we use the molecular axis to define the z direction, we are left with unhybridized p_x and p_y orbitals on each C atom. Combination of the two p_x orbitals yields one bonding and one antibonding π orbital, combination of the two p_y orbitals likewise. Each of the two π-bonding molecular orbitals contains two electrons, and all electrons are accounted for.

Problem 14.1 Do you expect butatriene, $H_2C{=}C{=}C{=}CH_2$, to be a planar molecule?

14.3 Single C–C bond distances and hybridization

The C–C single bond distance in normal butane is 153.1 pm. The single bond distance in *trans*-2-butene is 2.5 pm shorter. See Fig. 14.4. The difference is usually rationalized by noting that the C atoms in butane and the terminal C atoms in *trans*-2-butene are sp^3 hybridized while the two central C atoms in *trans*-2-butene are sp^2 hybridized. Since the s character of sp^2 orbitals is larger than the s character of sp^3 orbitals, the former are expected to form shorter and stronger bonds. The bonding radius of an sp^2 hybridized C atom should therefore be smaller than that of sp^3 hybridized C atom. The single bond distances in butane and *trans*-2-butene suggest that the bonding radii of sp^3 and sp^2 hybridized C atoms are about 76.6 and 74.1 pm respectively. The latter radius suggests that the single bond distance in butadiene (Fig. 14.4) should be about 148.2 pm, as compared to an observed distance of 146.5 pm.

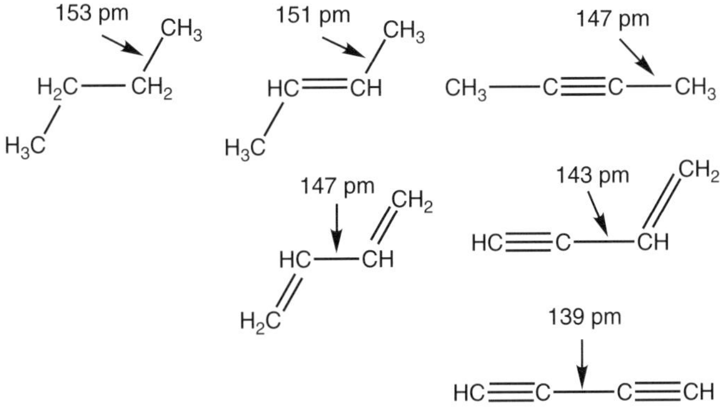

Fig. 14.4. C–C single bond distances in C_4 hydrocarbons with double and triple bonds.

The single bond distance in 2-butyne is 146.8 pm, suggesting that the bonding radius of sp^1 hybridized C atoms is about 6.3 pm smaller than the radius of sp^3 hybridized, i.e. about 70.3 pm. This estimate, in turn, implies that the single bond distance in 1,3-butadiyn should be 140.6 pm as compared to an observed value of 138.8 pm. Addition of the radii of sp^1 and sp^2 hybridized carbon atoms yields an estimated single C-C bond distance in 1-butene-3-yn equal to 144.4 pm as compared to an observed distance of 143.4 pm.

Taken together the bond distances in Fig. 14.4 suggest that the single bond radius of an sp^3 hybridized C atom is 77 pm, of an sp^2 hybridized C atom 73 pm and of an sp^1 hybridized C atom 70 pm.

14.4 Delocalized π orbitals in benzene

The benzene molecule is planar. All valence angles are 120°, all CC bond distances are equal to 140 pm, and all C-H bond distances equal to 109 pm.

The bonding may be described by assuming each C atom to be sp^2 hybridized, and to form three $2c, 2e$ σ bonds to its nearest neighbours. This leaves six p_z orbitals for the formation of π molecular orbitals, and six electrons to occupy them.

We begin by numbering the C atoms in the ring consecutively from 1 to 6, and denoting the p_z orbitals as Φ_1 to Φ_6. See Fig. 14.5. A general π molecular orbital of the ring may then be written as a linear combination of the atomic orbitals

$$\Psi = c_1\Phi_2 + c_2\Phi_2 + c_3\Phi_3 + c_4\Phi_4 + c_5\Phi_5 + c_6\Phi_6$$

The coefficients may be determined by simple Hückel calculations: the energy is calculated in terms of Coulomb, resonance and overlap integrals. We then take the derivative of the energy with respect to each of the six coefficients, and set the derivatives equal to zero in order to find the lowest possible value of ε. This procedure yields six secular equations, the first of which is

$$\begin{aligned}(H_{1,1} - \varepsilon)c_1 + (H_{1,2} - S_{1,2}\varepsilon)c_2 + (H_{1,3} - S_{1,3}\varepsilon)c_3 \\ + (H_{1,4} - S_{1,4}\varepsilon)c_4 + (H_{1,5} - S_{1,5}\varepsilon)c_5 + (H_{1,6} - S_{1,6}\varepsilon)c_6 = 0.\end{aligned}$$

We now simplify the secular equations by setting all overlap integrals equal to zero, and neglecting resonance integrals except those between nearest neighbors on the ring. Since

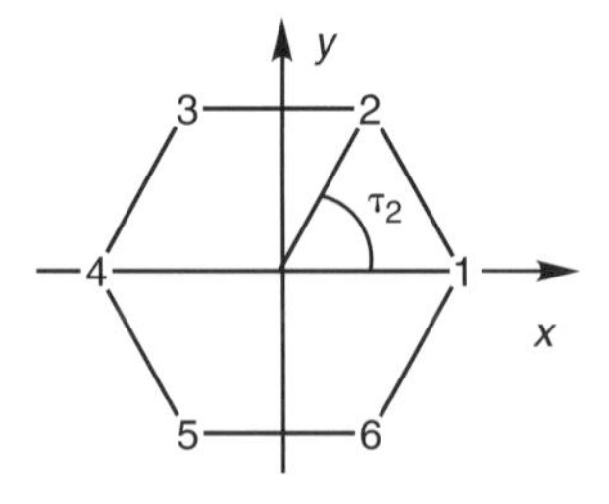

Fig. 14.5. Numbering of six p_z orbitals on the C atoms in benzene.

$H_{1,2} = H_{1,6}$, the first secular equation then becomes

$$(H_{1,1} - \varepsilon)c_1 + H_{1,2}c_2 + 0 + 0 + 0 + H_{1,2}c_6 = 0.$$

The other secular equations are simplified in a similar manner. The secular determinant, which must be equal to zero if the secular equations are to have non-trivial solutions, is

$$\begin{vmatrix} (H_{1,1}-\varepsilon) & H_{1,2} & 0 & 0 & 0 & H_{1,2} \\ H_{1,2} & (H_{1,1}-\varepsilon) & H_{1,2} & 0 & 0 & 0 \\ 0 & H_{1,2} & (H_{1,1}-\varepsilon) & H_{1,2} & 0 & 0 \\ 0 & 0 & H_{1,2} & (H_{1,1}-\varepsilon) & H_{1,2} & 0 \\ 0 & 0 & 0 & H_{1,2} & (H_{1,1}-\varepsilon) & H_{1,2} \\ H_{1,2} & 0 & 0 & 0 & H_{1,2} & (H_{1,1}-\varepsilon) \end{vmatrix} = 0$$

Expansion of the determinant yields a sixth-degree equation for ε. The roots, listed in order of increasing energy, are:

$$\varepsilon_1 = H_{1,1} + 2H_{1,2} \quad \varepsilon_2 = \varepsilon_3 = H_{1,1} + H_{1,2} \quad \varepsilon_4 = \varepsilon_5 = H_{1,1} - H_{1,2} \text{ and}$$
$$\varepsilon_6 = H_{1,1} - 2H_{1,2}$$

The lowest energy molecular orbital has the form

$$\Psi_1 = (\Phi_2 + \Phi_2 + \Phi_3 + \Phi_4 + \Phi_5 + \Phi_6)/\sqrt{D}$$

Since we have set all overlap integrals equal to zero,

$$D = c_1^2 + c_2^2 + c_3^2 + c_4^2 + c_5^2 + c_6^2 = 6 \tag{14.2}$$

The two next orbitals are degenerate. When orbitals are degenerate, we have a choice between several ways of describing them. The most useful forms are

$$\Psi_2 = [\Phi_1 + (1/2)\Phi_2 - (1/2)\Phi_3 - \Phi_4 - (1/2)\Phi_5 + (1/2)\Phi_6]/\sqrt{\mathrm{D}}$$
$$\Psi_3 = [(\sqrt{3}/2)\Phi_2 + (\sqrt{3}/2)\Phi_3 - (\sqrt{3}/2)\Phi_5 - (\sqrt{3}/2)\Phi_6]/\sqrt{\mathrm{D}}$$

The bonding orbitals Ψ_1, Ψ_2 and Ψ_3 are filled by six electrons.

The forms of the three bonding orbitals as well as of the three antibonding orbitals are indicated in Fig. 14.6. The lowest energy orbital is bonding between all neighbors in the ring. Two electrons in this orbital would create an electron density that consists of two doughnuts, one above the ring plane and the other below. The doughnuts would have lumps above and below each of the six C atoms.

Ψ_2 is bonding between the C atoms 6, 1 and 2, and between C atoms 3, 4 and 5. It is antibonding between C atoms 2 and 3 and between 5 and 6. The bonding interactions are, however, more important than the antibonding, and the net effect is a lowering of the orbital energy by $H_{1/2}$.

As we have seen, the energy of Ψ_3 is equal to that of Ψ_2. Ψ_3 is bonding between C atoms 2 and 3 and between 5 and 6. Since all CC bond distances in benzene are equal it is clear

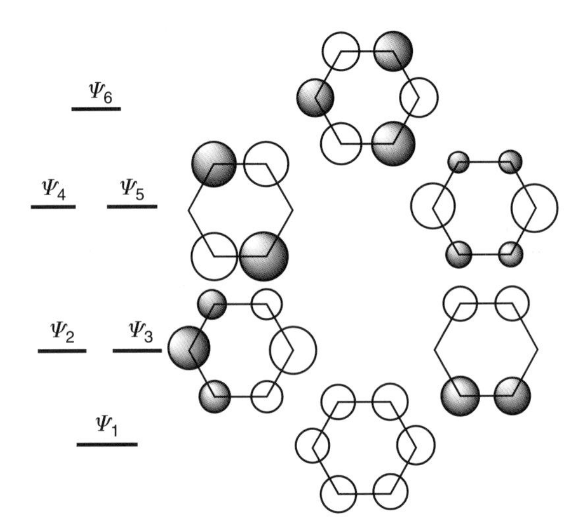

Fig. 14.6. The combination of p_z AOs on the carbon atoms in benzene to form π molecular orbitals.

that the net effect of two electrons in Ψ_2 and the two electrons in Ψ_3 is to make the six C–C bonds in the ring equally strong.

Problem 14.2 Use equation (14.2) to determine the normalization constants $\sqrt{D}$ for Ψ_2 and Ψ_3. Show that if each of the three π-bonding orbitals is filled with two electrons, then the electron densities at the six C atoms are equal.

Problem 14.3 Investigate bonding and antibonding interactions in the orbitals Ψ_4, Ψ_5 and Ψ_6.

Problem 14.4 What would be the effect on the molecular structure if one electron is removed from the Ψ_2 or Ψ_3 orbital to form the $C_6H_6^+$ cation?

The π orbital with the lowest energy, Ψ_1, has one nodal plane, namely the xy plane. Note that the molecular orbitals Ψ_2 and Ψ_3 have two nodal planes, viz. the xy plane and either the yz or xz plane respectively.

We may describe the position of each C atom on the ring by defining an angle τ as the angle between a line from the origin through the atom and the x-axis. The angle τ_2 is indicated in Fig. 14.5. Note that the six coefficients in Ψ_2 are Ψ_3 are given by $c_i(2) = \cos\tau_i$ and $c_i(3) = \sin\tau_i$ respectively.

Problem 14.5 How many nodal planes do you find in Ψ_4, Ψ_5 and Ψ_6?

Problem 14.6 The LCAO coefficients in Ψ_1, Ψ_4, Ψ_5 and Ψ_6 are given by $c_i = \cos(0\tau_i)$, $c_i = \cos(2\tau_i)$, $c_i = \cos(3\tau_i)$, $c_i = \sin(0\tau_i)$, $c_i = \sin(2\tau_i)$, or $c_i = \sin(3\tau_i)$. Calculate these coefficients and identify the corresponding molecular orbitals.

14.5 Ethane analogues of the heavier Group 14 elements

Ethane analogues of composition R_3EER_3, E = Si, Ge, Sn or Pb, where R is equal to a H atom or an organic radical, are referred to as disilanes, digermanes, distannanes,

Table 14.1. Hexamethyl-dimetallanes, Me_3EEMe_3, E = Si, Ge, Sn or Pb: experimental E–E bond distances and bond dissociation energies; and dimetallanes H_3EEH_3: calculated E–E bond distances and bond dissociation energies.

	Me_3EEMe_3 (exp)		H_3EEH_3 (calc)	
	R(E–E) (pm)	*D*(E–E)[a] (kJ mol^{-1})	*R*(E–E)[b] (pm)	D_0(E–E)[b] (kJ mol^{-1})
H_3CCH_3	153	375	151	363
$Me_3SiSiMe_3$	234[c]	339	232	293
$Me_3GeGeMe_3$	240[d]	305	241	257
$Me_3SnSnMe_3$	277	234	272	220
$Me_3PbPbMe_3$	–	230	293	185

[a] Data from reference [6]. [b] Data from reference [7]. [c] Estimated standard deviation equal to 1 pm. [d] Bond distance in H_3GeGeH_3.

or diplumbanes respectively. We shall also refer to them collectively as dimetallanes. Hexamethyl-disilanes, -digermanes, and -distannanes are relatively stable compounds and have been known for more than 50 years. Hexamethyl-diplumbane is also known, but is much less stable. Experimentally determined E–E bond distances and E–E bond dissociation energies in hexamethyl-dimetallanes are listed in Table 14.1. The table also includes the E–E bond distances and dissociation energies in the parent compounds H_3EEH_3 obtained by accurate quantum chemical calculations. There is reasonably good agreement between bond distances, but the experimental bond dissociation energies of the hexamethyl derivatives at 298 K are found to be significantly higher than those calculated for the parent molecules at zero K. This is in agreement with their greater thermal stability.

14.6 The molecular structures of ethene analogues of the heavier Group 14 elements

The ethene analogs of composition R_2SiSiR_2, R_2GeGeR_2, R_2SnSnR_2, or R_2PbPbR_2, are known as disilenes, digermenes, distannenes, and diplumbenes respectively and we shall refer to them collectively as dimetallenes. Early attempts to synthesize ethene analogs of the heavier Group 14 elements invariably led to the formation of rings, $(MR_2)_n$, or polymers. Such polymerization may, however, be avoided when the organic radicals R are very bulky, and during the last 25 years a number of dimetallenes have been synthesized [8].

The molecular structures of several disilenes in the solid state have been determined by X-ray crystallography. All of them are characterized by very bulky alkyl or aryl substituents, R^C. The structures of (*tert*-butyl)(Mes)SiSi(*tert*-butyl)(Mes), Mes = phenyl-2,4,6-trimethyl (**A**), and $(Mes)_2SiSi(Mes)_2$ (**B**) are shown in Fig. 14.7. The central C_2SiSiC_2 framework of the former is *planar* as expected for an ethene analogue. The SiSi bond distance is 214 pm, about 9% shorter than the single bond distance in $Me_3SiSiMe_3$, 234 pm. (By comparison the CC double bond in ethene is 13% shorter than the single bond in ethane.)

The shape of the central C_2SiSiC_2 framework in $(Mes)_2SiSi(Mes)_2$, on the other hand, is *non-planar*: The coordination geometries of the Si atoms are slightly pyramidal and the two

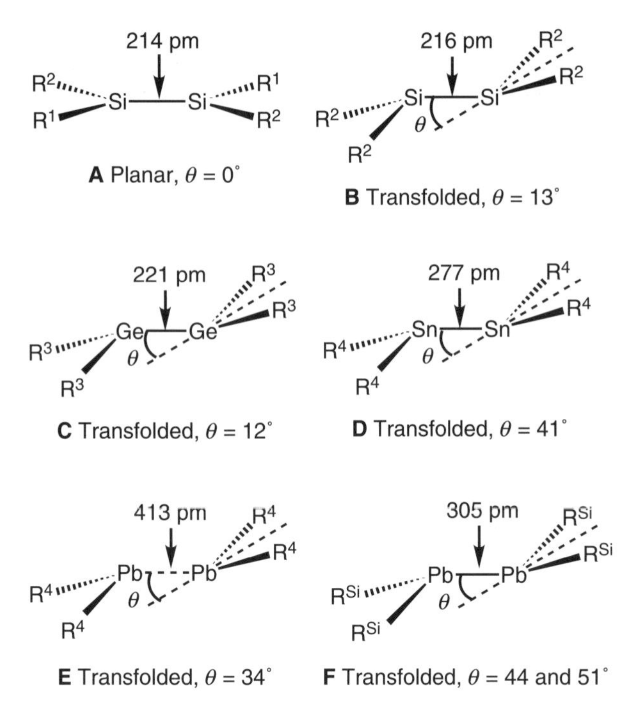

Fig. 14.7. The molecular structures of dimetallenes of the heavier Group 14 elements determined by X-ray crystallography. R^1 = *tert*-butyl; R^2 = mesityl = phenyl-2,4,6-trimethyl; R^3 = phenyl-2,6-diethyl; R^4 = $CH[Si(CH_3)_3]_2$; and R^{Si} = $Si(SiMe_3)_3$.

ends of the molecule are *trans*-folded as indicated in the figure. The folding angle θ, defined as the angle between the Si–Si bond and the plane through a Si atom and the two ligating C atoms, is 13°. The Si–Si bond distance is slightly elongated to 216 pm. The folding angles in other disilenes have been found to vary between 0 and 18°, the Si–Si bond distances have been found to vary between 214 and 225 pm, the latter is only 4% shorter than the single bond distance in $Me_3SiSiMe_3$. See Table 14.2. The great variation of the folding angles show that these molecules are easily deformed, and it is not immediately clear whether the unsubstituted disilene, H_2SiSiH_2, would be planar or not. Quantum chemical calculations on H_2SiSiH_2 have yielded folding angles ranging from 10 to 40°!

The molecular structure of the digermene $(R^3)_2GeGe(R^3)_2$, R^3 = phenyl-2,6-diethyl, is *trans*-folded with θ = 12° and a GeGe bond distance of 221 pm, about 8% shorter than the single bond distance in $Me_3GeGeMe_3$. The GeGe bond distances in other digermenes have been found to vary between 221 and 235, the upper limit is about 2% shorter than the single bond distance in H_3GeGeH_3. The folding angles substituted digermenes have been found to vary between 12 and 36°: the digermenes appear to be as flexible as the disilenes, but the folding angles are generally larger.

The molecular structure of the distannene **D** is *trans*-folded with θ equal to 41°. The magnitude of this folding angle may, perhaps, be better appreciated if it is recalled that if the ∠SnSnC and ∠CSnC valence angles were tetrahedral, then the folding angle would be

Table 14.2. Tetrasubstituted dimetallenes, $R^C_2EER^C_2$, carrying bulky alkyl or aryl substituents: E–E bond distances and folding angles determined by X-ray crystallography and experimental E–E bond dissociation energies in solution. Unsubstituted dimetallenes, H_2EEH_2: E–E bond distances, folding angles, E–E dissociation energies to yield EH_2 monomers in the electronic ground state; dissociation energies to yield EH_2 monomers in the excited triplet state; and singlet-triplet energy gaps of the EH_2 monomers, all determined by quantum chemical calculations.

	$R^C_2EER^C_2$ (exp)[a]			H_2EEH_2 (calc)[b]				
	R (E–E) (pm)	ϕ (°)	D (E–E) (kJ mol^{-1})	R (E–E) (pm)	ϕ (°)	D_0 (E–E) (kJ mol^{-1})	ΔE^* (kJ mol^{-1})	D^*_0(E–E) (kJ mol^{-1})
H_2CCH_2	134	0	716[c]	132	0	739	0	739
$R^C_2SiSiR^2_C$	214–225	0–18	265 ± 5[c]	215	36	250	74	398
$R^C_2GeGeR^C_2$	221–235	12–36	62[d]	225	47	180	96	372
$R^C_2SnSnR^C_2$	277–363	41–46	56[e]	257	51	121	97	315
$R^C_2PbPb^C_2$	299–413	41–51	–	282	54	42	143	328

[a] Data from reference [9] unless otherwise noted. [b] Data reference [7]. [c] Gas phase dissociation energies of for H_2CCH_2 and H_2SiSiH_2 respectively. [d] Dissociation energy of $R^2R^3GeGeR^2R^3$ in solution, reference [10]. [e] Dissociation energy of $R^3_2SnSnR^3_2$ in solution, reference [11].

55°. The SnSn bond distance in **D** (277 pm) is the shortest found in a distannene to date and indistinguishable from the single bond distance in the ethane analogue $Me_3SnSnMe_3$. Other distannenes exhibit bond distances up to 100 pm *longer* than in the ethane analogue. See Table 14.2.

The diplumbene **E** which carries the same alkyl ligands as the distannene **D**, exhibits a PbPb distance about 100 pm longer than estimated for a PbPb single bond distance by doubling the bonding radius of Pb (see Table 6.2). **D** is perhaps better described as a pair of weakly interacting PbR_2 monomers rather than a dimetallene molecule. The shortest PbPb bond distance in a diplumbene has been observed in **F** where the metal atoms carry Si-centered, rather than C-centered substituents. In this molecule the Pb–Pb distance is 305 pm, about 5 pm longer than estimated for a single Pb–Pb bond distance. The structure is *trans*-folded with two rather different folding angles of 44 and 51°.

Taken together the information presented in Fig. 14.7 and Table 14.2 show that the structures of the ethene analogues $R^C_2EER^C_2$ change gradually as the group is descended from the rigid planar structure of H_2CCH_2 to non-planar, flexible and increasingly *trans*-folded structures. At the same time the EE bond distances approach and, in the case of some Sn and Pb dimetallenes, surpass the single E–E bond distance in the ethane analogues R_3EER_3.

The E–E bonds in some dimetallenes are not only surprisingly long, they are also very weak. Both the digermene **C** and the distannene **D** are completely dissociated to ER^C_2 monomers in the gas phase. Some, but not all, of the known digermenes are partially dissociated in solution. The dissociation energy of $R^2R^4GeGeR^2R^4$ is only 62 kJ mol^{-1} as compared to a Ge–Ge bond dissociation energy of 305 kJ mol^{-1} in $Me_3GeGeMe_3$. All known distannenes are partially dissociated in solution, the dissociation energy of **D** is 56 kJ mol^{-1} as compared to a Sn–Sn bond dissociation energy of 234 kJ mol^{-1} in $Me_3SnSnMe_3$. All diplumbenes appear to be partially or completely dissociated in solution, as we have seen, **E** is only weakly associated even in the solid phase.

In Table 14.2 we compare the dissociation energies and structure parameters obtained by experimental studies of dimetallenes carrying bulky alkyl or aryl substituents with those obtained by quantum chemical calculations on the parent compounds H_2EEH_2. While there is reasonable agreement between the calculated and experimental gas phase dissociation energies of H_2CCH_2 and H_2SiSiH_2, the dissociation energies of $R^2R^3GeGeR^2R^3$ and $R^4_2SnSnR^4_2$ in solution are significant *smaller* than calculated for the hydrogen derivatives. The bulky groups in the former were introduced in order to destabilize oligomeric rings and polymers relative to the dimetallenes $R^C_2EER^C_2$. Unfortunately, it appears that steric repulsion between the bulky substituents also destabilize the dimetallenes relative to the monomeric ER^C_2 species. Comparison of E–E bond distances and folding angles θ shows that the bond distances tend to be *longer* and the folding angles *smaller* in the alkyl- or aryl-substituted compounds than calculated for the parent molecules. These differences are probably due to steric repulsion between the bulky substituents bonded to different metal atoms.

The most startling result of the quantum chemical calculations on the parent molecules is the steep decline of the dissociation energies as the group is descended. Comparison of the calculated dissociation energies of H_2GeGeH_2, H_2SnSnH_2, and H_2PbPbH_2 with the dissociation energies of the corresponding ethane analogues listed in Table 14.1 suggests that the putative EE double bonds in the dimetallenes are inherently weaker than the single bonds in the alkane analogues!

14.7 Molecular orbital descriptions of the ethene analogues of the heavier Group 14 elements

Quantum chemical calculations on planar models of H_2SiSiH_2 lead to a set of bonding molecular orbitals similar to those of ethene. The highest occupied molecular orbital (HOMO) is a 2*c* π orbital formed by combination of the two $3p_z$ orbitals, and the second highest MO (the HOMO-1) a 2*c* σ-bonding orbital formed by combination of two approximately sp^2 hybridized AOs on the Si atoms. Several quantum chemical calculations have nevertheless yielded *trans*-folded equilibrium structures with a folding angles ranging from 20 and 40°. The barrier to planarity, i.e. the difference between the optimal planar and the trans-folded equilibrium structure, is calculated to be 5–10 kJ mol^{-1}, i.e. smaller than the barrier to internal rotation in ethane! Perhaps surprisingly, examination of orbital energies shows that the energy of the SiSi π-bonding orbital is *lower* in the *trans*-folded structure while the energy of the 2*c* SiSi σ-bonding orbital is *higher*. This fact may be interpreted to mean that the *trans*-folding is caused by the π-electrons. Closer examination of the π MO reveals that folding is accompanied by a small tilting of the p_z orbital and a small hybridization with the 3*s* orbitals of the Si atom. See Fig. 14.8. The energy lowering accompanying the increase of the *s*-character of the atomic orbital presumably more than outweighs any loss of π-bonding energy.

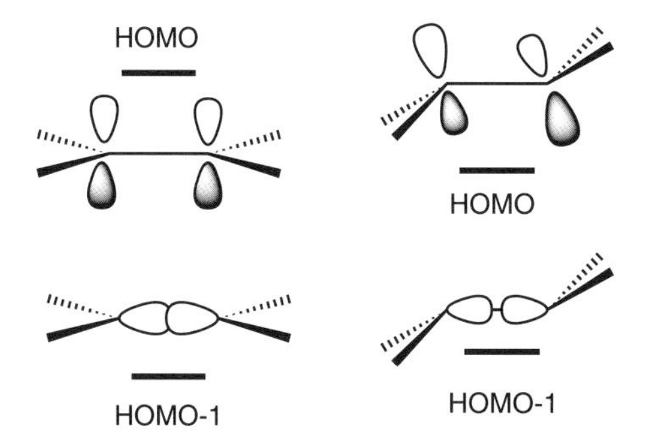

Fig. 14.8. A schematic representation of the highest molecular orbitals in planar and trans-folded disilene.

Calculations on the ethene analogues of Ge, Sn or Pb show that the folding angles become larger, and the barriers to linearity higher, as the group is descended. At the same time the *s* character of the hybrids formed from the former p_π orbitals continues to grow. Even if the folding reduces the overlap between the (approximate) p_π AOs on the Si atoms somewhat, there are good reasons to continue to regard the MO formed by their combination as a bonding molecular orbital. The molecular orbital description of the parent dimetellenes thus suggests that *four* electrons occupy two 2*c* E–E bonding orbitals. Why then, are these bonds so weak?

As we have seen in Section 13.7, the electronic ground state of the methylene radical CH_2 is a triplet, one non-bonding electron resides in an sp^λ hybrid atomic orbital in the plane of the molecule, the other in a p_π orbital perpendicular to the plane. When two methylene units are brought together the two sp^λ hybrid atomic orbitals combine to a σ-bonding MO, and the two p_π AOs to a π-bonding MO, each containing two electrons. After the H_2CCH_2 molecule has been formed, each of the four former non-bonding electrons resides in a molecular orbital with lower energy than before.

The electronic ground states of the methylene analogues of the heavier Group 14 elements are all diamagnetic: both nonbonding electrons reside in a non-bonding sp^λ hybrid σ atomic orbital. The energies (ΔE^*) required to excite an EH_2 fragment from the singlet ground state to the triplet state with one electron in the sp^λ AO and the other in the p_π are listed in Table 14.8. They are seen to increase as the group is descended from Si to Pb. The process of bond formation between the two EH_2 fragments may therefore be considered to occur in two steps. In the first step the two fragments are excited from the singlet ground state to the triplet state under absorption of the energy $2\Delta E^*$, in the second step the excited fragments combine and releases an energy equal to $2\Delta E^* + D$. The energy required to split the dimetallenes into EH_2 fragments in the triplet state at zero K is thus given by

$$D_0^* = D_0 + 2\Delta E^*$$

The dissociation energies calculated in this manner are listed in the last column in Table 14.1. Comparison with the electronic dissociation energies listed in Table 14.2 shows that the E–E bonds formed by combination of the fragments in their triplet state are significantly stronger than the single bonds in the alkane analogs. We conclude, therefore, that the E–E bonds

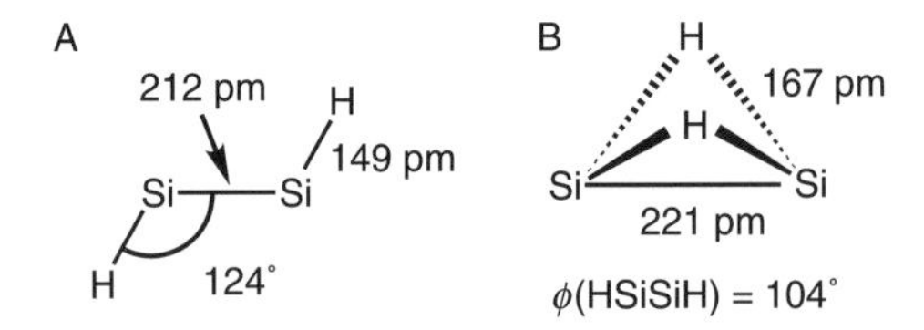

Fig. 14.9. A: The molecular structure of a high-energy form of Si_2H_2 found by quantum chemical calculations. B: The molecular structure of the most stable form of Si_2H_2 determined experimentally and by quantum chemical calculations.

in the Ge, Sn and Pb dimetallenes are weak because a large part of the energy released on formation of the bond has already been expended in exciting the EH_2 fragments from the singlet ground to the triplet valence state. Similarly the structures of the dimetallenes are *trans*-folded because the distortion from planarity allows the molecules to recoup a part of the singlet-triplet excitation energy.

Even though the nature of the bonding in the dimetallenes is well understood, the properties of digermenes and distannenes leave us in a terminological quandary: should we describe the E–E bonds in these molecules as double and accept that a double bond may be weaker than a single bond between the same atoms, or should we describe the bonds as single and accept that Lewis' electron pair model has failed?

14.8 The molecular structure of Si_2H_2

The similarity of the molecular formulae suggest that the structure of Si_2H_2 should be similar to that of C_2H_2, and high-level quantum chemical calculations taking a linear HSiSiH structure as a point of departure does lead to the planar, but trans-bent structure as shown in Fig. 14.9 A [12]. However, other calculations taking a structure with the two Si and the two H atoms occupying positions at alternating corners of a square leads to the identification of a second structure with energy more than 70 kJ mol^{-1} *below* that of the *trans*-bent structure.

The molecule Si_2H_2 has been detected in electric discharge tubes containing a mixture of SiH_4 and ethyne and its structure determined by microwave spectroscopy [13]. The experimental structure thus obtained is in excellent agreement with the calculations. See Fig. 14.9B. The structure may be described as containing two Si atoms bridged by the two H atoms. The ring formed by the four atoms is, however, not planar, the angle between the two Si_2H planes being about 104°.

Problem 14.7 Suggest a molecular orbital description of the *trans*-bent, high-energy structure of Si_2H_2.

Problem 14.8 Compare the bond distances in the most stable form of Si_2H_2 with the Si–H bond distance in SiH_4, 148 pm, and the Si–Si distance in Si_2H_6, 233 pm, and suggest a description of the bonding in terms of 2*c* and 3*c* molecular orbitals.

References

[1] N. Moazzen-Ahmadi, E. Kelly, J. Schroderus, and V.-M. Horneman, *J. Mol. Spect.*, 209 (2001) 228.
[2] F. M. Bickelhaupt and E. J. Baerends, *Angew. Chem. Int. Ed.*, 42 (2003) 4183 and references therein.
[3] V. Pophristic and L. Goodman, *Nature*, 411 (2001) 565 and F. Weinhold, Nature, 411 (2001) 539.
[4] F. Lattanzi, C. Di Lauro, and V.-M. Hornemann, *Mol. Phys.*, 104 (2006) 1795.
[5] M. W. Schmidt, P. N. Truong, and M. S. Gordon, *J. Am. Chem. Soc.*, 109 (1987) 5217.
[6] R. A. Jackson, *J. Organometal. Chem.*, 166 (1979) 17.
[7] H. Jacobsen and T. Ziegler, *J. Am. Chem. Soc.*, 116 (1994) 3667.
[8] For two recent reviews of structure and bonding in dimetallenes see P. P. Power, *J. Chem. Soc. Dalton Trans.*, (1998) 2939 or M. Weidenbruch, *Eur. J. Inorg. Chem.*, (1999) 373.
[9] H. Grützmacher and T. F. Fässler, *Chem. Eur. J.*, 6 (2000) 2317.
[10] K. Kishikawa, N. Tokitoh, and R. Okazaki, *Chem. Lett* (1998) 239.
[11] K. W. Zilm, G. A. Lawless, R. M. Merrill, J. M. Millar, and G. G. Webb, *J. Am. Chem. Soc.*, 1987 (109) 7236.
[12] R. S. Grev and H. F. Schaefer III, *J. Chem. Phys.*, 97 (1992) 7990.
[13] M. Bogey, H. Bolvin, M. Cordonnier, C. Demuynck, J. L. Destombes, and A. G. Császár, *J. Chem. Phys.*, 100 (1994) 8614.

Chapter 15

Structure and bonding in simple compounds of the Group 15 elements

Introduction

The first two elements in Group 15, nitrogen and phosphorus, have the electron configurations $[\mathrm{Ng}]ns^2np^3$ while the heavier elements, arsenic, antimony and bismuth, have the electron configurations $[\mathrm{Ng}](n-1)d^{10}ns^2np^3$. All the Group 15 elements form compounds of composition EH_3, $E(CH_3)_3$, ECl_3 and EF_3. These compounds are all monomeric in the gas phase and in solution. With the exception of NCl_3 and BiH_3, they are reasonably stable and have been extensively studied. We shall refer to a Group 15 element that forms three homolytic bonds as Lewis-valent.

Compounds of composition ECl are present in high temperature vapors, but have never been isolated: such subvalent compounds of Group 15 elements are therefore of little importance.

The heavier elements of the group, P, As, Sb and Bi also form compounds of composition EB_5, B = halogen, methyl or phenyl. In each of these compounds the Group 15 element is bonded to five atoms or groups. If the central atom is assumed to form a Lewis electron pair bond to each ligating atom, it is surrounded by ten electrons in the valence shell. Group 15 elements forming five homolytic bonds are described as *hypervalent*.

15.1 Compounds of Lewis-valent elements

All the Lewis-valent molecules of composition EH_3, ECl_3 or $E(CH_3)_3$ have C_{3v} symmetry. The bond distances, mean bond energies and valence angles are listed in Table 15.1. The bond distances are also displayed in Fig. 15.1. As usual bonds become longer and weaker as the group is descended. Note, however, that the MBE in NCl_3 is unknown. We have already seen that the C–Cl bonds in CCl_4 are weaker than the Si–Cl bonds in $SiCl_4$. Similarly the O–Cl bonds in OCl_2 are weaker than the S–Cl bonds in SCl_2. See Table 17.2. The fact that NCl_3 is highly explosive implies that the reaction

$$2NCl_3(g) = N_2(g) + 3Cl_2(g)$$

is highly exothermic, which in turn implies that

$$6\mathrm{MBE}(NCl_3) < D(N_2) + 3D(Cl_2)$$

or

$$\mathrm{MBE}(NCl_3) < 277\ \mathrm{kJ\ mol^{-1}}$$

as compared to a mean bond energy of 329 kJ mol^{-1} in PCl_3.

Table 15.1. Gaseous trihydrogen, trimethyl and trichloro derivatives of the Group 15 elements. E–H, E–C and E–Cl bond distances, *R* (in pm); mean bond energies, MBE (in kJ mol^{-1}); and valence angles ∠HEH, ∠CEC, and ∠ClECl (in degrees).

	EH_3			$E(CH_3)_3$			ECl_3		
E	*R*	MBE	∠HEH	*R*	MBE	∠CEC	*R*	MBE	∠ClECl
N	101	388	107	145	309	111	176	–	107
P	142	319	93	184	283	99	204	320	100
As	152	294	92	197	236	96	216	307	99
Sb	170	255	92	216	221	94	232	311	97
Bi	178	193	90	227	148	97	242	277	97

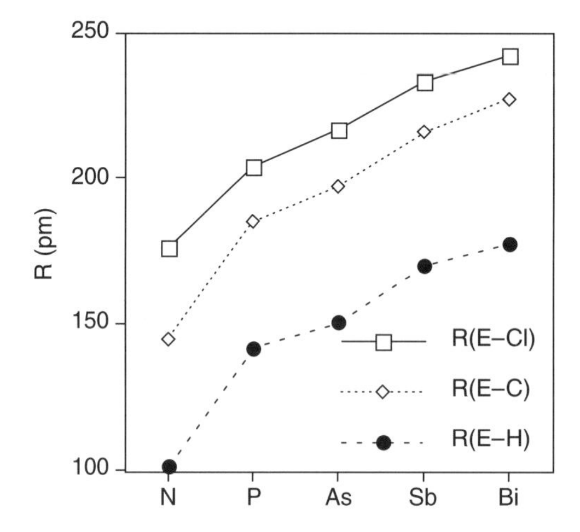

Fig. 15.1. Bond distances (in pm) in group valent hydrides, chlorides and methyl derivatives of the Group 15 elements.

The HNH valence angle in ammonia, ∠HNH = 106.7°, is three degrees smaller than tetrahedral. The valence angles in PH_3, AsH_3 or SbH_3, are only 92 or 93°, while the valence angle in BiH_3 is 90.3°. All these angles are in agreement with the VSEPR model: the central atom E is surrounded by four electron pairs. Repulsion between the four pairs would result in an approximately tetrahedral arrangement. Since the lone pair requires more space on the surface of the central atom than the bonding electron pairs, the valence angles subtended by the ligating atoms are expected to be smaller than tetrahedral. The relatively large valence angle in ammonia may be due to greater repulsion between the H atoms: because of the relatively short N–H bond distance, they are much closer to each other in this molecule than in the hydrogen derivatives of the heavier elements of the group.

The valence shell electron configuration of the Group 15 elements is ns^2np^3. The valence angles in PH_3, AsH_3, SbH_3 and BiH_3 indicate that 2c bonding MOs are formed by combination of nearly pure valence shell *p* orbitals on the central atom with 1s orbitals on the H atoms. The valence angles in ammonia indicate that the bonding MOs are formed from *sp* hybrid orbitals with less than 25% *s* character while the *s* character of the orbital containing the non-bonding electron pair is higher than 25%.

The valence angle in NCl_3 is very similar to that of NH_3. ∠ClECl angles decrease as the group is descended, but not below 97°, perhaps because the ligating atoms are larger and repulsion between them consequently greater than in the trihydrogen compounds. Valence angles in the trimethyl derivatives are very similar to those of the trichlorides, except that the valence angle in trimethylamine is 111° indicating that steric repulsion between the methyl groups is more important than repulsion between the valence shell electron pairs.

15.2 Inversion of the N atom in ammonia

In Fig. 15.2 we show the variation of the potential energy of the ammonia molecule as a function of the angle between the N–H bonds and the threefold symmetry axis of the molecule. The curve has two equally deep minima at $q_e = 68°$ and 112° separated by a maximum at 90°, 21.1 kJ mol^{-1} above the minima [1]. The energy required to deform an ammonia molecule from the pyramidal equilibrium structure to trigonal planar form is thus about twice as large as the energy required to convert ethane from a staggered to an eclipsed form. The lowest vibrational level in NH_3 is found about 6.0 kJ mol^{-1} above the minima. A molecule in this vibrational state undergoes deformation vibrations with frequency $\nu = 2.9 \times 10^{13}$ s^{-1} in which the molecule retains threefold symmetry while the angle q varies with up to 10 or 15° from one of the equilibrium values. Every once in a while, however, the molecule will invert, i.e. it will move from one minimum on the

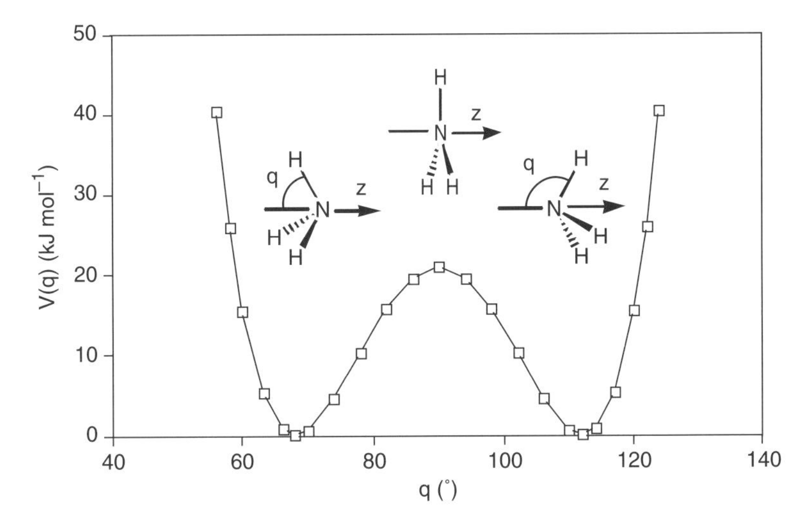

Fig. 15.2. Above: the inversion of NH_3. Below: the potential energy of the molecule (in kJ mol^{-1}) as a function of the inversion coordinate q which is defined as the angle between the N-H bonds and the threefold symmetry axis of the molecule. The inversion barrier $V(q = 90°) = 21$ kJ mol^{-1}.

potential energy curve to the other. This inversion will take place about 2.4×10^{10} times per second, i.e. after an average of about 1200 deformation vibrations [2].

How is this possible when the vibrational energy is less than one third of the energy required for the molecule to pass over the top of the barrier? When discussing the properties of one-electron atoms (Chapter 1) we noted that the total energy of the electron is given by the sum of the kinetic and potential energies and that the latter is given by

$$\mathrm{PE(r)} = -\frac{Ze^2}{4\pi \varepsilon_0 r}$$

If the total energy E is negative, then, according to classical mechanics, the maximum distance between electron and nucleus (r_{max}) is given by

$$E = -\frac{Ze^2}{4\pi \varepsilon_0 r_{\mathrm{max}}} \qquad (1.7)$$

Nevertheless examination of the 1s wavefunction of a one-electron atom showed that the probability of finding the electron outside a sphere with radius equal to r_{max} was greater than 20% (Problem 1.12). The phenomenon of the electron being in places where the potential energy is greater than the total energy is referred to as tunneling.

Similarly calculation of the wavefunction describing the deformation vibration of NH_3 shows that there is a small, but non-zero probability of finding the N atom in the plane of the H atoms ($q = 90°$), i.e. that it is possible for the molecule to invert through tunneling. The ability of atoms to tunnel through potential barriers depends strongly on their mass: while the rate of tunneling in NH_3 is 2.4×10^{10} inversions per second, doubling the mass of the hydrogen atoms reduces the rate by a factor of 0.06 to 1.5×10^9 inversions per second in ND_3 (D = deuterium).

The inversion barrier of phosphine has not been determined experimentally. The best quantum chemical calculations yield a barrier of 141 kJ mol^{-1}, i.e. more than five times higher than in ammonia [2]. Since the barrier to internal rotation in Si_2H_6 is smaller than in ethane, it may seem surprising that the inversion barrier of PH_3 is higher than in NH_3. The reason may be that the pyramidal structure of NH_3 is significantly destabilized by repulsion between the H atoms.

15.3 Some planar amines

The potential energy curve shown in Fig. 15.2 indicates that the energy required to open the HNH angle in ammonia from its equilibrium value to a value corresponding to the observed equilibrium CNC valence angle in trimethylamine is less than 3 kJ mol^{-1}. If the methyl groups were replaced by even larger ligands such as silyl or germyl groups, the valence angles would presumably become even larger, and if the energy of repulsion between the three ligands is reduced by more than 21 kJ mol^{-1} on going from a pyramidal to a planar N atom, then this repulsion would force the N atoms into a planar coordination geometry.

The coordination geometries of $N(SiH_3)_3$ and $N(GeH_3)_3$ are in fact found to be trigonal planar. The distance between the geminal Si atoms in $N(SiH_3)_3$ is 300 pm, the distance between geminal Ge atoms in $N(GeH_3)_3$ is 318 pm. See Fig. 15.3.

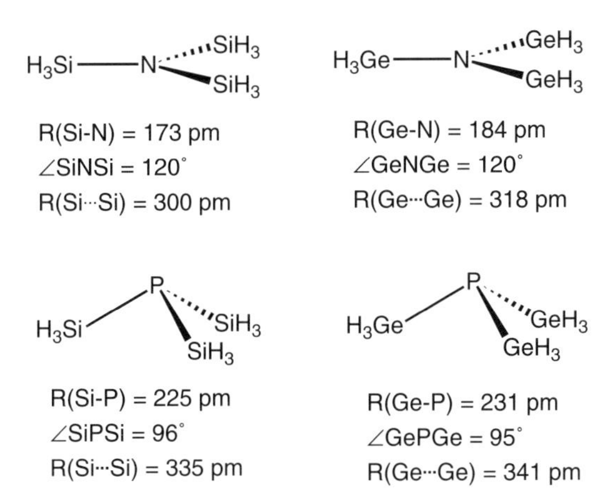

Fig. 15.3. Bond distances, valence angles and across-angle distances in $N(AH_3)_3$ and $P(AH_3)_3$, A = Si or Ge.

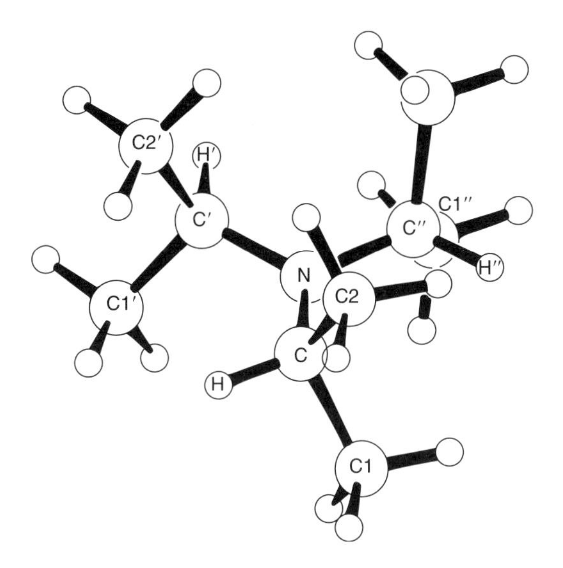

Fig. 15.4. The molecular structure of tri(*iso*-propyl)amine. Reproduced with permission from H. Bock, I. Goebel, Z. Havlas, S. Liedle, and H. Oberhammer, Angew. Chem. Int. Ed. Engl. 30 (1991) 187. Copyright (1991) Wiley-VCH Verlag GmbH & Co KGaA.

$P(SiH_3)_3$ and $P(GeH_3)_3$ on the other hand, are pyramidal with valence angles less than 100°. The phosphorus atom is larger than the nitrogen atom, and P–Si and P–Ge bond distances are about 50 pm longer than N–Si or N–Ge bond distances. As a geometrical consequence across-angle Si· · · Si and Ge· · · Ge distances are longer in the pyramidal phosphanes than in the planar amines.

Another simple amine with a planar (or nearly planar) N atom is tri(*iso*-propyl)amine. See Fig. 15.4. The molecule has a threefold symmetry axis through the N atom. In addition there is a symmetry plane through the N atom and the three ligating C atoms. Since we use the symmetry axis to define the vertical direction, the symmetry plane is horizontal. Note that there is no symmetry axis along the N–C bonds. Molecules with a threefold axis and a horizontal symmetry plane (and no more symmetry) are said to have C_{3h} symmetry.

Since trimethylamine is pyramidal, the planarity of tri(*iso*-propyl)amine cannot be due to repulsion between the geminal C atoms, but must be due to repulsion between the methyl groups on different ligands or between ligating C atoms and methyl groups on the neighbouring ligands. Note that the planarity of the N atom as well as the angle of rotation about the N–C bonds is such as to maximize the distance from ligating C atoms to methyl groups on the other ligands. The non-bonded C$\cdots$C1' and C$\cdots$C2' distances are equal to 250 pm. This is so much shorter than the van der Waals diameter of a methyl group, 400 pm, that there must be strong repulsion between them.

15.4 The gas phase molecular structures of phosphorus pentafluoride and pentachloride

Phosphorus forms a pentafluoride, PF_5, which is gaseous at room temperature, and solid compounds of composition PCl_5, PBr_5 and PI_5. Evaporation of the pentachloride yields a gas consisting of PCl_5, PCl_3 and Cl_2 in a temperature and pressure dependent equilibrium:

$$PCl_5(g) = PCl_3(g) + Cl_2(g).$$

The mole fraction of PCl_5 in the saturated vapor over a solid sample of PCl_5 at 90 °C is about 75%. Evaporation of the pentabromide yields a gas consisting largely of PBr_3 and Br_2: the mole fraction of PBr_5 in saturated vapor at 65 °C is less than 2%. The pentaiodide has never been observed in the gas phase.

The gas phase molecular structures of PF_5 and PCl_5 have been determined by gas electron diffraction. See Fig. 15.5.

In both molecules the coordination polyhedron of the P atom is a trigonal bipyramid or *tbp*. The phosphorus and the two axial halogen atoms define a vertical threefold symmetry axis, the three equatorial halogen atoms a horizontal symmetry plane. In addition to the threefold symmetry axis and the horizontal symmetry plane, each molecule has a twofold symmetry axes through each of the three equatorial P–X bonds. The symmetry of a molecules with one n-fold symmetry axis and n twofold symmetry axes perpendicular to it are denoted by D_n. The presence of a horizontal symmetry plane is indicated by the additional subscript h. The symmetry of PF_5 or PCl_5 then, is D_{3h}.

All valence angles in these compounds are determined by the symmetry. Thus all $\angle X_{eq}PX_{eq}$ angles are 120°, all $\angle X_{ax}PX_{eq}$ angles are 90°, and $\angle X_{ax}PX_{ax}$ is equal to 180°.

The two axial substituents are symmetry equivalent since they change places when the molecule is rotated 180° about one of the twofold axes, the three equatorial substituents are symmetry equivalent since they change places by rotation about the threefold or twofold symmetry axes. Since axial and equatorial atoms cannot be exchanged by any symmetry operation, they are symmetry inequivalent.

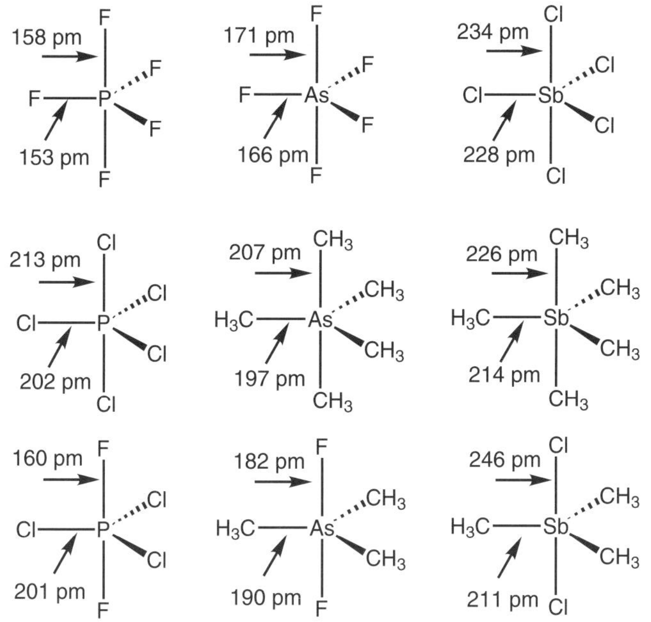

Fig. 15.5. Above: the six known molecular structures of gaseous, homoleptic, hypervalent compounds of Group 15 elements. Below: the structures of three heteroleptic compounds.

An important consequence of the symmetry inequivalence of axial and equatorial substituents is that we expect them to give separate peaks in the NMR spectra. We shall return to this point below.

15.5 The solid state structures of the phosphorus pentahalides

Investigation of solid PF_5 at very low temperature (109 K) by X-ray crystallography shows that it consists of monomeric *tbp* molecules very similar to those found in the gas phase. Investigation of PCl_5 by X-ray diffraction, on the other hand, shows that the crystal consists of tetrahedral $[PCl_4]^+$ cations and octahedral $[PCl_6]^-$ anions. We may regard these ions as formed by transfer of a Cl^- anion from one PCl_5 molecule to another. Such transfer is clearly energetically unfavorable in the gas phase, the Coulomb attraction energy between one cation and one anion is not large enough to compensate for the energy required for transfer a Cl^- anion between two PCl_5 molecules. The Coulomb attraction between a large number of oppositely charged ions in the crystal is, however, large enough to make an ionic crystal structure more stable than a non-ionic. The difference between the gas phase and solid state structures of PCl_5 provides an example of a more general effect: Coulomb interactions in the solid phase may stabilize a more ionic – or more polar – form than that found in the gas phase. Solid PBr_5 has been found to consist of $[PBr_4]^+$ cations and Br^- anions. The structure of solid PI_5 is unknown. It may be ionic, $[PI_4]^+ I^-$, but is probably similar to that of crystalline Ph_3PI_2 [3]. See Fig. 15.6 Pentaiodides of As, Sb, or Bi appear to be unknown.

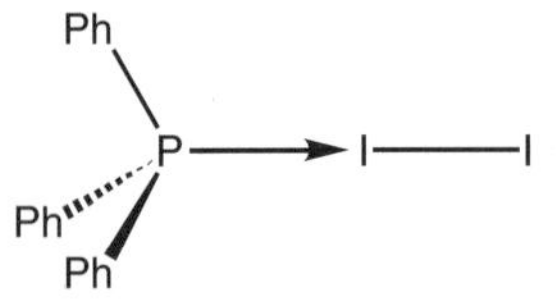

Fig. 15.6. The crystal structure of $Ph_3P{\rightarrow}I_2$, Ph = phenyl, with a dative bond between the P and I atoms. Dative bonding between electron donors and I_2 molecules will be discussed in Chapter 18.

Table 15.2. Homoleptic derivatives of hypervalent Group 15 elements EB_5. Me = methyl, Ph = phenyl. The stable phase at room temperature is indicated in parentheses. The compounds that are unstable at room temperature, are indicated by an asterisk.

B=	H	Me	Ph	F	Cl	Br
N	–	–	–	–	–	–
P	–	–	PPh_5(s)	PF_5(g)	PCl_5(s)	PBr_5(s)
As	–	$AsMe_5$(l)	$AsPh_5$(s)	AsF_5(g)	$AsCl_5$*	–
Sb	–	$SbMe_5$(l)	$SbPh_5$(s)	SbF_5(l)	$SbCl_5$(l)	–
Bi	–	$BiMe_5$*	$BiPh_5$(s)	BiF_5(s)	–	–

15.6 Other homoleptic derivatives of hypervalent Group 15 elements

Arsenic forms a gaseous pentafluoride and a solid pentachloride which decomposes at temperatures above −50°C. Antimony forms a pentachloride which may be vaporized without decomposition, and a pentafluoride which is tetrameric in the solid phase and trimeric in the gas phase. The structures of the trimer will be described in Chapter 16. The heaviest element in the group (Bi) forms a pentafluoride which is polymeric in the solid phase. The gas phase structure is unknown.

Pentamethylphospforane, $P(CH_3)_5$, is unknown, but both arsenic and antimony form stable pentamethyl derivatives. $Bi(CH_3)_5$ explodes on warming to room temperature. P, As, Sb and Bi all from solid pentaphenyl derivatives, the least stable of these compounds is $BiPh_5$.

The properties of known homoleptic compounds of hypervalent Group 15 elements, EB_5 where B is a singly bonded ligand, are summarized in Table 15.2. The properties of the homoleptic derivatives of hypervalent Group 16 and 17 elements are described in Sections 17.7 and 18.2 respectively. No hypervalent compound where the central atom carries only hydrogen or iodine substituents is known, and only one where it carries only Br atoms. Altogether there are five homoleptic hypervalent compounds with methyl substituents, five with phenyl and four with chlorine atoms. By far the largest number of hypervalent compounds (15) are fluorides. This tally suggests that the ability of singly bonded ligands to stabilize hypervalent molecules increases in the order

$$\mathrm{H} \approx \mathrm{I} < \mathrm{Br} < \mathrm{Cl} \approx \mathrm{Me} \approx Ph << F \tag{15.1}$$

It is noteworthy that in all these hypervalent molecules the ligating atoms are more electronegative than the central. As we shall see in the following, there is also considerable evidence to show that the bond distances in hypervalent compounds become shorter – and presumably stronger – when a ligating atom or group is replaced by one that is more electronegative. If such inductive effects are the dominating factor determining the stability of hypervalent compounds, the stabilizing influence of the ligands would be expected to increase with the electronegativity of the ligating atom:

$$H < I < Me \approx Ph < Br < Cl < F \qquad (15.2)$$

Comparison of sequences (15.1) and (15.2) shows that it is necessary to consider (at least) one more variable, perhaps destabilizing steric repulsion between halogen substituents, which would be expected to increase in the order

$$Cl < Br < I$$

Steric repulsion between the second period atoms C or F bonded to elements in periods 3 to 6 is expected to be negligible.

Formation of hypervalent compounds like NH_5 or NMe_5 is probably prevented by unfavorable bond polarities (both H and C are more electropositive than N) while formation of nitrogen pentahalides may be prevented by repulsion between the ligands.

Up to the present, the structure of six homoleptic, hypervalent compounds of Group 15 elements have been determined in the gas phase. All were found to be trigonal bipyramidal. The bond distances are displayed in Fig. 15.5. The equatorial bond distances are slightly (up to 4 pm) shorter than the bond distances in the analogous trivalent compounds. Thus the equatorial Sb–C distance in $Sb(CH_3)_5$ is about 2 pm shorter than the Sb–C bond distance in $Sb(CH_3)_3$. *The axial bond distances in the hypervalent compounds are 5–12 pm larger than the equatorial.*

15.7 The strength of the hypervalent bonds

We define the mean hypervalent bond energy, MHBE, as half the standard energy of the reaction

$$EX_5(g) \rightarrow EX_3(g) + 2X(g)$$

The thermodynamic data required for the calculation of the *MHBE*s are only available for PF_5, PCl_5, AsF_5 and $SbCl_5$. In Table 15.3 we compare these hypervalent bond energies with the mean bond energies of the corresponding trihalides.

In each of the four cases where data are available, the mean hypervalent bond energy is significantly lower than the mean bond energy in the corresponding trihalide:

$$HMBE(EX_5) < MBE(EX_3).$$

The axial bonds in the pentahalides are thus not only longer than the bonds in the trihalides, but also much weaker.

We also see that the standard enthalpy of the decomposition reaction

$$EX_5(g) \rightarrow EX_3(g) + X_2(g)$$

Table 15.3. Mean bond energies, MBE, in trihalides, and mean hypervalent bond energies, MHBE, of pentahalides of Group 15 elements. Standard enthalpies, ΔH_r°, of the reactions $EX_5(g) \rightarrow EX_3(g) + X_2(g)$. All energies in kJ mol^{-1}.

EX_3	MBE	EX_5	MHBE	ΔH°_r
PF_3	501	PF_5	395	636
PCl_3	320	PCl_5	162	86
AsF_3	439	AsF_5	302	451
$SbCl_3$	311	$SbCl_5$	157	76

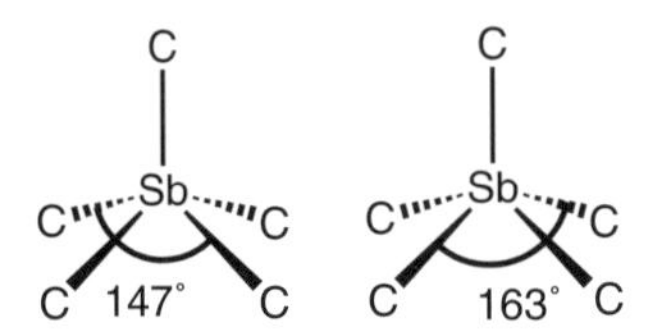

Fig. 15.7. The coordination polyhedron of the antimony atom in crystalline $SbPh_5$.

is much larger for X = F than for X = Cl. Since the only pentabromide observed in the gas phase, PBr_5, is nearly 100% dissociated to PBr_3 and Br_2, and since no pentaiodide appears to exist in the gas phase at all, we conclude that the stability of the hypervalent pentahalides increases in the order indicated by (15.1).

15.8 The pentaphenyl derivatives of Group 15 elements

The structures of these compounds have been determined by X-ray diffraction. The coordination geometries of the phosphorus atom in PPh_5 and the arsenic atom in $AsPh_5$ are both trigonal bipyramidal as expected, but the coordination geometry of the antimony atom in $SbPh_5$ is *approximately* square pyramidal (*sqp*). See Fig. 15.7. The symmetry of a *perfect* square pyramidal SbC_5 fragment is C_{4v}. The fragment possesses a fourfold symmetry axis that coincides with one of the Sb–C bonds. The C atom on the symmetry axis is referred to as *apical* (ap). The four symmetry equivalent C atoms are referred to as *basal* (ba).

While the valence angles of the trigonal bipyramidal molecules are fixed by the symmetry, the square pyramidal molecules contain one independent valence angle. This angle may be defined as the angle $\angle C_{ap}SbC_{ba}$ or as the angle $\angle C_{ba}SbC'_{ba}$ where C_{ba} and C'_{ba} are related through a rotation of 180° about the fourfold symmetry axis.

The two largest $\angle$CSbC angles in crystalline $SbPh_5$ are 147° and 163°. If the coordination geometry was perfectly square pyramidal, these angles should, of course, be equal. The reason why they are different is not clear, the deformation may be due to steric interaction between the ligand rings in the same molecule, or to steric interactions between neighboring molecules in the crystal. The apical Sb-C bond distance is 212 pm, considerably shorter than the average of the four basal distances 225 pm.

The crystals of $SbPh_5$ had been obtained by crystallization from isooctane. When $SbPh_5$ is crystallized from cyclohexane, solvent molecules are incorporated in cavities between the pentaphenylantimony molecules: the composition of the crystals is $SbPh_5$ 1/2 *cyclo*-C_6H_{12}.

Structure determination by X-ray crystallography shows that the Sb coordination geometry is now trigonal bipyramidal! *Clearly, not only the values adopted by the valence angles at the Sb atom, but the basic shape of the molecule is determined by intermolecular forces.* Investigations by Raman and IR absorption spectroscopy indicate that $SbPh_5$ is square pyramidal in CH_2Cl_2 or CH_2Br_2 solutions [4], but the coordination geometry in the gas phase is unknown.

Pentaphenyl bismuth, $BiPh_5$, is also approximately square pyramidal in the crystalline phase, the largest ∠CBiC angles being 150° and 164°. The apical Bi–C bond distance, 221 pm, is considerably shorter than the mean basal bond distance of 232 pm.

Problem 15.1 Calculate the mean values of the angles $\angle C_{ap}EC_{ba}$ in crystalline $SbPh_5$ and $BiPh_5$ respectively.

15.9 Structural non-rigidity and Berry pseudorotation

Phosphorus pentafluoride contains two sets of symmetry inequivalent fluorine atoms, and the ^{19}F NMR spectrum is expected to contain two peaks of relative areas 3:2. Experiments show, however, that the spectrum contains only *one line* (split to a doublet by P–F coupling). Barring the unlikely possibility that the chemical shifts of axial and equatorial F atoms happen to be virtually identical, this means that the molecule changes its shape in such a manner that each F atom change between axial equatorial positions several times per millisecond. A molecule that rearranges at such a high rate is referred to as *fluxional*. If the molecule is to rearrange at such a rate, the activation energy for the process must be very low.

In 1959 R.S. Berry postulated a general mechanism for the exchange of axial and equatorial ligands in trigonal bipyramidal molecules [5]. See Fig. 15.8.

When the process is complete, the ligands marked "a" occupy equatorial positions, while the ligands marked "e" occupy axial position. The process does not involve any rotation, but when it is complete, it looks as if the entire molecule has been rotated 90° about the p–E bond. The process is therefore referred to as a *pseudorotation*. A series of such pseudorotations with different pivotal ligands, will lead to complete scrambling of all ligands.

At some stage of the process the two angles ∠eEe and ∠aEa will be equal, say about 150°. At this point the coordination geometry of the central atom E is square pyramidal, and

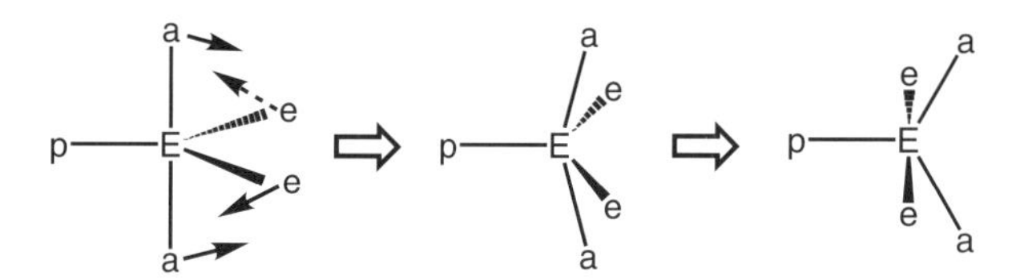

Fig. 15.8. Berry pseudorotation. One of the equatorial ligands, called the pivotal ligand (p), and the central atom E remain at rest during the process. The two other equatorial ligands (e) move in a plane perpendicular to the plane of the paper in such a way that the angle ∠eEe increases from 120° to 180°. At the same time the two axial ligands (a) move in the plane of the paper in such a manner that the angle ∠aEa is reduced from 180° to 120°.

the potential energy is at its maximum. The activation energy of the process, or the barrier to pseudorotation, is therefore equal to the energy difference between the square pyramidal and trigonal bipyramidal forms.

All the homoleptic, hypervalent compounds of the Group 15 elements described in this chapter are fluxional. The activation energies are not accurately known, but have been estimated to about 20 kJ mol^{-1} in PF_5 and PCl_5; 13 kJ mol^{-1} in AsF_5; and about 8 kJ mol^{-1} in $SbCl_5$ and $Sb(CH_3)_5$: the activation energies appear to decease as the group is descended. The magnitude of these barriers to pseudorotation is comparable to the magnitude of the rotational barrier in ethane, 12 kJ mol^{-1}, or the barrier to inversion in ammonia, 21 kJ mol^{-1}.

15.10 Pentacoordination and the VSEPR model

The central atom in a hypervalent compound is presumably surrounded by five bonding electron pairs. The application of the VSEPR model is not completely unequivocal as to whether repulsion between five electron pairs should lead to a trigonal bipyramidal *(tbp)* or a square pyramidal (*sqp*) arrangement. But if we assume that it is particularly important to avoid short distances between the electron pairs, i.e. to avoid small valence angles at the central atom, the *tbp* would be favored over an *sqp*; in the *tbp* there would be six electron pairs spanning valence angles of 90°, in the *sqp* there would be eight such pairs. In any event, $SbPh_5$ and $BiPh_5$ appear to be the only uncharged main group element molecules with five bond pairs (and no electron lone pairs) to adopt a square pyramidal arrangement.

The VSEPR model may also be used to rationalize the observation that the axial bonds in *tbp* molecules are longer than the equatorial: The axial bond electron pair is repelled by *three* (equatorial) bond pairs over a valence angle of 90°, while the equatorial bond electron pair is repelled by *two* (axial) bond pairs over the same angle: The additional repulsion leads to greater elongation of the axial bonds.

Problem 15.2 Suggest a rationalization for the observation that the apical bonds in the square pyramidal pentaphenyl bismuth is shorter than the basal.

15.11 Heteroleptic hypervalent compounds

When discussing the molecular structures of heteroleptic hypervalent compounds we shall seek answers to the following questions:

(i) Which ligands occupy the axial positions?
(ii) In what manner do the valence angles deviate from the ideal values of 90, 120 or 180°?
(iii) In what manner are bond distances changed when new substituents are introduced?

The molecular structures of four heteroleptic fluoro-methyl-phosphoranes $P(CH_3)_{5-n}F_n$, $n = 1, 2, 3$, or 4, are shown in Fig. 15.9. Note that the F atom in the monofluoro derivative occupies an axial position, so do both the F atoms in the difluoro compound. When more F atoms are introduced, they enter equatorial positions. The only member of the series exibiting an axial methyl group is thus the monofluoro compound. Studies of the mixed fluoro-chloro-phosphoranes $PCl_{5-n}F_n$, $n = 1, 2, 3$ or 4, reveal a similar pattern: the first two F atoms to be introduced occupy axial, additional F atoms enter equatorial positions. The molecular structures of $As(CH_3)_3F_2$ and $Sb(CH_3)_3Cl_2$ show the F atoms in the former and the Cl atoms in the latter also occupy axial positions (see Fig. 15.5). These observations may be generalized: *The axial positions in heteroleptic, trigonal bipyramidal, hypervalent compounds of the Group 15 elements are occupied by the more electronegative substituents.*

The monofluoro compound $P(CH_3)_4F$ has C_{3v} symmetry with the threefold symmetry axis running through the linear FPC_{eq} fragment. The $F_{ax}PC_{eq}$ valence angle is 85°, the $\angle C_{ax}PC_{eq}$ angle 95°. The P atom in thus lying below the plane defined by the three equatorial C atoms, and $\angle C_{eq}PC_{eq}$ is reduced to 119°. The coordination polyhedron of $P(CH_3)_3F_2$ has D_{3h} symmetry, though the symmetry of the molecule as a whole is lower (C_{3h}) because of the methyl group H atoms. The high symmetry implies that there are no deviations from the ideal *tbp* valence angles. The trifluoro compound has C_{2v} symmetry with the twofold axis running through the equatorial P–F bond. Any deviation of $\angle F_{ax}PF_{eq}$ from 90° is too small to be detected, but the angle spanned by the two equatorial C atoms has increased to about 124°. The coordination polyhedron of the tetrafluoro compound has C_{2v} symmetry with $\angle F_{ax}PF_{eq}$ increased to 92° and the angle spanned by the two equatorial F atoms decreased to 116°. All these observations may be summarized by the statement that the smallest valence angles are those spanned by the more electronegative substituents. (We have already drawn a similar conclusion regarding the valence angles in fluoro-methyl-silanes). The deviations from the ideal angles of 90, 120 and 180° are small, but they are all in a direction predicted by the VSEPR model.

Finally we turn our attention to the variation of bond distances. The structures shown in Fig. 15.9 show that

(i) Substitution of the four methyl groups in $P(CH_3)_4F$ by fluorine atoms to form PF_5 reduces the axial P–F distance from 176 to 158 pm, i.e. by 18 pm.

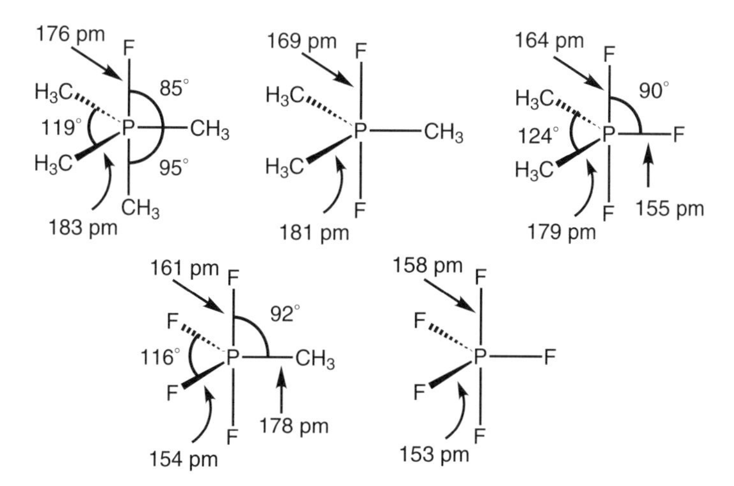

Fig. 15.9. The molecular structures of $PF_n(CH_3)_{5-n}$, $n = 1, 2, 3, 4$, or 5.

(ii) Substitution of the two methyl groups in $P(CH_3)_2F_3$ by F atoms reduces the equatorial P–F distance by 3 pm.
(iii) Substitution of three methyl groups in $P(CH_3)_4F$ by F atoms reduces the equatorial P–C distance by 5 pm.

Comparison of the molecular structures of $As(CH_3)_5$, $As(CH_3)_3F_2$ and AsF_5, and between $Sb(CH_3)_5$, $Sb(CH_3)_3Cl_2$ and $SbCl_5$ (Fig. 15.5) also show that replacement of methyl groups by more electronegative halogen atoms lead to shortening of both E–C and E–X bonds. (We have previously made the same observation regarding the bond distances the fluoro-methyl-silanes.) The structures of the phosphoranes also suggest that the axial bond distances are more sensitive to inductive effects than the equatorial.

Before going on, we pass to note that crystallization of $P(CH_3)_4F$ yields solid tetramethylphophonium fluoride, $[P(CH_3)_4]F$.

15.12 Molecular orbital descriptions of the hypervalent compounds

The molecular orbital description of bonding in the hypervalent molecules poses a problem: The valence shell orbitals of the Group 15 atoms are restricted to one *s* and three *p* orbitals: if we wish to describe the bonding in terms 2*c*,2*e* bonds we shall need *five* AOs on the central atom in order to form five orthogonal hybrid orbitals pointing towards the five ligands. (The reader may object that the 3*d* orbitals of an arsenic atom are part of the valence shell, but the 3*d* orbitals are filled and thus unavailable for formation of bonding MOs.)

So, if we wish to describe the molecules in terms of 2*c*,2*e* bonds we must take an AO from the next valence shell. Historically this was the solution that was first adopted: the valence shell s, p_x and p_y orbitals were assumed to be sp^2 hybridized, and the three hybrids combined with the appropriate AOs on the equatorial atoms to form three bonding 2*c* MOs. This description is consistent with the observation that the three equatorial bonds are slightly shorter than the bonds in the Lewis-valent compounds. The p_z orbital was then combined with the d_{z2} orbital of the next valence shell to produce two orthogonal hybrids pointing towards the two axial atoms, viz.

$$(p_z + d_{z2})/\sqrt{2} \quad \text{and} \quad (p_z - d_{z2})/\sqrt{2}$$

Finally these hybrids were combined with the appropriate AOs on the two axial atoms to form two 2*c* MOs.

It is however possible to describe the structure without taking recourse to *d* orbitals. Consider the molecular orbital

$$\Psi = (c_1\Phi_1 + c_2\Phi_2 + c_3\Phi_3)/\sqrt{D} \tag{15.3}$$

where Φ_2 is the valence shell p_z orbital on the central atom and Φ_1 and Φ_3 are suitable atomic orbitals on the axial atoms. If we are describing a pentahalide, Φ_1 and Φ_3 would be the valence shell p_z orbitals of the two axial halogen atoms. If we are describing a pentamethyl compound, they would be sp^3 hybrid orbitals, and if we are describing a pentaphenyl compound, sp^2 hybrid orbitals on the two axial C atoms. See Fig. 15.10. Note that we have chosen the signs of Φ_1 and Φ_3 in such a manner that both point a positive lobe in the direction of the central atom. This choice implies that the overlap integral $S_{2,3}$ will be positive, the resonance integral $H_{2,3}$ negative, that $S_{1,2} = -S_{2,3}$ and that $H_{1,2} = -H_{2,3}$.

We now use simplified Hückel calculations to obtain an estimate of the energy and the orbital coefficients c_1, c_2 and c_3. Calculation of the orbital energy and derivation with respect to the coefficients yields the three secular equations (see Section 12.2):

$$\begin{array}{lclclcl} (H_{1,1} - \varepsilon)c_1 & + & (H_{1,2} - S_{1,2}\varepsilon)c_2 & + & (H_{1,3} - S_{1,3}\varepsilon)c_3 & = & 0 \\ (H_{2,1} - S_{2,1}\varepsilon)c_1 & + & (H_{2,2} - \varepsilon)c_2 & + & (H_{2,3} - S_{2,3}\varepsilon)c_3 & = & 0 \\ (H_{3,1} - S_{3,1}\varepsilon)c_1 & + & (H_{3,2} - S_{3,2}\varepsilon)c_2 & + & (H_{3,3} - \varepsilon)c_3 & = & 0 \end{array} \quad (15.4)$$

We now simplify equations (15.2) by replacing all overlap integrals S by zero. Since the axial atoms are far apart, we also assume that $H_{1,3} = H_{3,1} = 0$. The resonance integrals $H_{2,3}$ and $H_{3,2}$ are equal, and the integrals $H_{1,2} = H_{2,1} = -H_{1,3}$. Finally we disregard the electronegativity difference between the centred and ligating atoms, and replace $H_{1,1}$ and $H_{3,3}$ by $H_{2,2}$. The secular equations may then be written as

$$\begin{array}{lclclcl} (H_{2,2} - \varepsilon)c_1 & - & H_{2,3}c_2 & + & 0 & = & 0 \\ -H_{2,3}c_1 & + & (H_{2,2} - \varepsilon)c_2 & + & H_{2,3}c_3 & = & 0 \\ 0 & + & H_{2,3}c_2 & + & (H_{2,2} - \varepsilon)c_3 & = & 0 \end{array} \quad (15.5)$$

These homogeneous linear equations will have nontrivial solutions if the secular determinant is equal to zero:

$$\begin{vmatrix} (H_{2,2} - \varepsilon) & -H_{2,3} & 0 \\ -H_{2,3} & (H_{2,2} - \varepsilon) & H_{2,3} \\ 0 & H_{2,3} & (H_{2,2} - \varepsilon) \end{vmatrix} = 0 \quad (15.6)$$

This equation has three roots:

$$\varepsilon_1 = H_{2,2} + \sqrt{2}H_{2,3} \quad \varepsilon_2 = H_{2,2} \quad \varepsilon_3 = H_{2,2} - \sqrt{2}H_{2,3}$$

Since the resonance integrals $H_{2,3}$ are negative, the first energy is that of a bonding orbital, the second that of a nonbonding, and the third that of an antibonding orbital. Insertion of $\varepsilon_1 = H_{2,2} + \sqrt{2}\, H_{2,3}$ and $c_2 = 1$ into the secular equations yields the coefficients $c_1 = -1/\sqrt{2}$ and $c_3 = 1/\sqrt{2}$. The normalization constant $D = c_1^2 + c_2^2 + c_3^2 = 2$, and

$$\Psi_b = \Psi_1 = -\frac{1}{2}\Phi_1 + \frac{1}{\sqrt{2}}\Phi_2 + \frac{1}{2}\Phi_3$$

Similarly the other molecular orbitals are

$$\Psi_{nb} = \Psi_2 = \frac{1}{\sqrt{2}}\Phi_1 + \frac{1}{\sqrt{2}}\Phi_3,$$

and

$$\Psi_a = \Psi_3 = \frac{1}{2}\Phi_1 + \frac{1}{\sqrt{2}}\Phi_2 - \frac{1}{2}\Phi_3$$

See Fig. 15.10.

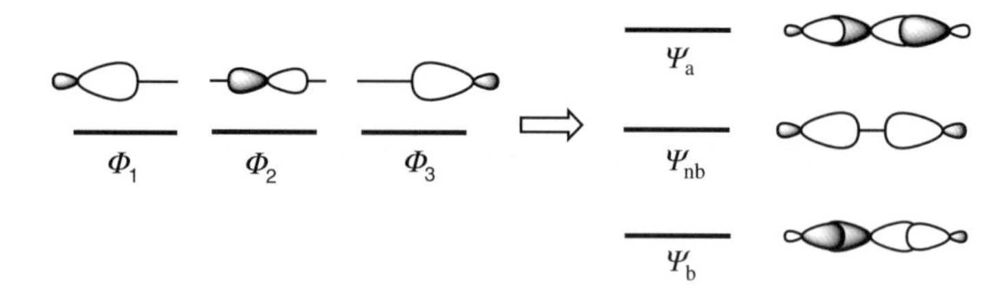

Fig. 15.10. Formation of 3c molecular orbitals from the p_z AO on the central atom and atomic orbitals, denoted by Φ_1 and Φ_3, on the two apical atoms.

As indicated by the Lewis formula these orbitals must accommodate four electrons. Two of these electrons come from the central atom, while the axial atoms contribute one electron each. Two electrons are assigned to the bonding orbital Ψ_b and two to the nonbonding orbital Ψ_{nb}.

The description of hypervalent bonding in terms of *pd* hybridization and two-center orbitals implies that the electron densities in the p_z and d_{z2} orbitals are approximately equal. The description in terms of three-center orbitals, Fig. 15.10, implies that the electron density in the d_{z2} orbital is negligible. Accurate molecular orbital calculations show that the latter description is closer to reality.

The Lewis structures of pentavalent compounds of Group 15 elements suggest that the central atom is surrounded by five electron pairs. This picture is modified at the level of the 3*c*,4*e* bonding model: one of the electron pairs occupies a non-bonding MO which confines these electrons to the axial ligands.

The axial XEX fragment is often described as a three-center, four-electron or 3*c*,4*e* bonding system. It should be kept in mind, however, that only two out of the four electrons reside in bonding orbitals. Since the axial XEX fragment is held together by only two bonding electrons, it is perhaps not surprising that the axial bonds are longer and weaker than the 2*c*,2*e* equatorial E–X bonds.

The two electrons in the non-bonding 3c orbital will give rise to an electron density corresponding to one electron on each axial atom. In addition there will be electron density on the axial atoms due to the two electrons in the bonding orbital. This means that the axial atoms will carry a net negative, and the central atom a net positive charge, even if there is no electronegativity difference between them. This result is consistent with our observation that the only stable homeleptic hypervalent compounds are those in which the ligating atoms are more electronegative than the central, and also with the observation that the axial positions in heteroleptic compounds are occupied by the more electronegative substituents.

References

[1] C. Léonard, S. Carter, and N. C. Handy, *Chem. Phys. Letters*, 370 (2003) 360.
[2] D. J. Grant and D. A. Dixon, *J. Phys. Chem. A*, 110 (2006) 12955.
[3] S. M. Godfrey, D. G. Kelly, C. A. McAuliffe, A. G. Mackie, R. G. Pritchard, and S. M. Watson, *J. Chem. Soc. Chem. Commun.*, 1991, 1163.
[4] G. L. Kok, *Spectrochim. Acta*, 30A (1974) 961.
[5] R. S. Berry, *J. Chem. Phys.*, 32 (1960) 933.

Chapter 16

Electron donor–acceptor complexes

Introduction

The amine borane molecule, H_3NBH_3, is isoelectronic with ethane, H_3CCH_3: the total number of electrons in each molecule is 18, and the number of valence electrons 14. The Lewis structures of both molecules indicate that one electron pair is shared between the two second-period elements:

$$H_3N{:}BH_3 \quad \text{or} \quad H_3C{:}CH_3.$$

As expected for isoelectronic molecules, they are found to have the same shape. See Fig. 16.1.

The CC bond distance in ethane is 153 pm, the NB bond distance in gaseous amine borane is 167 pm. This difference is surprisingly large since comparison of the B–H, C–H and N–H bond distances in the figure indicates that while the bonding radius of the B atom is about 10 pm larger than that of C, the bonding radius of N is 10 pm smaller. This would seem to imply that the NB bond distance in amine borane should be about equal to the CC bond distance in ethane, or perhaps a little shorter due to the electronegativity difference between the N and B atoms: the NB bond distance calculated from the MSS rule (Section 6.5) is 146 pm. We conclude that the NB bond distance in amine borane is surprisingly long.

The CC bond in ethane is not only shorter, but *three times stronger* than the NB bond in amine borane: Minimum energy rupture of the C–C bond in gaseous ethane yields two electrically neutral methyl radicals:

$$H_3C{:}CH_3(g) \rightarrow 2\,{\cdot}CH_3(g) \tag{16.1}$$

The dissociation energy at 298 K is $D(\mathrm{C-C}) = 373$ kJ mol^{-1}.

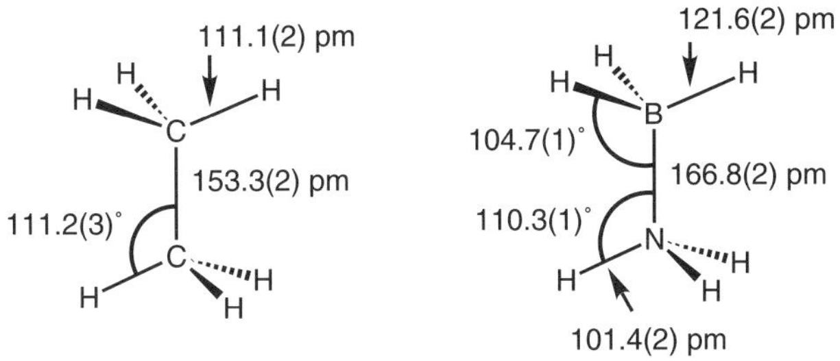

Fig. 16.1. The gas phase molecular structures of ethane and amine borane.

Minimum energy rupture of the NB bond in amine borane, on the other hand, yields the two electrically neutral, diamagnetic species H_3N and BH_3:

$$H_3N{:}BH_3(g) \rightarrow H_3N{:}(g) + BH_3(g) \tag{16.2}$$

and the dissociation energy at 298 K is D(N–B) $= 128$ kJ mol^{-1}.

The cubic crystalline form of boron nitride has a diamond-like structure, each boron atom is bonded to four nitrogen atoms at a bond distance of 156 pm. The mean NB bond energy calculated from the energy of atomization is 321 kJ mol^{-1}, i.e. more than 2.5 times larger than the NB bond dissociation energy of amine borane. We conclude that the NB bond in amine borane is not only unusually long, but also unexpectedly weak.

The dipole moment of ethane is zero by symmetry. Amine borane, on the other hand, is a very polar molecule with an electric dipole moment of 5.22 Debye, nearly as large as that of the ionic molecule LiF, 6.28 D.

16.1 Covalent and dative bonds

Most chemists would rationalize the existence of amine borane as due to the fact that the B atom in monomeric BH_3 has only six electrons in the valence shell. The formation of the NB bond is described as due to donation of the electron lone pair on N to the "electron poor" boron atom. A species formed in this manner from two relatively stable chemical entities, is referred to as a complex or coordination compound. The chemical species providing the electron pair is referred to as the electron donor or the Lewis base. The bonding partner is referred to as the electron acceptor or the Lewis acid. The new bond that has been formed between the donor and acceptor atoms, has been referred to as an electron donor–acceptor bond or as a dative bond.

We have previously defined a single bond as homolytic if minimum energy rupture proceeds to yield two electrically neutral radical species. The covalent CC bond in ethane then, is a homolytic bond. It has been objected that a distinction between covalent and dative bonds is artificial, since it seems to depend on the method of formation: if ethane is formed by combination of the methyl cation CH_3^+ with the methyl anion CH_3^-, both CC bonding electrons are provided by the anion. This formal difficulty is, however, overcome by defining covalent and dative bonds by the products formed on minimum energy rupture: *Minimum energy rupture of a single covalent bond in a neutral molecule proceeds homolytically to yield electrically neutral radicals, minimum energy rupture of a single dative bond in a neutral molecule proceeds heterolytically to yield electrically neutral closed shell (diamagnetic) species* [1].

Should dative bonds be counted when we define the valency of the boron and nitrogen atoms? There does not seem to be general agreement between chemists at this point. Some (including the author of this book) do not include dative bonds in the valency, others count both normal and dative bonds and describe both the B and the N atom as tetravalent. Some define the valency of an atom in terms of the number of electrons used for bond formation. This means that the N atom in amine borane is pentavalent while the B atom remains trivalent [2].

If the two electrons between the N and B atoms in the Lewis structure of $H_3N{:}BH_3$ were equally shared between the two atoms, the net electric charge on the acceptor (BH_3) would be equal to one elementary *negative* charge and the net charge on the donor to one elementary positive charge. If these net charges were located on the B and N atoms, the electric dipole moment of the complex would be

$$\mu_{el} = e \times R(\text{N–B}) = 8.0 \text{ Debye}.$$

Note that the negative pole would be on the B atom which is *less* electronegative than the N atom. The observed dipole moment of 5.22 Debye suggest that the two electrons *are not equally shared*, they continue to be closer to the N than to the B atom. In the following we shall nevertheless indicate a dative bond by an arrow pointing from the donor to the acceptor atom: N → B.

Ethane is a gas at room temperature. The melting point solid ethane is –183°C, the boiling point of the liquid −89°C. Amine borane, on the other hand, is a solid at room temperature, and the melting point is +124°C. The large difference (300°C) between the melting points of the two compounds is due to a large difference in the strength of intermolecular attractions: crystalline ethane is held together by relatively weak dispersion forces, amine borane by much stronger Coulomb attractions, in paticular by favourable electric dipole–dipole interaction between neighboring molecules [3]. The CC bond in crystalline ethane determined by X-ray diffraction is found to be 2 pm shorter than in the gas phase. This difference is, however, believed to be due to a systematic error in the X-ray study: thermal motion of the molecule in the crystal could not be properly corrected for. The NB bond in amine borane, on the other hand, is 11 pm shorter in a crystal at −73°C than in the gas phase!

The shorter bond distance in the solid phase may be interpreted as follows: The crystal is stabilized by favorable Coulomb interactions, particularly dipole–dipole interactions between neighboring molecules. The greater the transfer of negative charge from the electron donor to the electron acceptor, the greater the electric dipole moment. The crystalline environment will therefore favor larger electron transfer between donor and acceptor, and hence a stronger and shorter bond.

16.2 N→B bond distances and dissociation energies in substituted amine borane complexes

Inspection of Table 16.1 shows that replacement of the three H atoms on the acceptor atom (B) in the complexes H_3NBH_3 or $(CH_3)_3NBH_3$ by methyl groups *reduces* the dissociation energies of the complexes by about 70 kJ mol^{-1} or 50%. Replacement of the hydrogen atoms on the donor atom (N) in the complexes H_3NBH_3 and $H_3NB(CH_3)_3$ by methyl groups, on the other hand, *increases* the dissociation energies by about 15 kJ mol^{-1}. This represents a 12% increase of the dissociation energy of H_3NBH_3 and a 30% increase for $H_3NB(CH_3)_3$.

Comparison of the dissociation energies of the four compexes $(CH_3)_3NX_3$, X=H, CH_3, F or Cl show that the stength of the dative N→B bonds increases in the order

$$CH_3 < F < Cl < H$$

Table 16.1. Dissociation energies (at 298 K) and N → B bond distances in gaseous substituted amine borane complexes.

	$D(N \rightarrow B)^{a}$ (kJ mol^{-1})	$R(N \rightarrow B)$ (pm)
H_3NBH_3	128	167
$(CH_3)_3NBH_3$	143	166
$H_3NB(CH_3)_3$	55	–
$(CH_3)_3NB(CH_3)_3$	71	170
$(CH_3)_3NBF_3$	109	164
$(CH_3)_3NBCl_3$	125	165

[a] Data from references [1] and [4].

while consultation with Table 3.2 shows that the electronegativity coefficients vary in the order

$$H < C < Cl < F$$

This means that the observed variation of N→B bond strength cannot be explained in terms of inductive effects alone. Steric effects appear to be as important as electronic [5].

The N→B bond distances in the four strongest complexes in Table 16.1, with dissociation energies ranging from 109 to 143 kJ mol^{-1}, fall into a relatively narrow range from 164 to 167 pm. We note, however, that the N→B bond distances in the weaker complex, $(CH_3)_3NB(CH_3)_3$, $D(N \rightarrow B) = 71$ kJ mol^{-1}, is 170 pm. The N→B distance in $H_3NB(CH_3)_3$ may be even longer, but has not been determined.

We conclude that dative bonds are easily perturbed by a change of substituents at the acceptor atom, somewhat less so by a change at the donor atom. Since both steric and inductive effects are involved, it is difficult to predict whether a change of substituents in a given complex will strengthen or weaken the dative bond. However, it appears to be generally true that a dative bond is weakened if a hydrogen or halogen atom at the acceptor atom is replaced by a methyl- or alkyl-group.

Problem 16.1 Hydrogen cyanide, HCN, is isoelectronic with the N_2 molecule. The Lewis structure would presumably be H–C≡N:. Do you expect the molecule to be linear or bent? Give a description of the bonding in terms of 2*c* molecular orbitals.

16.3 Complexes of the electron donors HCN and H_3CCN with the electron acceptor BF_3

A particularly dramatic demonstration of the sensitivity of complexes to change of substituents, or to intermolecular forces in the solid phase, is provided by the complexes of hydrogen cyanide, HCN, or acetonitril, H_3CCN, with BF_3. See Fig. 16.2.

The Lewis structure of acetonitril is H_3C–C≡N:, the structure is linear and the bonding similar to that of hydrogen cyanide. The energy required to rupture the dative bond in

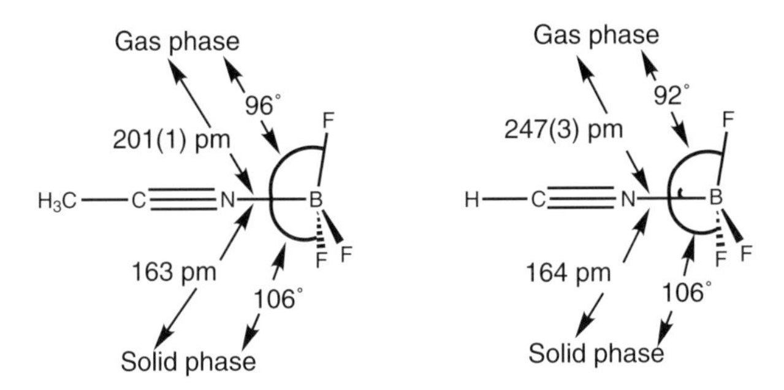

Fig. 16.2. The gas phase (above) and solid state structures (below) of the complexes $CH_3CN \rightarrow BF_3$ and $HCN \rightarrow BF_3$.

gaseous $H_3CCN \rightarrow BF_3$ has not been determined experimentally, quantum chemical calculations yields a dissociation energy of about 24 kJ mol^{-1}, less than a quarter of the dissociation energy of $(CH_3)_3N \rightarrow BF_3$, and just above the threshold we have suggested for "chemical bonds". The molecular structure in the crystalline phase was determined by X-ray diffraction about 40 years ago, the dative N→B bond distance was found to be 163 pm, i.e. indistinguishable from that found in gaseous $(CH_3)_3N \rightarrow BF_3$. Ten years ago the structure of the complex was determined in the gas phase, and the N→B bond distance was found to be 201 pm, nearly 40 pm longer than in the crystalline solid!

We have seen that replacement of the three methyl groups at the N atom in $(CH_3)_3N \rightarrow BH_3$ by H atoms decreases the N→B dissociation energy by 15 kJ mol^{-1}. Replacement the methyl group in $H_3CCN \rightarrow BF_3$ by a H atom might therefore be expected to lead to a weakening of the dative bond. The N→B bond dissociation energy of $HCN \rightarrow BF_3$ has never been determined, but indirect evidence suggests that is less than 15 kJ mol^{-1}: this bond is so weak that the reader may decide to classify it as a van der Waals interaction rather than a chemical bond.

The N→B bond distance in gaseous $HCN \rightarrow BF_3$ is 247 pm, more than 40 pm longer than in the acetonitril complex. In the solid state, however, the N→B bond distance is unexceptional: 164 pm.

16.4 Some complexes of trimethylaluminum

Trimethylaluminum forms stable complexes with a large number of electron donors such as trimethylamine, trimethylphosphine, dimethylether and dimethyl sulphide. The dative bond dissociation energies and bond distances are listed in Table 16.2.

The dative bond dissociation energies follow a fairly general trend for electrically neutral complexes: the strongest dative bonds to main group elements are formed by N centered electron donors, followed by P or O centered donors. There is also a general tendency for the strength of the dative bond to decrease when the donor atom is exchanged for a heavier element in the same group.

Table 16.2. Complexes of trimethylaluminum with trimethylamine, trimethylphosphine, dimethylether and dimethylsulphide: dative bond dissociation energies $D(\mathrm{D}\rightarrow\mathrm{Al})$ (D = donor atom); dative bond distances $R(\mathrm{D}\rightarrow\mathrm{Al})$; and single, polar covalent bond distances $R(\mathrm{D{-}Al})$ estimated by the MSS rule.

	$D(\mathrm{D}\rightarrow\mathrm{Al})$ (kJ mol^{-1})	$R(\mathrm{D}\rightarrow\mathrm{Al})$ (pm)	$R(\mathrm{D{-}Al})$ (pm)
$(CH_3)_3N\rightarrow Al(CH_3)_3$	126	210	182
$(CH_3)_3P\rightarrow Al(CH_3)_3$	85	253(4)	230
$(CH_3)_3O\rightarrow Al(CH_3)_3$	89	201	174
$(CH_3)_3S\rightarrow Al(CH_3)_3$	73	255	220

In these complexes of $Al(CH_3)_3$ the dative D→Al bond distances are 20–30 pm longer than the single, polar covalent bond distances between the same atom pair predicted by the MSS rule.

Like the borane complexes described previously, trimethylalane complexes are sensitive both the change of state from gas to crystal and to substitution effects: The N→Al bond distance in $(CH_3)_3N\rightarrow Al(CH_3)_3$ decreases to from about 210 pm in the gas phase to 205 pm in the solid phase. Exchange of the methyl groups on Al by more electronegative Cl atoms to give $(CH_3)_3NAlCl_3$ increases the dative bond energy from about 126 to about 196 kJ mol^{-1} and decreases the bond distance in the solid phase to 195 pm.

16.5 Two complexes containing dative and covalent bond distances between identical atom pairs

The structures of the two complexes in Fig. 16.3 allow us to compare dative and covalent bond distances between identical, AlO or AlN, atom pairs. In order to determine the character

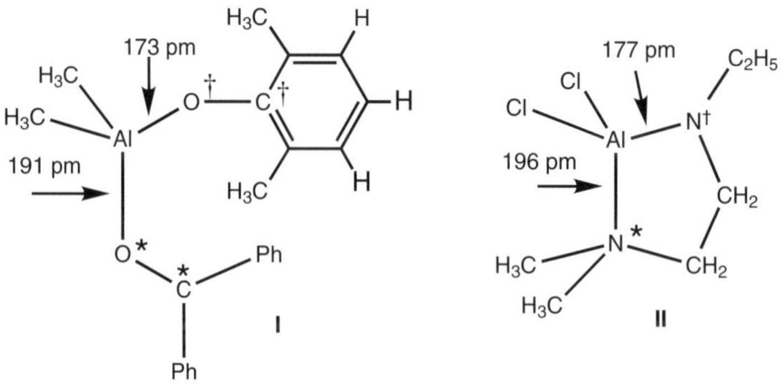

Fig. 16.3. Comparison of normal and dative bond distances between Al and O and between Al and N. Ph = phenyl.

of the two Al–O bonds in complex **I**, we begin at the periphery of the molecule, describe each bond as dative or normal and move in towards the Al and O atoms:

We know that H atoms, which have only one electron in the valence shell, can form only one covalent bond, and that carbon atoms with four valence electrons may form four single covalent bonds or alternatively two single and are double bond. Each carbon atom in the three phenyl rings of **I** must form one covalent bond to a H atom or methyl group outside the ring, and one CC double bond and one CC single bond in the ring. The C atom marked by a dagger (†) must therefore form a single, covalent bond to $O^{\dagger}$. An O atom with six valence electrons will form two single covalent bonds. It follows that the bond from $O^{\dagger}$ to Al must be covalent. Indeed, the observed bond distance is just 1 pm shorter than the MSS estimate for a single, polar covalent Al-O bond distance. The Al atom forms covalent bonds to the C atoms of two methyl groups. It has then used all its valence electrons: the O*–Al bond must therefore be dative.

Since the C atom marked by an asterisk must form single, covalent bonds to one C atom in each of the two phenyl rings, it follows that the C*O* bond must be double: the donor molecule in this complex is $Ph_2C{=}O$ and the Lewis structure of the free donor would place two non-bonding electron pairs on the O atom. It is thus able to form a dative bond to the Al atom. The dative bond O*→Al bond distance is 18 pm longer than the covalent Al–$O^{\dagger}$ bond.

Problem 16.2 Carry out a similar analysis of the bonding in **II** and show that the Al–$N^{\dagger}$ bond must be covalent while the Al–N* bond must be dative. Compare the Al–N* bond distance with the dative bond distance in $(CH_3)_3N{\rightarrow}Al(CH_3)_3$. Suggest a reason why the dative bond distance in **II** is shorter.

Problem 16.3 Suggest Lewis structures for the following compounds: $(CH_3)_3GeGe(CH_3)_3$, $(CH_3)_3GaAs(CH_3)_3$, $(CH_3)_3GeAs(CH_3)_2$, $(CH_3)_3GeGa(CH_3)_2$, $(CH_3)_2GaAs(CH_3)_2$, $(CH_3)_4GeAs(CH_3)_3$, and $(CH_3)_3GeAs(CH_3)_4$. What will be the products if the central bonds between the fourth-period elements are broken to yield neutral products? Classify these bonds as polar covalent or dative.

16.6 Dative bonds and the VSEPR model

In Fig. 16.4 we compare the structures of the two complexes $(CH_3)_3N{\rightarrow}Al(CH_3)_3$ and $(CH_3)_3P{\rightarrow}Al(CH_3)_3$ with the structure of the free acceptor $Al(CH_3)_3$. It is seen that the Al–C bonds are elongated and pushed away from the electron donor as the complex is formed. These changes may be interpreted as due to repulsion between the dative D→Al bond electron pair and the three covalent Al–C bond electron pairs, but may be augmented by steric repulsion between the entire donor molecule and the methyl groups bonded to the Al atom. It should be noted, however, that in both complexes the angle ∠DAlC is smaller than tetrahedral, indicating that *the accepted electron pair requires less space on the acceptor atom than the three covalent bond pairs.*

While valence shell electron pair repulsion and steric repulsion between ligands may act in a synergetic manner on the acceptor molecule, they are expected to be opposed on the donor molecule: the VSEPR model predicts that partial removal of a lone pair on the donor atom should lead to a reduction of D–C bond distances and an increase of the CDC valence angles. Steric repulsion between the methyl groups bonded to D and the entire acceptor

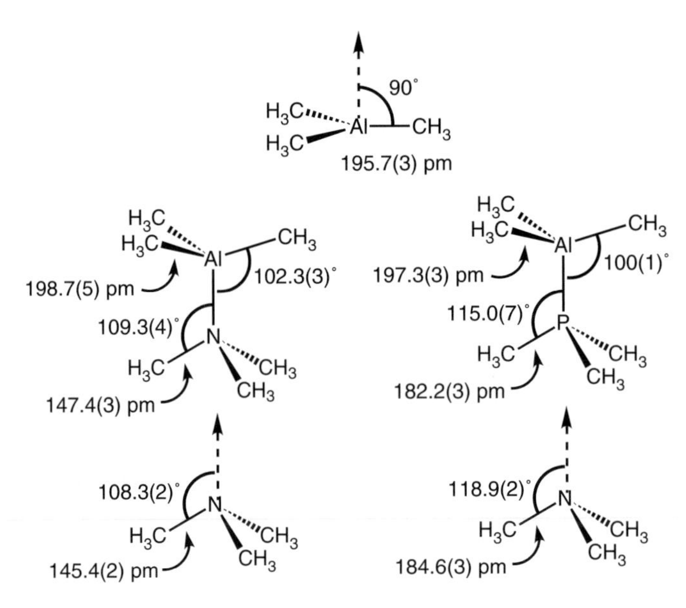

Fig. 16.4. The molecular structures of the complexes $(CH_3)_3NAl(CH_3)_3$ and $(CH_3)_3PAl(CH_3)_3$, the free donors $(CH_3)_3N$ and $(CH_3)_3P$ and the free acceptor $Al(CH_3)_3$. Threefold symmetry axes of the isolated donor and acceptor molecules are indicated by stippled arrows.

molecule, on the other hand, would be expected to increase D–C bond distances and reduce CDC valence angles.

Comparison of the structure of $(CH_3)_3P{\rightarrow}Al(CH_3)_3$ with the structure of the free donor $(CH_3)_3P$ shows that as the complex is formed the angle between the threefold symmetry axis and the P–C bonds is reduced from 118.9(2)° to 115.0(7)°, presumably because the spatial requirements of the lone pair becomes smaller as the electrons are moving closer to the acceptor atom. The small decrease in the P–C bond distances is also consistent with reduced repulsion between the dative electron pair and the three covalent bond pairs.

The distortion of the donor molecule in the trimethylamine complex is, however, in the opposite direction: the N–C bonds become longer and the angle between the threefold symmetry axis and the N–C bonds slightly larger on complex formation. These changes which are in the opposite direction to those predicted by the VSEPR model, are probably due to steric repulsion between the methyl groups bonded to N and the entire acceptor molecule $Al(CH_3)_3$.

Investigations of a large number of complexes indicate that the structure changes of the acceptor molecule generally are in the direction predicted by the VSEPR model while the structure changes of the donor are due to a combination of VSEPR and steric repulsion effects.

Problem 16.4 Use the VSEPR model to predict the coordination geometry around the O atom in $(CH_3)_2O{\rightarrow}Al(CH_3)_3$ and the F atom in the complex $HFAlF_3$.

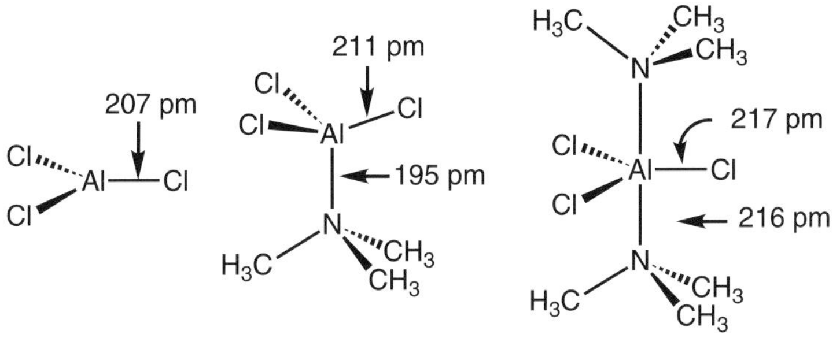

Fig. 16.5. The molecular structure of $AlCl_3$ in the gas phase and of the complexes Me_3NAlCl_3 and $(Me_3N)_2AlCl_3$ in the solid phase.

Problem 16.5 Consider the structures of the two complexes in Fig. 16.2. Are the NBF valence angles in the gaseous complexes in agreement with the VSEPR model? How would you explain the changes of these angles when the complex is studied in the crystalline phase?

Problem 16.6 Use the VSEPR model to predict the coordination geometries around the fourth period atoms in the following compounds: $(CH_3)_3GeGe(CH_3)_3$, $(CH_3)_3GaAs(CH_3)_3$, $(CH_3)_3GeAs(CH_3)_2$, $(CH_3)_3GeGa(CH_3)_2$, $(CH_3)_2GaAs(CH_3)_2$, $(CH_3)_4GeAs(CH_3)_3$, and $(CH_3)_3GeAs(CH_3)_4$.

16.7 The molecular structure of $(Me_3N)_2AlCl_3$, a complex with two dative bonds

The complex Me_3NAlCl_3 will add a second mole of trimethylamine to form $(Me_3N)_2AlCl_3$. The structure of this 2:1 complex is shown in Fig. 16.5. It is clear that both amines are bonded to Al through dative bonds. The energy of the reaction

$$(Me_3N)_2AlCl_3 \rightarrow Me_3NAlCl_3(g) + Me_3N(g)$$

is not known, but it is certain that the second trimethylamine donor is less strongly bonded than the first. This is to be expected, there must be a limit to how much electron density the $AlCl_3$ unit will accept.

The Al atom is surrounded by three covalent bond electron pairs and two dative bond electron pairs. According to the VSEPR model the most stable arrangement of five electron pairs in the valence shell of the central atom is trigonal bipyramidal, and this is indeed the structure observed. The VSEPR model may also be used to rationalize the observation that the donor atoms occupy axial positions: an axial bond electron pair is repelled by *three* (equatorial) bond pairs, while an equatorial bond electron is repelled by *two* axial bond pairs across the same angle. Since a covalent bond electron pair requires more space at the aluminum atom than a dative bond electron pair, the covalently bonded atoms occupy the equatorial positions.

It is seen that introduction of a second trimethylamine donor molecule leads to elongation of the first dative Al→N bond as well as further elongation of the three Al–Cl bond distances.

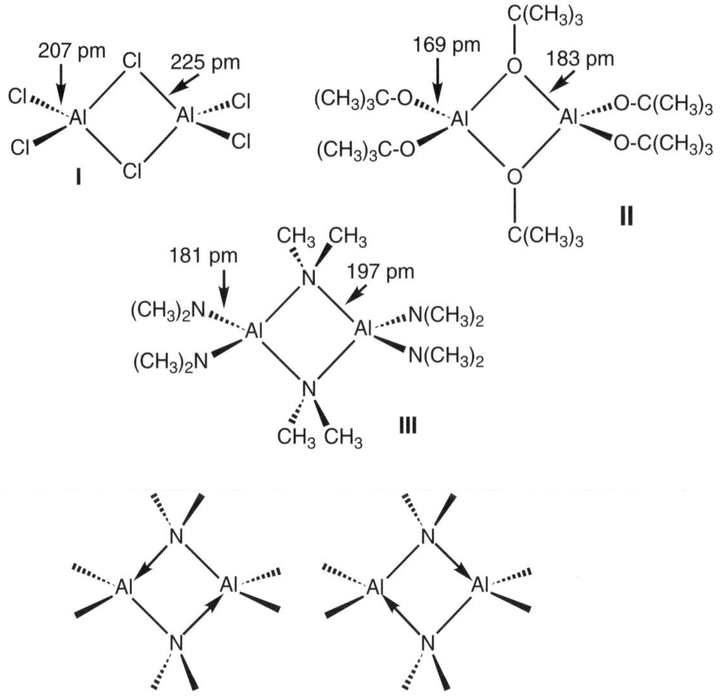

Fig. 16.6. Above: the molecular structures of $Cl_4Al_2(\mu\text{-}Cl)_2$, $(Me_2N)_4Al_2(\mu\text{-}NMe_2)_2$ and $(Me_3CO)_4Al_2(\mu\text{-}OCMe_3)_2$. Below: the two canonical forms of $(Me_2N)_4Al_2(\mu\text{-}NMe_2)_2$.

16.8 Bridge bonds: bonds with equal covalent and dative contributions

The aluminum trisamide $Al(NMe_2)_3$, the tris-*tert*-butoxide, $Al(OCMe_3)_3$, and the trichloride, $AlCl_3$, are all dimeric in hydrocarbon solution and in the gas phase at low temperatures. (The structure of monomeric $AlCl_3$ described in Chapter 11 had been determined in the gas phase at 530°C). The molecular structures of the dimers are shown in Fig. 16.6. The structure of Al_2Cl_6 is similar to that of diborane: the central part of the molecule consists of a planar, rhombohedral four-membered ring of two Al and two bridging Cl atoms. In addition each Al atom is bonded to two terminal Cl atoms. The structures of the Al_2N_6 and A_2O_6 skeletons of the other two molecules in Fig. 16.6 are similar and the formulae of the three compounds in the figure may be written as $Cl_4Al_2(\mu\text{-}Cl)_2$, $(Me_2N)_4Al_2(\mu\text{-}NMe_2)_2$ and $(Me_3CO)_4Al_2(\mu\text{-}OCMe_3)_2$ respectively.

N atoms must form three covalent bonds in order to reach an electron octet. The terminal N atoms in compound **III** in Fig. 16.6 are three-coordinate, we therefore assume that the terminal Al–N bonds are covalent. Each Al atom is able to form three covalent bonds. It follows that two of the four AlN bonds in the central Al_2N_2 ring must be described as covalent and two as dative. Just as we describe the bonding in benzene in terms of resonance between two canonical forms, each with alternating double and single bonds around the ring, we may describe the bonding in each of the three molecules in Fig. 16.5 in terms of two canonical forms with alternating covalent and dative bonds in the four-membered

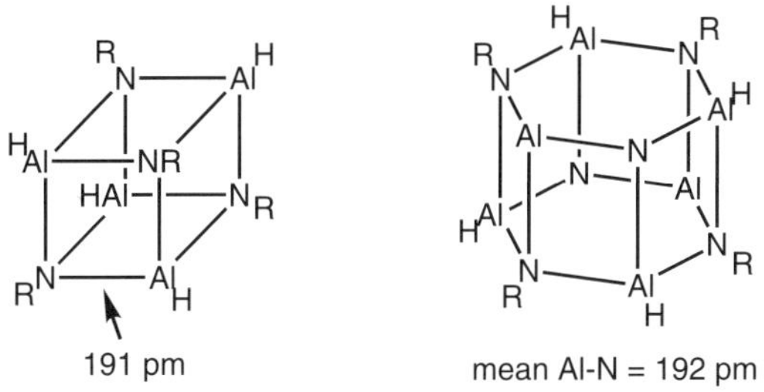

191 pm

mean Al-N = 192 pm

Fig. 16.7. The molecular structures of $H_4Al_4(\mu^3\text{-N-}i\text{-Pr})_4$ and $H_6Al_6(\mu^3\text{-N-}i\text{-Pr})_6$, R = *i*-Pr = *iso*-propyl. Two terminal H atoms and two terminal *iso*-propyl groups in the latter have been omitted for clarity.

rings. This description implies that the terminal bond distances should correspond to single covalent bond distances while the bridging bond distances should be intermediate between covalent and dative bond distances. This is indeed found to be the case.

Similarly we describe the Al–Cl bridge bonds in **I** as bonds with equal covalent and dative contributions. Thus, even though the structure of Al_2Cl_6 is similar to that of diborane, the bonding in the two molecules is quite different: Diborane is an electron deficient molecule where the central four-membered ring is held together by just two electron pairs, the central ring in Al_2Cl_6 on the other hand, is held together by four electron pairs.

Problem 16.7 Compare the bridging Al–O bond distances in **II** with the average of the terminal (covalent) bond distance and the dative O→Al distance from Table 16.2. Make a rough estimate of a dative Cl→Al bond distance.

16.9 Triple bridges and cage compounds

A monomeric compound of composition RAlNR′ would be referred to as an aluminum imide or an iminoalane. Compounds of such composition are known, but are invariably found to form oligomers, most often tetramers or hexamers. The structures of tetrameric and hexameric HAlN–*i*-Pr, *i*–Pr = *iso*-propyl, are shown in Fig. 16.7. The tetramer forms a cubic cage with Al and N atoms occupying alternating corners. The imino groups are said to be triply bridging and the formulae of the oligomers written as $H_4Al_4(\mu^3\text{–}NC_3H_7)_4$ and $H_6Al_6(\mu^3\text{–}NC_3H_7)_6$ respectively. Each Al or N atom presumably forms a covalent bond to a terminal H atom or *iso*-propyl group respectively. Bonding within the cubic cage may accordingly be described in terms of canonical forms in which each Al or N atom is bonded to its three neighbors in the cube through two covalent and one dative AlN bond. Each bond may thus be described as 2/3 covalent and 1/3 dative.

Problem 16.8 Discuss the bonding in the hexamer in terms of dative and covalent bonding contributions. Is the Al–N bond distance consistent with a description as 1/3 dative and 2/3 normal?

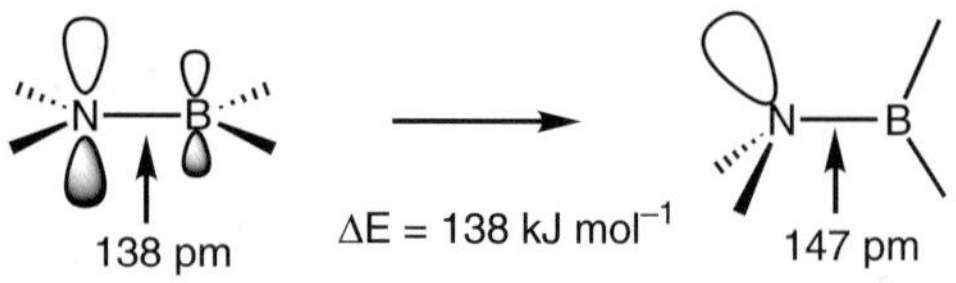

138 pm   ΔE = 138 kJ mol^{-1} 147 pm

Fig. 16.8. Dative π-bonding in aminoborane, H_2NBH_2.

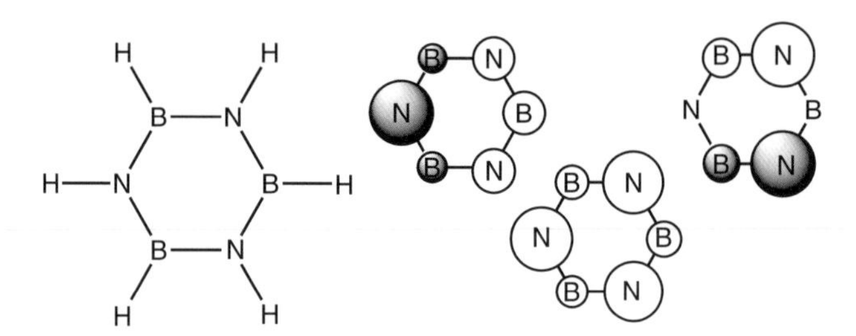

Fig. 16.9. Left: the molecular structure of borazine. Right: the three occupied π-bonding molecular orbitals.

16.10 Dative π-bonding

The aminoborane molecule H_2NBH_2 is isoelectronic with ethene and has a similar planar structure with an NB bond distance of 139 pm, as compared to the single, polar covalent NB bond distance of 146 pm estimated by the MSS rule. A Lewis structure for this molecule in which the N atom is surrounded by three bonding electron pairs and one lone pair, indicates that the coordination geometry should be pyramidal as in NH_3: both the planarity of the N atom and the relatively short NB distance indicate that the NB bond in aminoborane has partial double bond character. Indeed molecular orbital calculations yield a NB π-bonding orbital as indicated in Fig. 16.8. If the π-bond is broken by rotating the BH_2 fragment 90° into an orthogonal orientation, the N atom becomes pyramidal and the NB bond distance increases by 8 pm to 147 pm. Comparison with Fig. 14.2 shows that the energy required to break the dative π-bond in H_2NBH_2 is about half the energy required to break the covalent π-bond in ethene.

Borazine, $N_3B_3H_6$, is isoelectronic and isostructural with benzene. See Section 14.4. The NB bond distance, 144 pm, is intermediate between the bond distances in the planar and orthogonal forms of aminoborane. The bonding may be described in terms of molecular orbitals similar to those of benzene: each N or B atom may be regarded as sp^2 hybridised and to form three $2c, 2e\sigma$-bonds to its nearest neighbors. Six valence electrons occupy the three π-bonding orbitals formed by combination of the six $2p_\pi$ AOs as indicated in the Fig. 16.9. Compare with the corresponding orbitals in benzene, Fig. 14.6.

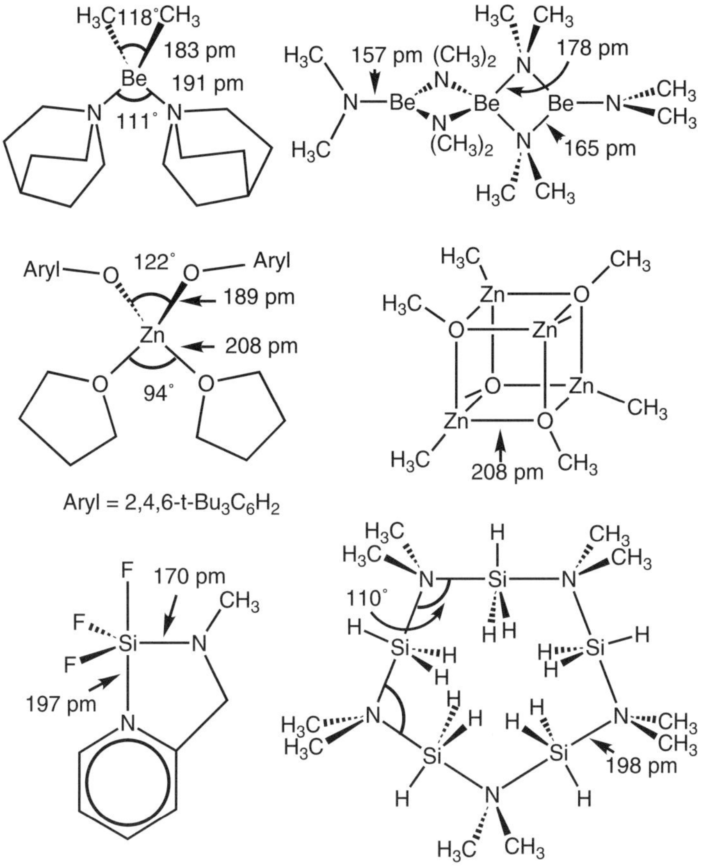

Fig. 16.10. The solid state structures of the complexes $(HC(C_2H_4)_3N)_2Be(CH_3)_2$, $[Be(N(CH_3)_2)_2]_3$, $((CH_2)_4O)_2Zn(OAr)_2$, OAr = phenyl-2,4,6-tri-*tert*-butyl, $[CH_3ZnOCH_3]_4$, $(F_3SiN(CH_3)(CH_2)C_5H_4N$, and $[H_3SiN(CH_3)_2]_5$.

16.11 Some complexes of beryllium, zinc and silicon

The ability to act as electron acceptors is not limited to the elements in Group 13. In Fig. 16.10 we sketch the structures of some complexes of a Group 2 metal (Be), a Group 12 metal (Zn) and a Group 14 metalloid (Si). The reader should be able to classify the bonds as covalent, dative or as bonds with both covalent and dative character, and to examine whether the coordination geometries are consistent with the VSEPR model.

16.12 Electron donor–acceptor complexes with Group 15 acceptor atoms

Both Lewis-valent and hypervalent compounds of the Group 15 elements may react with electron donors to form electron donor–acceptor complexes. Most of these complexes are

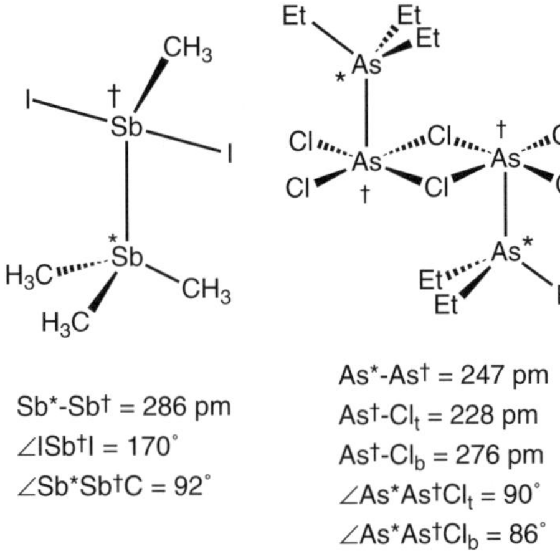

Sb*-Sb† = 286 pm
∠ISb†I = 170°
∠Sb*Sb†C = 92°
As*-As† = 247 pm
As†-Cl_t = 228 pm
As†-Cl_b = 276 pm
∠As*As†Cl_t = 90°
∠As*As†Cl_b = 86°

Fig. 16.11. The molecular structures of the complexes $(CH_3)_3SbSbCH_3I_2$ and $[Et_3AsAsCl_3)]_2$.

Sb-F_a = 181 pm
Sb-F_t = 181 pm
Sb-F_b = 204 pm
F_bSbF_b = 81°
F_tSbF_t = 98°
F_aSbF_a = 163°

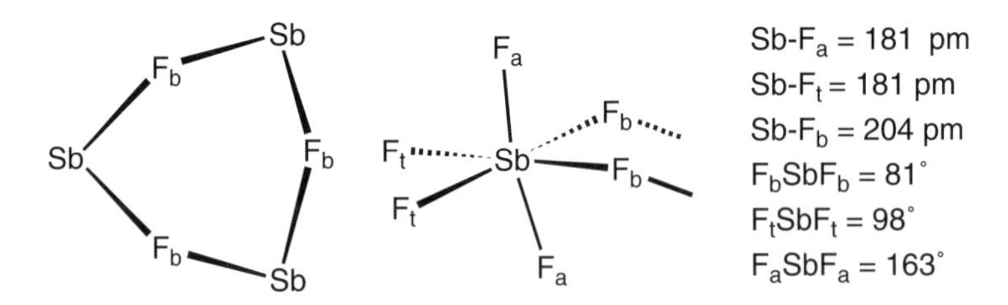

Fig. 16.12. The molecular structure of $(SbF_5)_3$. Left: the central Sb_3F_3 ring. Right: the coordination geometry of the metal atoms.

observed in the crystalline phase since they are too weakly bonded to survive transfer to the gas phase.

Thus crystallization from a solution containing equal amounts of $(CH_3)_3Sb$ and CH_3SbI_2 yields the complex shown in Fig. 16.11. The coordination polyhedron of the Sb* atom is distorted tetrahedral, that of the $Sb^\dagger$ atom a distorted trigonal bipyramid with the iodine atoms in axial positions and a vacant equatorial site.

Crystallization from a solution containing equal amounts of triethyl- and trichloro-arsane (Et_3As and $AsCl_3$), on the other hand, yields the dimeric complex $[Et_3AsAsCl_3)]_2$. See Fig. 16.11. The coordination geometries of the two As* atoms are distorted tetrahedral, the coordination geometries of the $As^\dagger$ atoms distorted square pyramidal. The two $As^\dagger$ atoms and the six chlorine atoms are nearly, but not completely, coplanar.

Problem 16.9 How would you describe the complexes $(CH_3)_3SbSbCH_3I_2$ and $[Et_3AsAsCl_3)]_2$in terms of covalent and dative bonding? Are the coordination geometries and valence angles in agreement with the VSEPR model? Would you expect $AsCl_3$ to be a better electron acceptor than $AsEt_3$? Is the structure consistent with this expectation?

Antimony pentafluoride is tetrameric in the solid phase and trimeric in the gas phase. The trimer consists of a somewhat buckled ring of alternating Sb and bridging fluorine atoms. See Fig. 16.12. The coordination geometries of the Sb atoms are distorted octahedral.

Problem 16.10 How would you describe the bonding in $(SbF_5)_3$ in terms of covalent and dative bonding? Are the coordination geometries and valence angles at the Sb atoms in agreement with with the VSEPR model? How do you explain the wide valence angles at the bridging fluorine atoms?

Problem 16.11 PF_5 forms a solid donor acceptor complex with NH_3. Use the VSEPR model to predict the coordination geometry of the phosphorus atom.

References

[1] A. Haaland, *Angew. Chem. Int. Ed. Engl.*, 28 (1989) 992.
[2] G. Parkin, *J. Chem. Ed.*, 83 (2006) 791.
[3] G. Merino, V. I. Bakmutov, and A. Vela, *J. Phys. Chem.*, A 106 (2002) 8491. See also W. T. Klooster, T. F. Koetzle, P. E. M. Siegbahn, T. B. Richardson, and R. H. Crabtree, *J. Am. Chem. Soc.*, 121 (1999) 6337.
[4] E. N. Guryanova, I. P. Goldshtein, and I. P. Romm, *The Donor–Acceptor Bond*, John Wiley & Sons, New York, 1975.
[5] F. Bessac and G. Frenking, *Inorg. Chem.*, 45 (2006) 6956.

Chapter 17

Structure and bonding in simple compounds of the Group 16 elements

Introduction

The first two elements in Group 16, oxygen and sulfur, have the electron configurations [Ng]ns^2np^4 while the heavier elements, selenium and tellurium, have the electron configurations [Ng]$(n-1)d^{10}ns^2np^4$. All the Group 16 elements form thermally stable compounds of composition EH_2 and $E(CH_3)_2$. Oxygen and sulfur also forms relatively stable dichlorides and difluorides, but the properties of the corresponding dihalides of selenium and tellurium are more complex: SeF_2 decomposes at low temperature and TeF_2 does not seem to exist at all; selenium or tellurium dichlorides are perfectly stable in the gas phase, but *not* in the solid state. (Gaseous $SeCl_2$ may be obtained from solid $SeCl_4$ which decomposes on evaporation to yield $SeCl_2$ and Cl_2, gaseous $TeCl_2$ may be obtained by evaporation from a melt formed from a 1:1 solid mixture of $TeCl_4$ and elemental tellurium.) Condensation of gaseous $TeCl_2$ returns the solid mixture of Te and $TeCl_4$, condensation of $SeCl_2$ yields a mixture of solid $SeCl_4$, solid selenium and a liquid consisting of diselinium dichloride, ClSeSeCl).

The central atom in the molecules of composition EH_2, $E(CH_3)_2$, ECl_2 or EF_2 are presumably surrounded by four electron pairs, two bonding electron pairs and two lone pairs. We refer to Group 16 elements that form two homolytic bonds, and thus are surrounded by an electron octet, as Lewis-valent.

The heavier elements of the group, S, Se and Te all form tetrafluorides, EF_4 and hexafluorides, EF_6. If the central atom in these compounds form Lewis electron pair bonds to all the ligating fluorine atoms, it must accommodate five or six electron pairs in the valence shell, and we refer to the atoms as hypervalent. Sulfur forms no further homoleptic hypervalent derivatives. Selenium forms a solid tetrachloride, but as mentioned in the last paragraph, it decomposes on evaporation. Tellurium forms a solid tetrachloride, which may be evaporated without decomposition, as well as tetraphenyl-, tetramethyl- and hexamethyl- derivatives.

17.1 Compounds of the Lewis-valent elements

The bond distances, mean bond energies and valence angles of the compounds EH_2, $E(CH_3)_2$ and ECl_2 are listed in Table 17.1. The bond distances are also displayed in Fig. 17.1. As usual, bonds become longer as the group is descended from O to Te, while mean bond energies decrease from S to Te. Again the first element in the group, O, displays irregular behavior so long as the MBE in OCl_2 is smaller than in SCl_2. The unexpected weakness of the O–Cl bonds may, at least in part, be due to repulsion between the Cl atoms.

The central atom in each of these Lewis-valent compounds is surrounded by two bond electron pairs and two lone pairs. According to the VSEPR model, the lone pairs require more space at the central atom than the bond pairs. The valence angles are therefore expected

Table 17.1. Gaseous dihydrogen, dimethyl, and dichloro derivatives of the Group 16 elements: E–H, E–C and E–Cl bond distances (in pm); mean bond energies (in kJ mol^{-1}); and ∠HEH, ∠CEC and ∠ClECl valence angles (in degrees).

	EH_2			$E(CH_3)_2$			ECl_2		
E	R	MBE	∠HEH	R	MBE	∠CEC	R	MBE	∠ClECl
O	96	461	104	142	358	112	169	199	111
S	134	364	92	182	297	99	201	266	103
Se	146	319	91	194	249	96	216	253	100
Te	166	272	90	214	220[a]	93	233	–	97

[a] Reference [1]

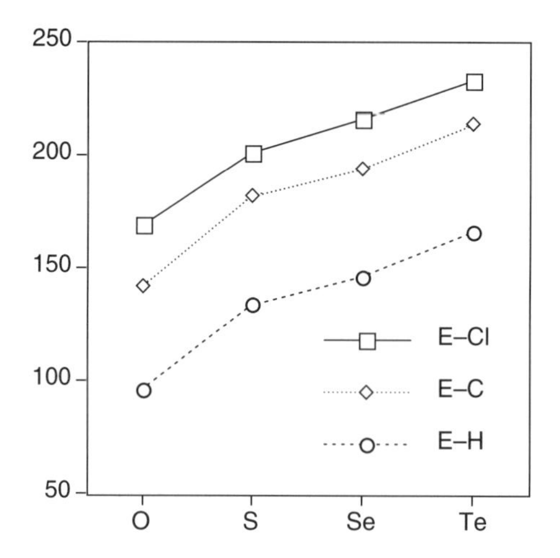

Fig. 17.1. The bond distances in hydrogen, methyl and chlorine derivatives of Lewis-valent Group 16 elements.

to be less than tetrahedral. The valence angles in $O(CH_3)_2$ and OCl_2 are, however, slightly larger than 109.5°. All the other valence angles in the table are less than tetrahedral and decrease as the group is descended.

If the methyl groups in dimethylether are replaced by larger groups, the valence angle increases: thus the SiOSi valence angle in gaseous disilylether, $O(SiH_3)_2$, is ∠SiOSi = 144°. This angle is not only very large, it is also very soft: the energy of the linear configuration is only 1.4 kJ mol^{-1} above that of the equilibrium geometry [2]. The SiOSi valence angle in gaseous $O(SiMe_3)_2$, Me = methyl, is ∠SiOSi = 148°, and in solid $O(SiPh_3)_2$, ∠SiOSi = 180°!

17.2 Across angle radii

When the planar structure of trisilylamine was first discovered, it was assumed that the planar N atom must be engaged in double bonding to the three Si atoms, just as the central sulfur atom in SO_3 is engaged in double bonding to each of the three O atoms. See Section 20.2.

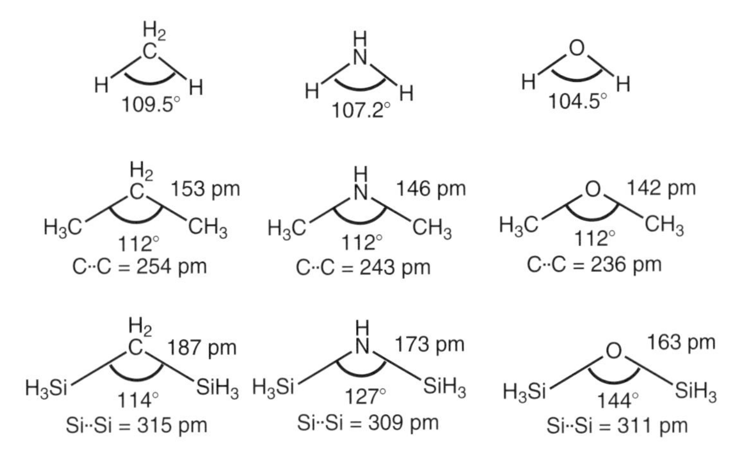

Fig. 17.2. Valence angles, bond distances and across-angle distances in some simple derivatives of C, N and O.

Table 17.2. Across-angle radii $r_<$ (in pm) of some main group elements.

C: 118			
Si: 155	P: 143	S: 141	Cl: 140
Ge: 158	As: 158	Se: 155	Br: 153
Sn	Sb: 173	Te: 170	I: 168

Similarly it was assumed that the wide ∠SiOSi angle in disilylether implied that the O atom is engaged in double bonding to the two Si atoms. This view is, however, not in agreement with the result of high-level quantum chemical calculations. We now believe that these wide angles are due to steric repulsion between the Si atoms. See Fig. 17.2.

The ∠SiESi angles in disilylmethane, disilylamine and disilylether vary from 114 to 144°, while the Si··· Si distance remains relatively constant at 315, 309 and 311 pm respectively. It has therefore been suggested that repulsion between geminal silyl groups becomes prohibitively large if the Si··· Si distance is less than 310 pm. If the central atom is small (C, N, or O), repulsion between silyl groups will override repulsion between valence shell electron pairs, and the ∠SiESi angle will open until the across angle Si··· Si distance is at least 310 pm. Half this critical distance has been defined as the across-angle radius ($r_<$) of a Si atom (carrying only H substituents).

Similarly half the C···C distance in dimethyl ether and half the Ge··· Ge distance in digermyl ether are taken as the across-angle radii of C and Ge atoms respectively. The smallest POC valence angle in gaseous $P(OCH_3)_3$ is equal to 117°. An estimate of the across-angle radius of the P atom is obtained by subtracting the across-angle radius of C from the observed across angle P··· C distance. Across angle radii for these and other elements in Goups 14–17 are listed in Table 17.2[3]

We have seen that the VSEPR model fails to predict the angular equilibrium structure of BaF_2 and other dihalides of the heavier Group 2 metals or the square pyramidal equilibrium structures of $SbPh_5$ and $BiPh_5$. In the next chapter we shall see that the model is of little help for the prediction of structures when the central atom is surrounded by seven electron pairs. By far the larger number of failures of the VSEPR model is, however, ascribed to steric repulsion between large ligating atoms or groups.

The most rational approach to the prediction of coordination geometries is probably to begin with the VSEPR model and then use across-angle radii or other means to see if valence shell electron pair repulsion may be overridden by steric repulsions: The prediction of the unknown structure of oxygen diiodide would begin with the calculation of the non-bonded I···I distance under the assumption that the valence angle is tetrahedral. (Since there is no reference value for the O-I bond distance, it would have to be estimated with the MSS rule). If the I···I distance thus obtained is larger than twice the across-angle radius of I, electron pair repulsion is expected to dominate and reduce the valence angle to some value less than tetrahedral. If the I···I distance obtained is smaller than twice the across-angle radius of the I atom, steric repulsion is expected to prevail and open the valence angle until the distance reaches $2r_<$.

Repulsion between large ligating *groups* as in $O(SiPh_3)_2$ or triisopropylamine and the effect of such repulsion on the molecular structure may be estimated by so-called molecular mechanics calculations.[4]

Problem 17.1 Predict the IOI valence angle in OI_2 and the GeCGe valence angle in digermyl methane, $H_2C(GeH_3)_2$.

Problem 17.2 While the ClOCl angle in OCl_2 is greater than tetrahedral, the ClNCl angle in NCl_3 is found to be 107°. Is this observation consistent with the across-angle radius of N listed in Table 17.2?

Problem 17.3 The ∠SiOC valence angle in methyl(silyl)ether is 120°. Could this angle be determined by across-angle repulsion? Would you expect the coordination geometry of the N atom in methyl(disilyl)amine to be planar? What about dimethyl(silyl)amine? (The former is found to be planar, the second nearly so.)

Problem 17.4 A Si–S bond is distance is expected to be 214 pm, a Ge–S bond distance to be 221 pm. Do you expect the valence angles at S in $S(SiH_3)_2$ or $S(GeH_3)_2$ to be larger than tetrahedral?

17.3 MO description of bonding in the Lewis-valent compounds

In order to describe the Lewis-valent compounds, EH_2, $E(CH_3)_2$ or ECl_2, in terms of $2c$, $2e$ bonds, we must form appropriate hybrid orbitals on the central atom E pointing towards the two ligating atoms. Each of these hybrids may then be combined with an appropriate orbital on the ligating atom to form a $2c$ MO which accommodates two electrons.

The two nonbonding electron pairs may be described in two ways. We may choose to accommodate two electrons in a third hybrid orbital h_3 with maximum electron density in the plane of the molecule corresponding to the h_3 hybrid orbital defined by equation (13.5) and the last two electrons in an uhybridized p_π orbital perpendicular to the molecular plane. The other possibility is to combine h_3 and p_π to form two new equivalent hybrids, h_4 and h_5 pointing in approximately tetrahedral directions:

$$h_4 = (h_3 + p_\pi)/\sqrt{2} \quad \text{and} \quad h_5 = (h_3 - p_\pi)/\sqrt{2}$$

See Fig. 17.3. The *total* electron density calculated for the electron configuration $h_4^2 h_5^2$ is identical to the electron density calculated for the electron configuration $h_3^2 p_\pi^2$: the states

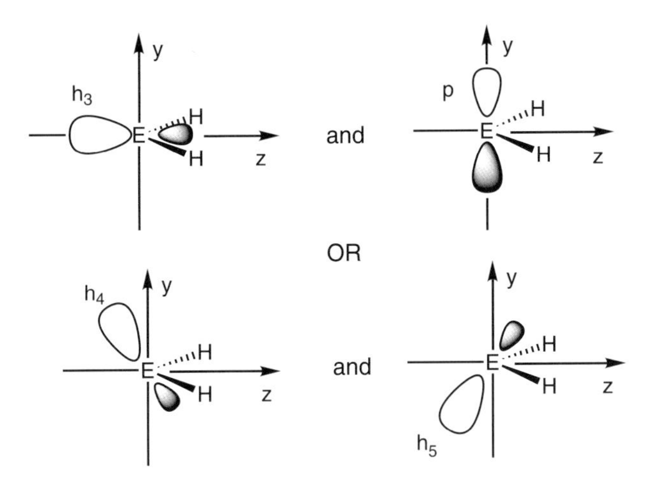

Fig. 17.3. Two alternative descriptions of the AOs containing the two nonbonding electron pairs in EH_2. The molecule is lying in the xz plane and the direction of the z-axis has been chosen to coincide with the twofold symmetry axis.

are in fact identical from quantum mechanical point of view. We may therefore choose one or the other depending on the problem we are discussing. A description in terms of h_4 and h_5 is in accord with the VSEPR model, a description in terms of h_3 and p_π provides a better starting point for the discussion of anomeric effects (see Section 17.6).

17.4 Ions and radicals formed from H_2O

The gas phase structures of the OH radical, the hydroxide ion, and the oxonium ion H_3O^+ have all been determined by spectroscopic methods. Bond distances and valence angles are listed in Table 17.3. The near-constant length of the OH bonds is striking.

H_3O^+ is isoelectronic with the ammonia molecule NH_3 and has a similar trigonal pyramidal shape. The HOH valence angle is slightly larger than the corresponding angle in NH_3 ($\angle$HNH $= 107°$) and the inversion barrier (11 kJ mol^{-1}) in less than half the barrier in ammonia, perhaps because O–H bond distance is about 5 pm shorter than the N–H bond distance.

Problem 17.5 Suggest a molecular orbital description of the bonding in the hydroxide ion.

17.5 The molecular structure of hydrogen peroxide, HOOH

The molecular structure of gaseous HOOH is shown in Fig. 17.4. The molecule has a twofold symmetry axis through the midpoints of the O–O bond and the line connecting the two H atoms. The molecule is nonplanar, the angle between the two OOH planes τ(HOOH) $= 118°$. The O–H bond distance is similar to that in the water molecule, the O–O

Table 17.3. O–H bond distances (in pm), and ∠HOH valence angles (in degrees) of H_2O, OH, OH^- and H_3O^+ in the gas phase.

	R(O-H)	∠HOH
H_2O	95.8	104.5
•OH	97.1	–
OH^-	96.4	–
H_3O^+	97.6	110.7

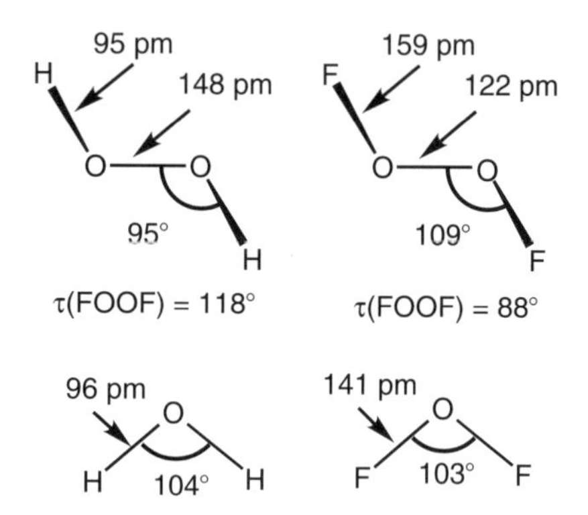

Fig. 17.4. The molecular structures of HOOH and FOOF. Both molecules have a twofold symmetry axis through the midpoint of the O–O bond and perpendicular to the plane of the paper.

distance one or two pm longer than the estimate obtained by the MSS rule. The ∠OOH valence angle is unexceptional.

The structure of HOOH is chiral, the molecule cannot be superposed on its mirror image. The structure can, however, be converted into its mirror image through internal rotation about the O–O bond in such a way that both H atoms first move up from the paper towards the viewer and then down and away. See Fig. 17.4. The energy barrier to the interconversion, i.e. the energy of a planar *syn*-structure with $\tau = 0°$ relative to the equilibrium structure is about 4.5 kJ mol^{-1}. The interconversion by internal rotation in the opposite direction through a planar *anti*-structure with $\tau = 180°$ requires much more energy: the energy of the planar *anti* form is about 30 kJ mol^{-1} above that of the equilibrium structure[5].

The mean O–H bond dissociation energy is hydrogen peroxide, defined as half the standard energy of the reaction

$$\text{HOOH(g)} \rightarrow 2\text{H(g)} + \text{OO(g)} \tag{17.1}$$

is only 284 kJ mol^{-1} as compared to 461 kJ mol^{-1} in the water molecule! A moments reflection should be sufficient to show that this is not a significant comparison: when the two O–H bonds in the peroxide molecule are broken the OO distance is reduced to 121 pm and an O=O double bond is formed. The energy released during the double bond formation reduces energy of reaction (17.1).

What does make sense, is to calculate the energy of atomization

$$HOOH(g) \rightarrow 2H(g) + 2O(g)$$

Then transfer the mean O–H bond energy from the water molecule, and calculate the thermochemical O–O bond energy, TBE(O–O) = 141 kJ mol^{-1} as compared to an O=O double bond dissociation energy of 494 kJ mol^{-1} in O_2.

The analogous dihydrogen derivatives of the heavier Group 16 elements are known. The molecular structures of HSSH and HSeSeH have been shown to be similar to that of HOOH, but the molecular structure of the tellurium analogue has not been determined.

17.6 The peculiar structure of dioxygen difluoride, FOOF: anomeric delocalization

Replacement of the two H atoms in HOOH by F atoms, does not change the shape or symmetry of the molecule. See Fig. 17.4. Note, however, that the O–O bond distance is 26 pm shorter than in HOOH and indistinguishable for the OO double bond distance in the O_2 molecule, while the O–F bond distance is 18 pm longer than in OF_2! Note, also, that the ∠OOF valence angle is 14° smaller than the ∠OOH valence angle in HOOH, and that the dihedral angle τ(XOOX) has been reduced by 30° to a value close to 90°.

The energy of atomization of FOOF is 637 kJ mol^{-1}, that is more than 100 kJ mol^{-1} higher than estimated by combining the mean O–F bond energy in OF_2 (191 kJ mol^{-1}) with the thermochemical O–O bond energy in HOOH (121 kJ mol^{-1}). Taken together, the O–O and O–F bond distances and the observed energy of atomization, suggest that the FOOF molecule has been stabilized by a stronger bond between the two O atoms, though this stabilization is accompanied by a weakening of the two O–F bonds.

Quantum chemical calculations indicate that the barriers to internal rotation in FOOF are much higher than in HOOH: the energies of the planar *syn* and *anti* forms are both calculated to be about 100 kJ mol^{-1} above that of the equilibrium structure[6]. Interestingly the O–O bond distances in the planar forms are similar to the O–O bond distance in HOOH while the O-F bond distances are similar to the O–F bond distance in OF_2.

The unexpected properties of FOOF are best explained by assuming that the equilibrium structure is stabilized by an "anomeris effect", i.e. by a favorable interaction between the p_π electron lone pair on each O atom and the O–F bond on the other. This stabilizing interaction may in turn be rationalized as due to delocalization of the electron lone pair on one O atom into the O–F antibonding orbital on the other O, i.e. that the electron pair occupies a molecular orbital formed by linear combination of the p_π orbital on one O atom with the antibonding O–F σ^*-orbital on the other. See Fig. 17.5. Such conjugation is expected to lead to a reduction of the O–O distance, an elongation of the O–F distance and an opening of the OOF valence angle. The interaction would be at its maximum when the relative orientation of the OF fragments is such that the overlap between the p_π orbital and the O–F antibonding orbital is maximized, i.e. when the dihedral angle τ(FOOF) is 90°, and to vanish when τ(FOOF) = 0 or 180°. The presence of such delocalization has been confirmed by quantum chemical calculations[6].

The results for FOOF may be generalized in the following manner: Consider a molecular fragment AOEX where the O atom forms a single, covalent bond to an atom E from Group

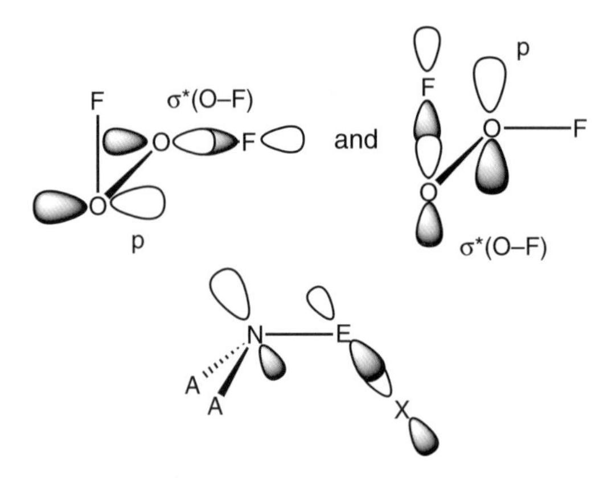

Fig. 17.5. Above: delocalization of the two p_π electron pairs on the oxygen atoms in FOOF into the antibonding O–F σ^*-orbitals. Below: anomeric delocalization of a nonbonding electron pair on a N atom into an antibonding E-X σ^*-orbital.

14, 15 or 16, which in turn forms a single bond to an electronegative element X such as F, Cl, O or N. Structure determination of a large number of molecules containing such fragments show that they tend to adopt structures where the dihedral angle τ(AOEX) is close to 90°, i.e. a structure which allows for maximum interaction between the p_π electrons on the O atom and the E–X antibonding orbital, and that such structures are characterized by unexpectedly short O–E bonds, unexpectedly long E–X bonds and large OEX angles. FOOF is unique in regard to the magnitude of the anomeric effect, but bond distance changes of several picometers and angle changes of several degrees are quite common.

Similar anomeric effects are also observed in molecules containing A_2NEX fragments. The available evidence shows that in such molecules the anomeric stabilization is at its maximum when the electron lone pair orbital on the N atom is *anti* the E–X bond, i.e. when the NEX plane bisects the ANA angle. See Fig. 17.5.

17.7 The molecular structures of compounds of hypervalent Group 14 elements

The properties of homoleptic compounds of hypervalent Group 16 elements are summarized in Table 17.4. Sulfur and selenium are significantly more electronegative than the third and fourth period elements in Group 15 (P and As), and the only homoleptic hypervalent derivatives that are sufficiently stable to be studied in the gas phase are the tetrafluorides EF_4 and the hexafluorides EF_6. Methyl and chloro ligands are presumably not sufficiently electronegative to stabilize hypervalent compounds of these elements. Tellurium, on the other hand, has an electronegativity coefficient between P and As, and $Te(CH_3)_4$, $Te(CH_3)_6$, $TeCl_4$, and TeF_6 are all sufficiently stable to have their structures determined in the gas phase. See Fig. 17.6. The molecular structure of TeF_4 is, however, unknown.

The molecular symmetry of the derivatives of the tetravalent elements is C_{2v}. The shape may be described as derived from a trigonal bipyramid with a missing equatorial ligand. Both axial and equatorial ligands are bent away from the vacant site. The axial bond distances

Table 17.4. Homoleptic derivatives of hypervalent Group 16 elements. Ph = phenyl. The stable phase at room temperature is indicated in parentheses.

	EH_4	EPh_4	$E(CH_3)_4$	$E(CH_3)_6$	ECl_4	EF_4	EF_6
O	–	–	–	–	–	–	–
S	–	–	–	–	–	SF_4(g)	SF_6(g)
Se	–	–	–	–	$SeCl_4$(s)*	SeF_4(l)	SeF_6(g)
Te	–	$TePh_4$(s)	$TeMe_4$(l)	$TeMe_6$(s)	$TeCl_4$(s)	TeF_4(s)	TeF_6(g)

* Decomposes on evaporation.

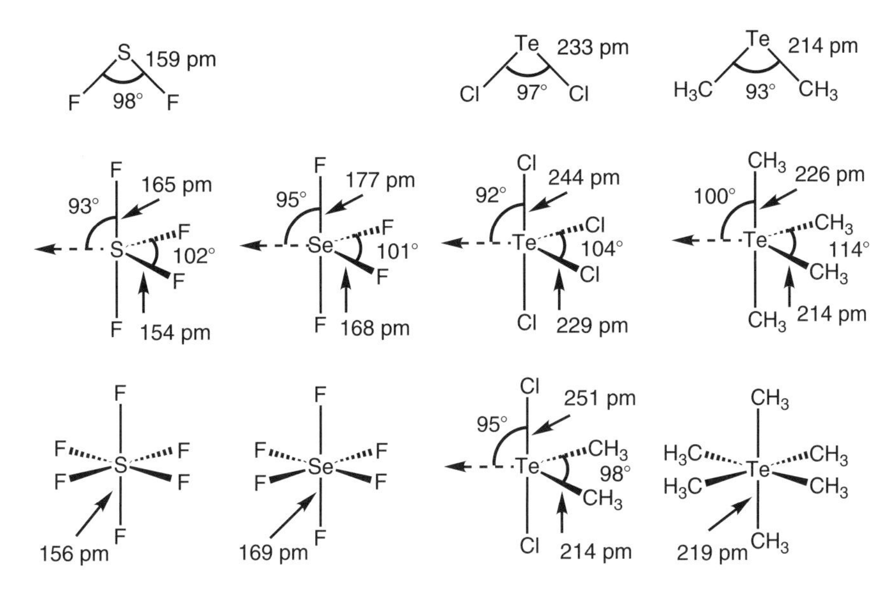

Fig. 17.6. Gas phase molecular structures of compounds of hypervalent Group 16 elements. The twofold symmetry axes of the derivatives of tetravalent elements are indicated by stippled arrows. The structure of TiF_6 (not shown) is octahedral with a Te–F band distance of 182 pm.

are 9 to 15 pm longer than the equatorial. The coordination geometries of the hexavalent elements are octahedral.

We now turn our attention to the variation of bond distances, and since Te–C bonds are expected to be less polar than Te–Cl or E–F bonds, we begin by comparing bond distances in $Te(CH_3)_2$, $Te(CH_3)_4$ and $Te(CH_3)_6$. The equatorial Te–C bond distances in $Te(CH_3)_4$ are indistinguishable from the Te–C bond distances in $Te(CH_3)_2$, the axial bonds are about 12 pm longer. Comparison with the structure of $(CH_3)_2TeCl_2$ shows that the replacement of two axial methyl groups by Cl atoms leaves the equatorial Te–C distances essentially unchanged. The Te–C bond distance in the hexamethyl derivative is 5 pm longer than in the dimethyl, and about 1 pm shorter than the average of axial and equatorial bond distances in the tetramethyl compound.

Since a Cl atom is significantly more electronegative than a C, Te–Cl bonds are expected to be more polar than Te–C bonds, and hence more sensitive to inductive effects. Indeed, comparison of the structures of $(CH_3)_2TeCl_2$ and $TeCl_4$ shows that replacement of two equatorial methyl groups in the former by Cl atoms reduces the axial Te–Cl bond distances by about 7 pm. Similarly, comparison of the structures of $TeCl_2$ and $TeCl_4$ shows that the introduction of two axial Cl atoms where previously there were none, has reduced the (equatorial) Te–Cl bond distances by about 4 pm.

The electronegativity coefficients of S and F indicate that S–F bonds should be more polar than Te–Cl bonds. Comparison of the structures of SF_2 and SF_4 shows that introduction of two axial F atoms where previously there were none, has reduced the (equatorial) S-F bond distance by about 5 pm. The axial S–F bonds are about 11 pm longer than the equatorial. The S–F bond distance in the hexafluoride is found to be 3 pm *shorter* than the bond distance in the difluoride or the average bond distance in the tetrafluoride.

The axial and equatorial substituents in these compounds are symmetry inequivalent, and are therefore expected give separate absorption peaks in NMR spectra. SF_4, SeF_4 and $Te(CH_3)_4$ are, however, all found to be fluxional. Exchange of axial and equatorial ligands probably occurs by Berry pseudorotation with the electron lone pair as "pivotal ligand."

Reliable data for the calculation of bond energies are only available for the series SF_2, SF_4 and SF_6: the MBE of SF_2 defined as half the energy of the reaction

$$SF_2(g) \rightarrow S(g) + 2\,F(g) \qquad\qquad MBE = 363 \text{ kJ mol}^{-1}.$$

The hypervalent mean bond energy in SF_4 is defined as half the energy of the reaction

$$SF_4(g) \rightarrow SF_2(g) + 2\,F(g) \qquad\qquad HMBE1 = 310 \text{ kJ mol}^{-1},$$

and the second HMBE defined as half the energy of the reaction

$$SF_6(g) \rightarrow SF_4(g) + 2\,F(g) \qquad\qquad HMBE2 = 305 \text{ kJ mol}^{-1}.$$

The hypervalent bonds are thus significantly weaker than the bonds in the Lewis-valent compound, but the difference in bond strength is much smaller than observed for compounds of the Group 15 elements.

17.8 Compounds of hypervalent Group 16 elements: model considerations

A tetravalent Group 16 atom is presumably surrounded by five electron pairs, viz. four bond pairs and one non-bonding electron pair. According to the VSEPR model the five electron pairs should arrange themselves in a trigonal bipyramidal manner. Since the non-bonding electron pair requires more space around the central atom than the bond pairs, it is expected to occupy an equatorial position, and deform the molecule in such a way that the bond electron pairs are pushed away. This is indeed the kind of distortion that is observed. The difference between axial and equatorial bond distances may be explained in the same manner as for PF_5 and the other compounds of hypervalent Group 15 elements.

A hexavalent Group 16 atom is presumably surrounded by six bond electron pairs. The octahedral coordination geometry is in accord with the VSEPR model. The observation that the bond distances in the hexavalent compounds are 8–10 pm shorter than the axial bonds in

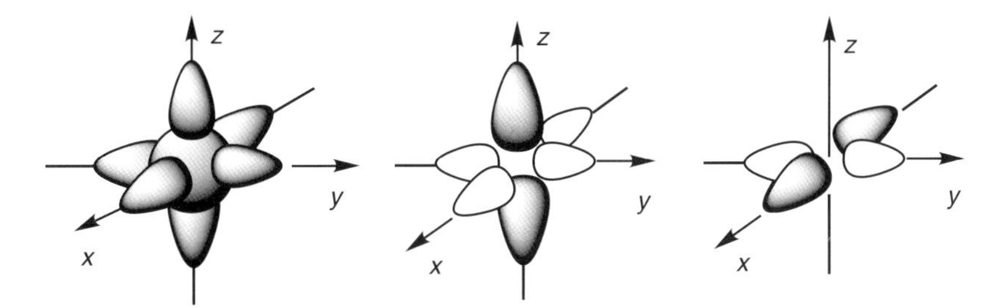

Fig. 17.7. Left: a seven-center bonding molecular orbital in $Te(CH_3)_6$ or SF_6 formed by combination of the valence shell *s* orbital of the central atom with a suitable AO on each of the six ligating atoms. Center and right: two degenerate, non-bonding molecular orbitals formed by linear combination of atomic orbitals on the ligating atoms.

the tetravalent compounds is not easily rationalized in terms of the VSEPR model: since the bond electrons in the octahedron are repelled by four other pairs over an angle of 90°, and the axial bonds in the *tbp* by two bond pairs and one lone pair, the bonds in the octahedron should, if anything, be the longer.

We now turn our attention to molecular orbital descriptions. The compounds of tetravalent elements may be described in terms of two $2c, 2e$ bonds to the equatorial ligands. Such a description is in agreement with the observation that the equatorial bond distances in $Te(CH_3)_4$ is equal to the bond distances in $Te(CH_3)_2$. The non-bonding electron pair occupies a third hybrid atomic orbital in the equatorial plane. The axial bonds, like the axial bonds in the pentavalent Group 15 element compounds, be described in terms of a $3c, 4e$ bonding system. The observation that the axial bonds are longer than the equatorial is, of course, in accord with the molecular orbital description of the bonding.

The description of bonding in compounds like $Te(CH_3)_6$ or SF_6 is less straight-forward: an adequate description in terms of molecular orbitals is only possible if one includes the *seven center* bonding orbital obtained by combination of the valence shell *s* orbital on the central atom with an appropriate atomic orbital on each ligating C or F atom. See Fig. 17.7. This molecular orbital, which provides bonding between the central atom and all of the six ligands, contains two electrons. Six electrons occupy the three $3c$ bonding molecular orbitals obtained by combination of each of the three valence shell *p* orbitals on the central atom with the appropriate atomic orbitals on two C or F atoms as indicated by the orbital Ψ_b in Fig. 15.10. Finally four electrons are accommodated in two degenerate, nonbonding molecular orbitals obtained by combination of AOs on the ligating C or F atoms. Se Fig. 17.7. Though the Lewis structures of $Te(CH_3)_6$ or SF_6 suggest that the central atom is surrounded by six electron pairs, this picture is again modified at the MO level: two of the electron pairs occupy non-bonding MOs which confine these electrons to the ligands.

The observed variation of Te–C bond distances in the series $Te(CH_3)_2$, $Te(CH_3)_4$ and $Te(CH_3)_6$ is entirely consistent with the MO description of the bonding. The observed variation of S–F bond distances in the series SF_2, SF_4, and SF_6, on the other hand, suggests that inductive effects leads to progressive shortening of the bonds with increasing number of attached F atoms.

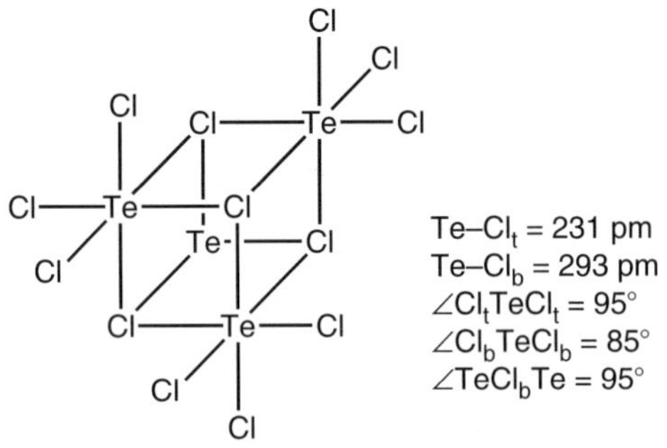

Te–Cl_t = 231 pm
Te–Cl_b = 293 pm
$\angle Cl_tTeCl_t$ = 95°
$\angle Cl_bTeCl_b$ = 85°
$\angle TeCl_bTe$ = 95°

Fig. 17.8. The molecular structure of $Cl_{12}Te_4(\mu^3\text{-}Cl)_4$. The terminal Cl atoms on one of the Te atoms have been omitted for clarity.

17.9 The solid state structure of tellurium tetrachloride

$TeCl_4$ is tetrameric in the solid phase. The structure of the tetramer, $Cl_{12}Te_4(\mu^3\text{-}Cl)_4$, has been determined by X-ray diffraction. See Fig. 17.8. The inner part of the molecular unit consists of four tellurium and four triply bridging chlorine atoms at alternating corners of a distorted cube. Each Te atom is surrounded by three terminal and three bridging Cl atoms at the corners of a distorted octahedron. The terminal Te–Cl bond distance is similar to the bond distance in $TeCl_2$ (233 pm) and to the equatorial bond distance in monomeric $TeCl_4$ (229 pm), the bridging Te–Cl bond distance is 62 pm longer.

The structure of crystalline $SeCl_4$ is similar to that of $TeCl_4$. The latter evaporates as a monomer, while $SeCl_4$ decomposes to the dichloride and Cl_2.

The terminal Te–Cl bonds may be regarded as normal $2c, 2e$ single bonds, the bridging bonds as 1/3 normal and 2/3 dative. This description implies that each bridging Cl atom is surrounded by four electron pairs; i.e. one non-bonding pair and three bonding pairs with mixed dative/covalent character. Since the nonbonding pair requires more space than each of the bonding pairs, such a description is consistent with the observed $TeCl_bTe$ angle of 95°.

The description of Te–Cl_b bonding as 1/3 normal and 2/3 dative implies that each Te atom is surrounded by seven electron pairs; one nonbonding electron pair, three normal bond pairs and three bond pairs with partial dative character. In the next chapter we shall discuss the application of the VSEPR model to compounds with seven electron pairs in the valence shell of the central atom. In the case of tetrameric $TeCl_4$ or $SeCl_4$ the lone pair does not seem to have any effect on the structure; if the lone pair is disregarded the six bond pairs would be expected to arrange themselves in octahedral directions around the Te atom, and since normal bond pairs are expected to require more space than the bond pairs with partial dative character, the $\angle Cl_tTeCl_t$ valence angles would be expected to be greater than $\angle Cl_bTeCl_b$ angles as observed.

At the MO level, the "stereochemical inactivity" of the lone pair may be rationalized if it is assumed to occupy the spherically symmetrical $5s$ atomic orbital, while each nearly linear Cl_tTeCl_b fragment is described as a $3c, 4e$ bonding system.

References

[1] V. I. Tel'noi and M.S. Sheiman, *Russ. Chem. Revs.*, 64 (1995) 309.
[2] J. Koput and A. Wierbicki, *J. Mol. Spectrosc.*, 99 (1983) 116.
[3] The across-angle radius of As has been estimated manner from the shortest As···C distance in $As(NMe_2)_3$ (spanning an AsNC valence angle of 115°), that of Sb from the Sb···Sb distance in $O(SbMe_2)_2$ ($\angle$SbOSb = 122°). The across-angle radius of S is estimated from the across-angle S··C distance in $S(OCH_3)_2$ ($\angle$SOC = 116°). $r_<$ for Cl and Br have been estimated from the X···X distances in OCl_2 ($\angle$ClOCl = 111°) and OBr_2 ($\angle$BrOBr = 112°). The remaining across angle radii listed in the table have been estimated by interpolation.
[4] See, for example N. A. Allinger, Molecular Mechanics, in *Accurate Molecular Structures*, A. Dominicano and I. Hargittai, Eds., International Union of Crystallography, Oxford University Press, 1992.
[5] J. M. Flaud, C, Camy-Peyret, J. W. C. Johns, and B. Carli, *J. Chem. Phys.*, 91 (1989) 1504.
[6] E. Kraka, Y. He, and D. Cremer, *J. Phys. Chem.*, 105 (2001) 3269. See also D. Feller and D. A. Dixon, *J. Phys. Chem.*, 107 (2003) 9641.

Chapter 18

Structure and bonding in simple compounds of the Group 17 elements

Introduction

The first two elements in Group 17, fluorine and chlorine, have the electron configurations $[Ng]ns^2np^5$ while the heavier elements, bromine and iodine, have the electron configurations $[Ng](n-1)d^{10}ns^2np^5$. All of them form stable diatomic molecules. F_2 and Cl_2 are gaseous at room temperature, Br_2 is a liquid and I_2 a solid. We have discussed the bond energies, bond distances and force constants of the dihalogen molecules F_2, Cl_2, Br_2 and I_2 in Chapter 4, the molecular orbitals of F_2 were described Chapter 9. The ionic bonding in monomeric gaseous alkali metal halides was described in Chapter 5, and the properties of the hydrogen halides and of diatomic interhalogen compounds in Chapter 6. In later chapters we have discussed the structures and bonding in gaseous dichlorides of the Group 2 and 12 metals, in the gaseous Lewis-valent and subvalent chlorides of the Group 13 and 14 elements, and in the Lewis-valent and hypervalent chlorides and fluorides of the elements in Groups 15 and 16. We begin this chapter with a short discussion of C–X bond distances and bond dissociation energies of the methyl halides, CH_3X, and then go on to describe some hypervalent derivatives of chlorine, bromine and iodine.

Cl_2, Br_2, I_2 and dihalogen compounds XX' react with electron donors like amines or ethers to form complexes where the dihalogen molecules act as electron acceptors. The structure of some such complexes will be discussed in Section 18.6. Similarly the hydrogen halides HX react with electron donors to form hydrogen-bonded complexes, and the structures of some of them will be described in Section 18.7. The chapter ends with descriptions of the structures of the hydrogen-bonded H_2O dimer in the gas phase, of solid ice and liquid water, and a brief account of the "polywater" episode.

18.1 Compounds of Lewis-valent halogens

In Table 18.1 we list the C–X bond distances and dissociation energies of the gaseous methyl halides as well as bond distances and bond dissociation energies of the gaseous HX and XCl diatomic molecules. Bond energies decrease in the expected way as the group is descended distances increase in an approximately linear fashion which bond. See Fig. 18.1. Comparison with similar plot of bond distances to Group 13 elements (Fig. 11.1) shows that the effect of the d-block contraction between the third and fourth period elements has been greatly reduced.

Table 18.1. Gaseous hydrogen halides, methyl halides or dihalogens XCl : X–H, X–C and X–Cl bond distances (in pm) and bond dissociation energies (in kJ mol^{-1}).

	HX		CH_3X		XCl	
X	*R*	*D*	*R*C–X)	*D*(C–X)	*R*	*D*
F	92	568	139	459	163	249
Cl	128	429	179	342	199	240
Br	141	364	193	287	214	216
I	161	296	213	232	232	208

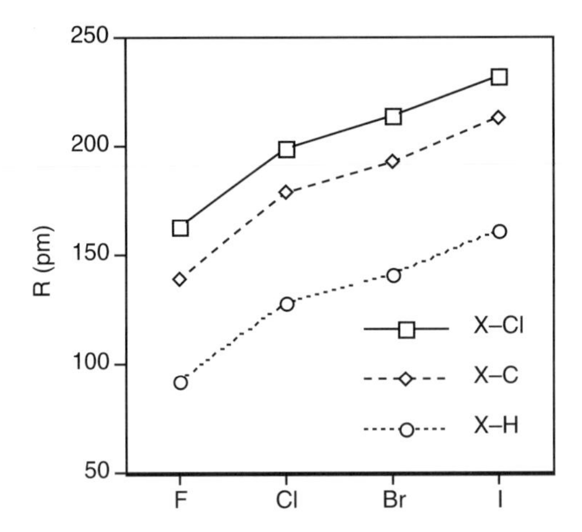

Fig. 18.1. Bond distances in hydrogen, methyl and chlorine derivatives of Lewis-valent Group 17 elements.

18.2 Compounds of hypervalent halogens

As we have seen, the second-period element, fluorine, induces hypervalency in many other *p*-block elements, but it does not itself display hypervalency. The third and fourth period halogens, i.e. chlorine and bromine, form trifluorides and pentafluorides, XF_3 and XF_5. The fifth-period element, iodine, forms an unstable trifluoride which decomposes below room temperature, a pentafluoride and a heptafluoride. Iodine also forms a solid trichloride which decomposes on evaporation: as in Groups 15 and 16 we find that the greater number of hypervalent derivatives are formed by the fifth-period element.

The gas phase structures of the fluorides are shown in Fig. 18.2. ClF_3 and BrF_3 are planar T-shaped molecules. Alternatively their shape may be described as a trigonal bipyramid with two equatorial ligands missing. In each of the two trifluorides the "equatorial" X–F bond is shorter than in the bond in the corresponding monofluoride, while the two "axial" bonds are significantly longer.

The three pentafluorides are all square pyramidal with C_{4v} symmetry. Note that the valence angles spanned by the apical and one of the basal F atoms, are less than 90°. The

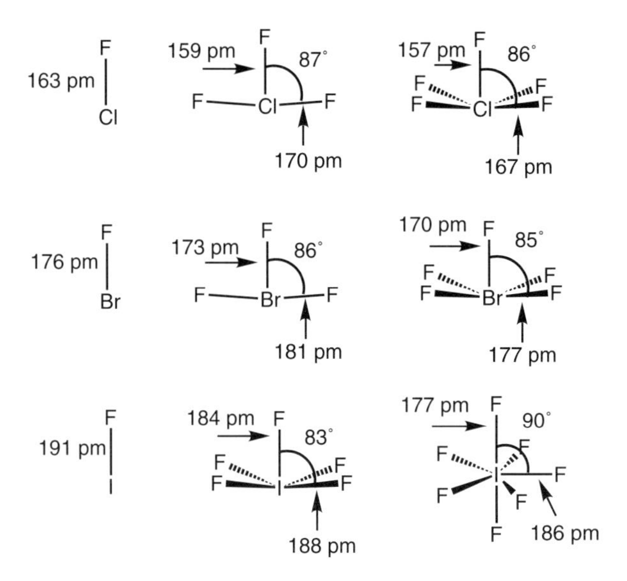

Fig. 18.2. Gas phase molecular structures of compounds of hypervalent Group 17 elements.

Table 18.2. The mean bond energies (in kJ mol^{-1}) of Lewis-valent and hypervalent chlorine, bromine and iodine fluorides.

X	XF	XF_3	XF_5	XF_7
Cl	249	170	–	–
Br	247	200	185	–
I	279	–	266	230

apical X–F bond is considerably shorter than in the corresponding monofluoride, while the four basal bonds are slightly longer.

Iodine heptafluoride forms a pentagonal bipyramid with five fluorine atoms in the equatorial plane [1,2]. The valence angles spanned by neighboring equatorial bonds are only 72°. The two axial bonds are longer than the five equatorial, all of them are shorter than the bond in the monofluoride.

Accurate enthalpies of formation for calculation of X-F bond energies are available for all the compounds in Fig. 18.2 except ClF_5. See Table 18.2. Somewhat surprisingly, the bond in IF is significantly stronger than the bonds in ClF or BrF. Part of the explanation may be sought in the fact that the I–F molecule is more polar. For each of the three elements the mean bond energies indicate that the X–F bonds become progressively weaker with increasing valency of X.

18.3 Compounds of hypervalent halogens: model considerations

If the hypervalent halogen atoms are assumed to form a 2*c*,2*e* bond to each ligating atom, the central atoms in ClF_3 and BrF_3 are surrounded by 10 valence electrons, viz. three bond pairs and two lone pairs. The central atoms in XF_5, X = Cl, Br or I, are surrounded by 12 electrons, viz. five bond pairs and one lone pair. Finally the iodine atom in IF_7 is surrounded by 14 electrons, or seven bond pairs.

According to the VSEPR model the five electron pairs in ClF_3 or BrF_3 should arrange themselves in a trigonal bipyramidal manner with the two lone pairs in equatorial positions. Since the electron lone pairs require more space around the central atom they are expected to deform the molecule is such a way that the valence angles spanned by one axial and one equatorial F atom are reduced to values below 90°. This is indeed what is observed.

The equatorial bonds in the trifluorides may be may be described as 2c,2e bonds, and the two axial bonds as a 3*c*,4*e* bonding system. This description is consistent with the observation that the equatorial bonds are shorter than the axial.

Problem 18.1 Draw a Lewis structure of the pentafluorides and discuss structure and bonding in terms of the VSEPR and MO models.

The pentagonal bipyramidal geometry in IF_7 is difficult to rationalize in terms of the VSEPR model since the smallest FIF valence angles could be larger than 72° if the coordination polyhedron was a distorted capped octahedron [3]. See Fig. 18.3.

Problem 18.2 Could the difference between axial and equatorial bond distances in IF_7 be rationalized in terms of the VSEPR model?

The observation that the axial bonds are shorter than the equatorial suggests an MO description in terms of sp^1 hybridization of the iodine atom and the formation of 2*c*,2*e* bonds to the axial ligands. This leaves 10 electrons for bonding in the equatorial plane. Four of these electrons occupy two degenerate six- or five-center bonding orbitals formed by combination of the valence shell p_x and p_y orbitals on iodine with the appropriate *p*orbitals on equatorial F atoms. See Fig. 18.4. Three electron pairs are left to occupy non-bonding orbitals formed by combination of *p* orbitals on the equatorial F atoms.

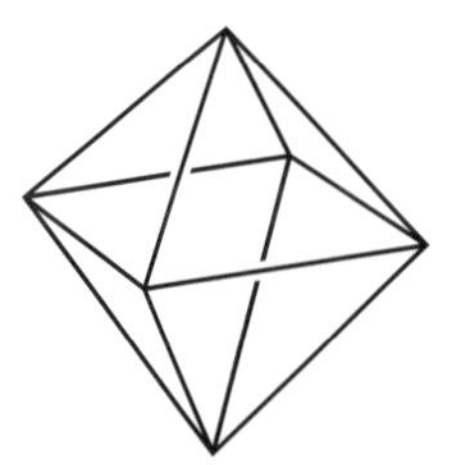

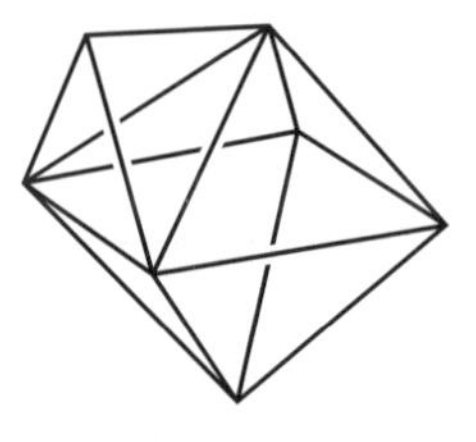

Fig. 18.3. Left: an octahedron, right: a capped octahedron. The symmetry of a capped octahedron is C_{3v} with the symmetry axis running through the capping atom and the center of the octahedron. A more realistic geometric model for an atom surrounded by seven electron pairs may be obtained by allowing the capped octahedron to relax in such a manner that the smallest valence angles are maximized. The smallest valence angles in the model thus obtained are larger longer than in a pentagonal bipyramid.

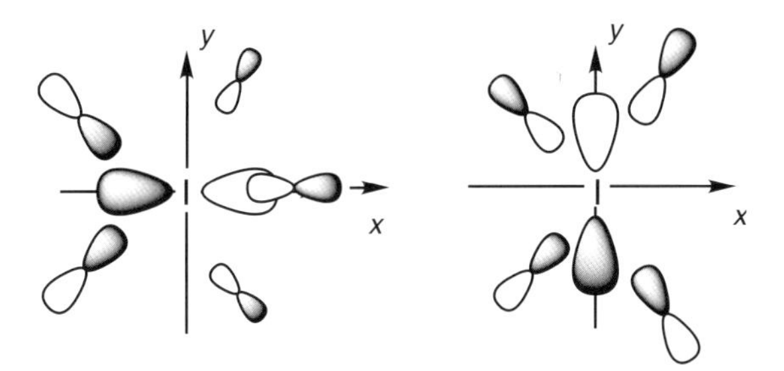

Fig. 18.4. The six- and five-center bonding orbitals in the equatorial plane of IF_7.

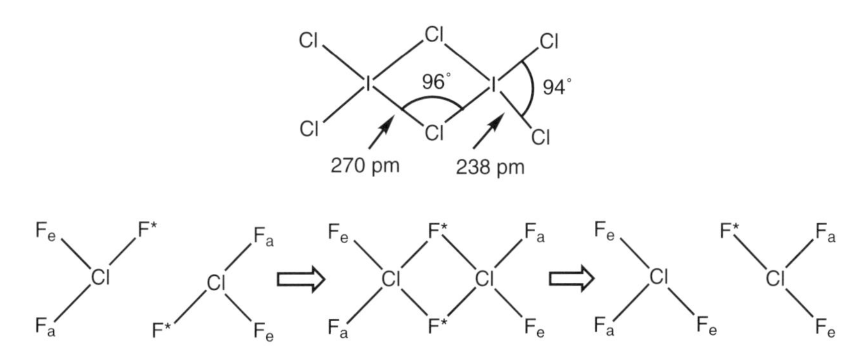

Fig. 18.5. Above: the structure of I_2Cl_6. Below: exhange of F atoms between axial and equatorial positions in monomeric ClF_3.

The reader may note that all the so-called hypervalent compounds that have been mentioned in this or earlier chapters, have been described in such a way that the four AOs in the valence shell of the central atom are combined by appropriate ligand AOs to form *four* bonding molecular orbitals. One or more of these bonding MOs extend over more than two atoms, and have lower orbital energies than two-center orbitals. The remaining electrons are lodged in non-bonding MOs on the ligands.

18.4 The crystal structure of iodine trichloride

ICl_3 is dimeric in the solid phase as indicated in Fig. 18.5. Like the dimer of aluminum trichloride the molecule contains two bridging and four terminal Cl atoms, and the bridging I–Cl distances are considerably longer than the terminal. Note, however, that the I_2Cl_6 molecule is *planar*. As in the case of Al_2Cl_6, the bonding in the four-membered ring may be rationalized in terms two canonical forms each with two covalent and two dative bonds. This means that each iodine atom is surrounded by twelve electrons, viz. three covalent bond electron pairs, one dative bond electron pair, and two non-bonding electron pairs. According to the VSEPR model their domains will be directed towards the corners of an octahedron. Since the molecule is found to be planar, the two nonbonding electron pairs must occupy domains above and below the plane of the paper.

18.5 ^{19}F NMR spectra of ClF_3 and BrF_3

The ^{19}F NMR spectra of ClF_3 or BrF_3 consist of one peak only, showing that the F atoms must undergo rapid exchange between axial and equatorial positions. The exchange probably occurs through formation of short-lived dimers as indicated in Fig. 18.5. One axial fluorine atom from each monomeric unit (denoted by F* in the figure) enters a bridging position in the dimer. The other axial fluorine atom F_a and the equatorial atom F_e enter terminal positions. The dimer may dissociate to yield the original monomers, but as indicated in the figure, it may also split in such a manner that the two F* atoms have been transferred from an axial position of one Cl atom to an axial position on the other. The other axial F_a atom remains bonded to the same Cl atom, but at the end of the process it is found to occupy an equatorial position. Similarly the F_e atom is found to occupy an axial site at the same Cl atom as before.

18.6 Some electron donor–acceptor complexes of dihalogens

Cl_2, Br_2, I_2, and some interhalogen compounds XX' form complexes with ammonia that are sufficiently stable to allow the structures to be determined in the gas phase. The structure of the complex H_3NICl is shown in Fig. 18.6 [4]. The ICl molecule is situated on the threefold symmetry axis of the NH_3 molecule, i.e. in the direction of the electron lone pair on the N atom. The N–I distance is 271 pm, 71 pm longer than the estimated single, covalent bond distance obtained from the MSS rule, but still 82 pm shorter than the sum of the van der Waals radii of N and I. The experimental information needed for calculation of the dissociation energy of the complex is unfortunately not available, quantum chemical calculations have yielded dissociation energies ranging from 42 to 52 kJ mol^{-1} [5,6].

The structure of H_3NICl is consistent with the formation of a weak dative bond N→I. This description implies that the iodine atom in the complex is surrounded by one covalent bond pair, one dative bond pair, and three lone pairs. According to the VSEPR model the five electron pairs will be directed towards the corners of a trigonal bipyramid, and since the three non-bonding electron pairs require more space, they will occupy equatorial positions.

The structure of the complex HFClF (see Fig. 18.6) is consistent with the formation of an dative bond corresponding to HF→ClF [7]. Quantum chemical calculations have yielded dissociation energies ranging from 6 to 10 kJ mol^{-1} [8]. The dative F→Cl bond is nearly

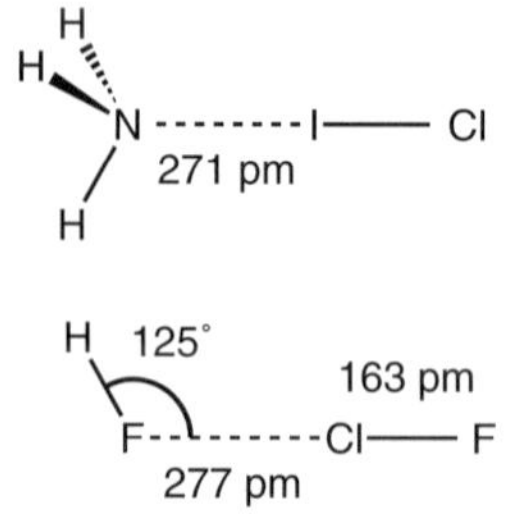

Fig. 18.6. The structures of two gaseous dihalogen complexes.

100 pm longer than the covalent Cl–F bond. The coordination geometry of the acceptor atom (Cl) is linear as in $H_3N{\rightarrow}ICl$, while the coordination geometry of the donor atom (F) is angular. The donor atom is surrounded by one covalent bond electron pair, one dative bond electron pair, and two non-bonding pairs. The observed coordination geometry is in agreement with the VSEPR model in so far as it is angular, but in disagreement in so far as the valence angle is larger than tetrahedral.

The conclusions reached on the basis of the two complexes H_3NICl and HFClF may be generalized: if a monovalent halogen atom acts as an electron donor, its coordination geometry is angular; if the halogen atom acts as an electron acceptor, the coordination geometry is linear [9].

In Fig. 18.7 we show a section of the structure of the most stable crystal modification of ICl [10]. This rather intricate structure contains two types of ICl molecules. Half the molecules which we denote by $I^{\dagger}$–$Cl^{\dagger}$, are characterized by a bond distance 3 pm longer than in the gas phase, the other half, which we denote by I^*–C^*, are characterized by a bond distances that is 13 pm longer. The molecules are joined by weak bonds to form planar zigzag $\cdots I^*\text{-}Cl^*\cdots I^{\dagger}\cdots I^*\text{–}Cl^*\cdots I^{\dagger}\cdots$ chains running through the crystal.

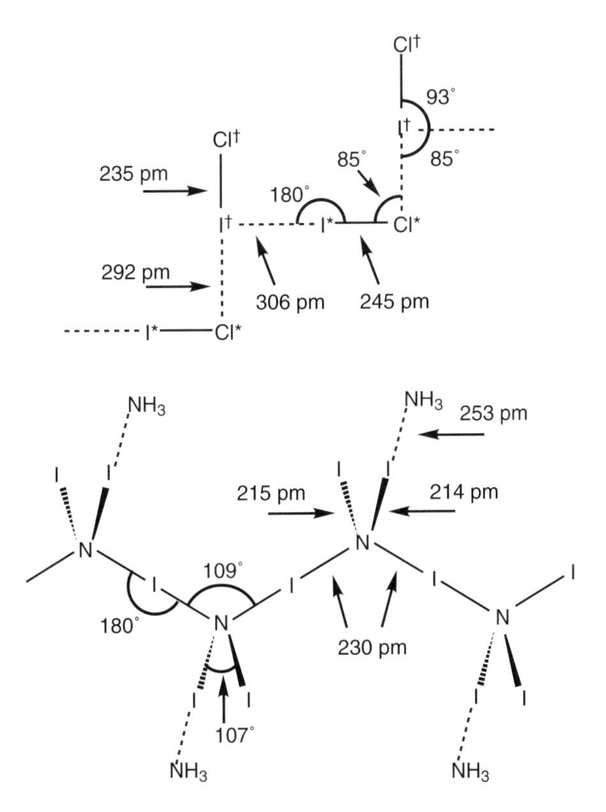

Fig. 18.7. Fragments of the crystal structures of α-ICl and NH_3NI_3. Interatomic distances that are more than 30 pm longer than single bond distances between the same elements and but substantially shorter than the sum of their van der Waals radii are indicated by stippled lines.

Both atoms in the I*–Cl* molecules are incorporated in the chain, in I†–Cl† molecules only the iodine atom is incorporated. The Cl*···I† distance in the chain is 60 pm longer than the covalent ICl bond distance in the gaseous molecule, but 80 pm shorter than the sum of the van der Waals radii. The I†···I* distance in the chain is 40 pm longer than in gaseous I_2, but nearly 100 pm shorter than the sum of the van der Waals radii of the two atoms.

The structure is easily interpreted in terms of dative Cl*→I† and I†→I* bonding. The angular coordination geometry of Cl* and the linear coordination geometry of I* are clearly consistent with such a description. The I† atom is surrounded by five electron pairs: two nonbonding electron pairs and the I†→I* dative electron pair occupy equatorial positions, the covalent bond pair and the Cl*→I† dative electron pair occupy equatorial positions. Remember that a dative bond electron pair is expected to occupy more space than a covalent bond electron pair at the donor atom, but less space than a covalent bond electron pair at the acceptor. See Section 16.6.

Nitrogen triiodide, NI_3, is a light-, temperature-, and extremely shock-sensitive substance. No wonder then, that it has never been isolated. Mixed crystals of composition $NI_3 \cdot NH_3$ have, however, been prepared, and a fragment of the crystal structure is shown in Fig. 18.8 [11]. The nitrogen triiodide molecules form an infinite chain of alternating N and bridging iodine atoms, $[I_2N\text{-}\mu\text{-}I]_\infty$. Each nitrogen atom in the chain is surrounded by two bridging and two terminal iodine atoms at the corners of a distorted tetrahedron. The coordination geometry of the bridging iodine atoms is linear. The ammonia molecules are coordinated to one terminal iodine atom at each N atom in the chain, with a N→I distance of 253 pm.

Problem 18.3 Rationalize the coordination geometries and N-I bond distances in solid $NI_3 \cdot NH_3$ in terms of covalent bonds, dative bonds and bonds of 50% covalent and 50% dative character.

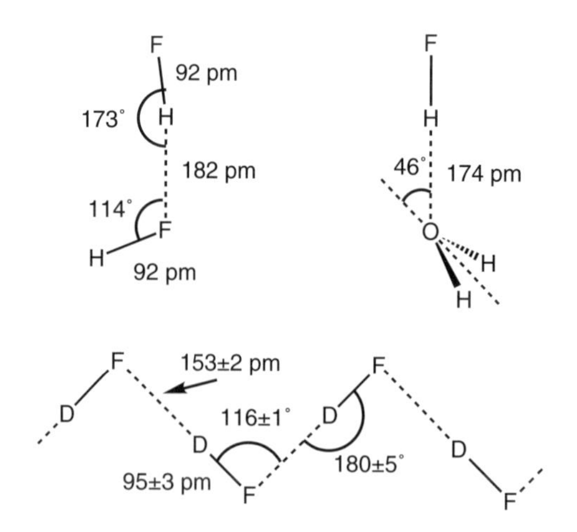

Fig. 18.8. Above: the "recommended" molecular structures of the gaseous hydrogen-bonded complexes $(HF)_2$ and H_2OHF in the gas phase. Below: the crystal structure of deuterium fluoride at 4.2 K.

18.7 Hydrogen bonded complexes

Hydrogen fluoride vapors contain monomeric HF molecules, dimers $(HF)_2$ and polymeric species, $(HF)_n$, with n varying from 3 to 8. The structures of the polymers remain unknown, but the structure of the dimer has been determined by spectroscopic measurements and by high-level quantum chemical calculations. The structure indicated in Fig. 18.8 represents a "recommended" structure based on information from both experiments and calculations [12]. The complex consists of two monomeric units with essentially the same H–F bond distance as in an isolated monomer. The monomers are joined through a H···F bond which is twice as long as bond distance in the monomer, but nearly 100 pm shorter than the sum of the van der Waals radii of H and F (Table 8.4). Note the similarity to the structure of HFICl (Fig. 18.6). The dissociation energy of the dimer at zero kelvin is $D_0 = 13$ kJ mol^{-1} [13].

The attraction between an "acidic H-atom" (a H-atom which is believed to carry a net positive charge) in one molecule and an atom with a pair of nonbonding electrons in another, is referred to as a "hydrogen bond." There is no doubt that the hydrogen bond is significantly stabilized by Coulomb attraction between the electron lone pair and the acidic hydrogen atom, but there is also ample evidence for transfer of negative charge between the two molecules, particularly in strongly hydrogen-bonded complexes. Hydrogen bonds have traditionally been described as a bond type of its own, and the two molecules have been referred to as the proton donor and acceptor respectively. We shall prefer to describe them as weak electron donor–acceptor complexes similar to the dihalogen complexes described in the preceding section [14].

Problem 18.4 Use the VSEPR model to rationalize the structure of $(HF)_2$.

Crystal structures are normally determined by X-ray diffraction. The X-rays are diffracted by the electrons in each atom, and the coordinates of the atom determined as the center of gravity of the electron cloud. The X-ray scattering power of an atom increases with the number of electrons. This means that the position of the center of gravity of the electron cloud of a hydrogen atom will be determined with less accuracy than that of a heavier atom. It is well known, moreover, that in the case of H atoms, polarization of the electron cloud may lead to discrepancies of several picometers between the center of gravity of the negative charge and the position of the nucleus. More accurate determinations of the positions of hydrogen atoms are obtained by neutron diffraction. Since the neutrons are scattered by the nuclei, the experiment leads directly to nuclear positions. Since the neutron scattering power of a deuteron is greater than that of a proton, the best results are obtained using deuterated samples.

The crystal structure of deuterium fluoride determined by neutron diffraction at 4.2 K is shown if Fig. 18.8 [15]. The DF molecules in the crystal form infinite chains which are approximately linear at the hydrogen and angular at the fluorine atoms. The hydrogen bond distance, F→D = 153 ± 2 pm, is 30 pm shorter than in the gaseous HF dimer. The shortening may be interpreted as a synergetic effect: when a dative F→D bond is formed, part of the negative charge transferred from the donor to the acceptor atom is passed on to the substituent on of the latter, i.e. to the next F atom in the chain. This atom thus acquires an additional negative charge, which increases its donor capability and enables it to form a stronger dative bond. When doing so, it withdraws more negative charge from the hydrogen atom, which in turn becomes a better acceptor.

In Fig. 18.8 we also display the structure of the gaseous complex $H_2O \rightarrow HF$, where the water molecule acts as an electron donor [16]. Note that the coordination geometry of the oxygen atom is pyramidal as predicted from the VSEPR model. The dissociation energy at zero kelvin, $D_0 = 34$ kJ mol^{-1}, [17] is more than twice as large as for $HF \rightarrow HF$. Since an oxygen atom is less electronegative than a fluorine atom, it is perhaps not surprising that it should be a better electron donor.

We now turn our attention of hydrogen-bonded complexes of HCl with ammonia and trimethylamine. Direct combination of gaseous ammonia and hydrogen chloride results in the formation of solid ammonium chloride, $(NH_4)^+Cl^-$. The crystal structure of $(ND_4)Cl$ at 5 K has been determined by neutron diffraction [18]. The ammonium ions are found to be perfectly tetrahedral with N–D bond distances of 103 pm. Each of them is surrounded by eight chloride ions at the corners of a cube, with $N \cdots Cl$ distances of 382 pm. The cations are oriented in such a manner that the four N–D bonds are pointing towards alternating corners of the cube formed by the anions. This means that the coordination geometry of the H atoms is linear with $H \cdots Cl$ distances of 279 pm.

Ammonium chloride is relatively volatile for an ionic compound, the vapor pressure reaches 1 torr at 160 °C. The vapor is found to consist of hydrogen chloride, ammonia and a 1:1 complex NH_3HCl. The structure of the complex shows that it is held together by a hydrogen bond, $H_3N \rightarrow HCl$. See Fig. 18.9. The complex has C_{3v} symmetry, the H–Cl bond distance is 130 pm, perhaps a couple of pm longer than in the free molecule, the $N \rightarrow H$ distance is 183 pm, 80 pm longer than the N–D bonds in crystalline ND_4Cl [19,20]. The dissociation energy of the complex is $D_0 = 22$ kJ mol^{-1} [20].

Formation of solid ammonium chloride from the gaseous complexes may be divided into three steps. In the first, endothermic, step the complexes are dissociated to form neutral NH_3 and HCl molecules; in the second, undoubtedly also endothermic, protons are transferred from HCl to NH_3 molecules to form the separated NH_4^+ and Cl^- ions; in the third, *exothermic* step the ions are combined to form a crystal which is stabilized by strong Coulomb attractions between cations and anions. The system $NH_z{\cdot}HCl$, then, provides another demonstration of how intermolecular interactions is the solid phase may lead to more polar or ionic structure than in the gas phase.

We have seen that replacement of H atoms on the donor atom by methyl groups tends to increase the strength of a dative bond. One might therefore expect the dissociation energy of the complex $(CH_3)_3N \rightarrow HCl$ to be significantly higher than that of $H_3N \rightarrow HCl$, and

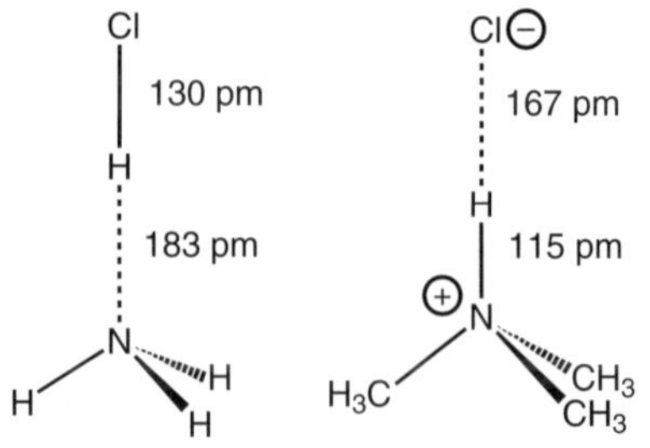

Fig. 18.9. "Recommended" molecular structures of the gaseous hydrogen-bonded complexes H_3NHCl and $(CH_3)_3NHCl$.

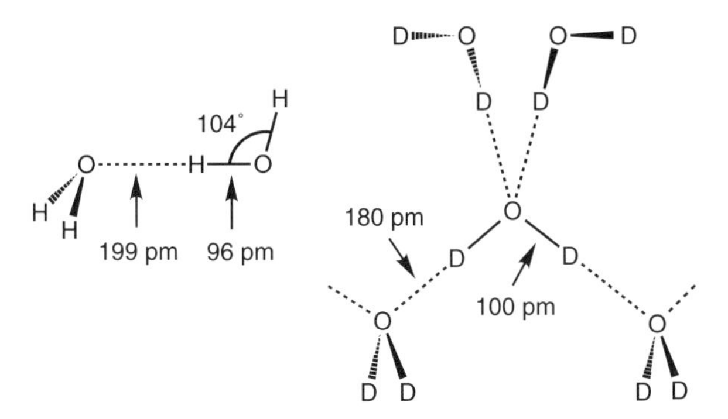

Fig. 18.10. Left: the structure of the water dimer in the gas phase. Right: a fragment of the structure of crystalline ice.

that the dative N→H distance should be significantly shorter. Consultation with Fig. 18.9 shows that introduction of the three methyl groups alters the structure radically: the N→H distance is reduced from 183 to 115 pm which is only about 12 pm longer than the N–H distance in the ammonium ion, while the H–Cl bond distance is increased from 130 to 167 pm [21,22]! These results indicate that the complex is best described as a gaseous ion pair: $[(CH_3)_3NH]^+Cl^-$. An accurate value for the dissociation energy to give $(CH_3)_3N$ and HCl is unfortunately not available, but quantum chemical calculations indicate that it is about 40 kJ mol^{-1} or nearly twice as large as that of H_3N→HCl [23].

The gas-phase structure of the water dimer is presented in Fig. 18.10 [24]. The three atoms of the acceptor molecule and the O atom of the donor are lying in a symmetry plane, and the O···H–O fragment appears to be linear. Both H–O–H valence angles are close to 104°. The dissociation energy at zero K is 15 kJ mol^{-1} [25,26,27].

18.8 The structures of ice and liquid water

Freezing water under atmospheric pressure yields hexagonal ice crystals held together by a three-dimensional network of hydrogen bonds. A neutron diffraction study of such crystals at 60 K shows that each water molecule interacts with four neighbors through linear O–H···O and O···H–O bonds [28]. See Fig. 18.10. The four oxygen atoms of the neighboring molecules form a nearly perfect tetrahedron. The covalent O–H bonds are a few picometers longer than in the gaseous dimer, the dative H→O bonds are 20 pm shorter than in the dimer. The sublimation energy of ice at 273 K is 48 kJ mol^{-1}. Since the number of hydrogen bonds is twice as large as the number of molecules in the crystal, this sublimation energy corresponds to a mean hydrogen bond dissociation energy of about 24 kJ mol^{-1}, about 50 % higher than in the gaseous dimer.

The structure of ice represents an interesting contrast to that of crystalline methane. See Section 9.12. Methane crystals are held together by weak dispersion forces, which are largely independent of the relative orientation of nearest neighbor molecules. The lattice is close-packed cubic, each CH_4 molecule is surrounded by 12 nearest neighbors at C···C distances of 316 pm. Ice crystals, on the other hand, are held together by hydrogen bonds.

Since the oxygen atom in each water molecule is covalently bonded to two hydrogen atoms and carries only two non-bonding electron pairs, the number of nearest neighbor interactions is reduced to four (there is no reason to doubt that there would have been room for more), and each molecule is oriented in such a manner that the strength of the hydrogen bonds is maximized. Since the hydrogen bonds are much stronger than dispersion attraction in crystalline methane, the nearest neighbor O···O distance of 280 pm is shorter than the corresponding distance in crystalline methane.

The energy of fusion of ice is 6.0 kJ mol^{-1}. Comparison with the energy of sublimation suggests that hydrogen bonds in the liquid state are about 10% weaker than in the solid state, alternatively that about 10% of the hydrogen bonds present in the ice are broken when the solid melts. We have seen that in liquid methane a few degrees above the melting point the average number of nearest neighbors has been reduced from 12 to about 9.5. X-ray investigations of liquid water at 4 °C, on the other hand, shows that the number of nearest neighbour molecules has *increased* from 4 to 4.4. Even though the average water molecule now interacts with a larger number of neighbors than before, the relative orientations of the molecules are so much less favorable for the formation of strong hydrogen bonds, that the total energy of the liquid is higher than that of the solid [29].

Problem 18.5 Are the structures of the gaseous water dimer and hexagonal ice crystals in agreement with the VSEPR model? Why are the O-H hydrogen bonds shorter and stronger in ice than in the gaseous dimer?

18.9 Polywater

In the early 1960s Russian scientists discovered that condensation of water vapor in narrow quartz capillary tubes with diameters less than 100 μm produced water with anomalous properties: the density was 20 or 30% higher than for normal water; the freezing point was at least 30° lower; and the boiling point at least 30° higher. Particularly startling was the observation that the vapor pressure was lower than for normal water at the same temperature; this implies that the "anomalous water" is thermodynamically more stable. Hence, with time, all water should be converted to the anomalous form. The unusual properties could, of course, be explained by impurities. Chemical analysis was made very difficult by the fact that less than 100 μg (10^{-4} g) of the mysterious liquid could be produced in a process which lasted for days. Anyway, no impurities were found.

After the leader of the Russian research group had given lectures on anomalous water in Britain during the summer of 1966, the experiments were repeated in several British research laboratories with the same results. From Britain the information about and interest in anomalous water spread to the US. Most chemists appear to have been deeply skeptical about the existence of a previously unknown, thermodynamically more stable form of water. On the other hand, *if* the existence of such a form could be established, it might prove to be the "the most important physical-chemical discovery of the century."

In 1969 a group of American spectroscopists published the infrared (IR) and Raman spectra of anomalous water [30]. The observed spectra did *not* correspond to those of normal H_2O molecules. The spectra were therefore interpreted as evidence for the presence of water *polymers* held together by hydrogen bonds with linear O–H–O fragments; two equal O–H bond distances of about 115 pm; and O···O distances of about 230 pm, as compared to 280 pm in ice. The strength of such short, symmetric hydrogen bonds was estimated to be at least 20 kJ mol^{-1} greater than that of the asymmetric O-H···O bonds found in normal water or ice.

The apex of research on polymeric water or "polywater" was reached in the following year when a group of quantum chemists published calculations that seemed to indicate that polywater consists of strands of fused, planar and hexagonal O_6H_6 rings held together by short, symmetric O–H–O bonds [31]. See Fig. 18.11. Each oxygen atoms in one strand forms one or two symmetric hydrogen bonds to oxygen atoms in other strands.

Two years later polywater was dead. Several groups had published studies showing that polywater was indeed solutions of silicates and/or organic materials. The spectroscopists had published a declaration to the effect that their samples had been impure, and the quantum chemists had found that more accurate calculations showed that the polymeric structure with symmetric O-H-O bonds had *higher* energy than normal water [32].

Was the polywater episode a "scandal?" This author does not think so. A difficult experiment had led to an unexpected result that was be reproduced in other laboratories. The original interpretation was unorthodox, but not impossible. A sufficient number of scientists accepted the challenge, and the error was rectified in less than a dozen years.

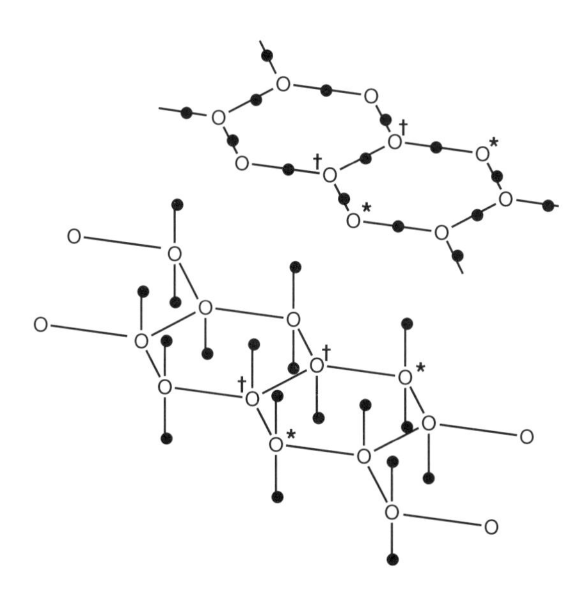

Fig. 18.11. Suggested structure of polywater [31]. Above: fused O_6H_6 rings in a polymeric strand. Bridging H atoms are indicated by black dots. Oxygen atoms marked with a dagger (†) form three hydrogen bonds to O atoms in the strand, oxygen atoms marked by an asterisk (*) form two. Below: hydrogen bonds between strands. The oxygen atoms in other strands have been omitted for clarity. Oxygen atoms marked with a dagger form one hydrogen bond to an O atom in another strand, oxygen atoms marked by an asterisk form two. All O–H distances within a strand are about 115 pm, O–H bonds connecting strands are a few pm longer.

References

[1] K. O. Christe, E. C. Curtis, and D. A. Dixon, *J. Am. Chem. Soc.*, 115 (1993) 1520.
[2] R. Marx, A. R. Mahjoub, K. Seppelt, and R. M. Ibberson, *J. Chem. Phys.*, 101 (1994) 585.
[3] R. J. Gillespie and I. Hargittai, *The VSEPR Model of Molecular Geometry*, Allyn and Bacon, Boston, 1991.
[4] E. R. Waclawik and A. C. Legon, *Phys. Chem. Chem. Phys.*, 1 (1999) 4695.
[5] J.-W. Zou, Y.-J. Jiang, M. Guo, G.-X. Hu, B. Zhang, H.-C. Liu, and Q.-S. Yu, *Chem. Eur. J.*, 11 (2005) 740.
[6] O. Kh. Poleshchuk, V. Branchadell, B. Brycki, A. V. Fateev, and A. C. Legon, *J. Mol. Struct., THEOCHEM*, 760 (2006) 175.
[7] S. E. Novick, K. C. Janda, and W. Klemperer, *J. Chem. Phys.*, 65 (1976) 5115.
[8] J. R. Pliego, S. M. Resende, and W. B. De Almeido, *Theor. Chim. Acta*, 93 (1996) 333 and references therein.
[9] For a review of the gas phase structures of complexes of dihalogen molecules with electron donors see A. C. Legon, *Angew. Chem. Int. Ed*, 38 (1999) 2686.
[10] R. Minkwitz and M. Berkei, *Z. Naturforsch.*, 54b (1999) 1615.
[11] H. Hartl, H. Bärnighausen, and J. Jander, *Z. anorg. allg. Chemie*, 357 (1968) 225.
[12] K. A. Peterson and T. H. Dunning, J. Chem. Phys., 102 (1995) 2032.
[13] E. J. Bohac, M. D. Marshall, and R. E. Miller, *J. Chem. Phys.*, 96 (1992) 6681.
[14] For a review of the gas phase structures hydrogen-bonded complexes of hydrogen halides see A. C. Legon, *Angew. Chem. Int. Ed*, 38 (1999) 2686.
[15] M. W. Johnson, E. Sándor, and E. Arzi, *Acta Cryst.*, B31 (1975) 1998.
[16] Z. Kisiel, A. C. Legon, and D. C. Millen, *Proc. Roy. Soc. (London)*, A 381 (1982) 419.
[17] A. C. Legon, D. J. Millen, and H. M. North, *Chem. Phys. Lett.*, 135 (1987) 303.
[18] A. Alavi, R. M. Lynden-Bell, P. A. Willis, I. P. Swainson, and R. J. C. Brown, *Can. J. Chem.*, 76 (1998) 1581.
[19] N. W. Howard and A. C. Legon, *J. Chem. Phys.*, 88 (1988) 4694.
[20] G. Corongiu, D. Estrin, G. Murgia, L. Paglieri, L. Pisani, G. S. Valli, J. D. Watts, and E. Clementi, *Int. J. Quant. Chem.*, 59 (1996) 119.
[21] A. C. Legon and C. A. Rego, *J. Chem. Phys.*, 90 (1989) 6867.
[22] G. B. Bacskay and J. S. Craw, *Chem. Phys.* Lett., 221 (1994) 167.
[23] C. S. Brauer, M. B. Craddock, J. Kilian, E. M. Grumstrup, M. C. Orilall, Y. Mo, J. Gao, and K. R. Leopold, *J. Phys. Chem.* A, 110 (2006) 10025.
[24] J. A. Odutola and T. R. Dyke, *J. Chem. Phys.*, 72 (1980) 5062.
[25] S. S. Xantheas and T. H. Dunning, *J. Chem. Phys.*, 99 (1993) 8774.
[26] L. A. Curtiss, D. J. Frurip, and M. Blander, *J. Chem. Phys.*, 71 (1979) 2703.
[27] J. Reimers, R. Watts, and M. Klein, *Chem. Phys.*, 64 (1982) 95.
[28] W. F. Kuhs and M. S. Lehmann, *J. Phys. Chem.*, 87 (1983) 4312.
[29] For a fuller discussion of the structure of liquid water see F. Franks, *Water*, The Royal Society of Chemistry, London, 1983.
[30] E. R. Lippincott, R. R. Stromberg, W. H. Grant, and G. L. Cessac, *Science*, 164 (1969) 1482.
[31] L. C. Allen and P. A. Kollman, *Science*, 167 (1970) 1443.
[32] For a fuller account of the polywater affair see F. Franks, *Polywater*, The MIT Press, Cambridge, Mass., 1981.

Chapter 19

Structure and bonding in simple compounds of the noble gases

Introduction

The first element in Group 18, helium, has the electron configuration $1s^2$, the second, neon, the configuration $[He]2s^22p^6$, while the heavier elements, argon, krypton and xenon, have the electron configurations $[Ng](n-1)d^{10}ns^2np^6$. According to Lewis' electron octet model, the noble gases should not form compounds at all, and after unsuccessful attempts to prepare halides of Ar, Kr or Xe had been reported in the early 1930s, the failure of the noble gases to form chemical compounds was taken for granted by most chemists. They were referred to collectively as the "inert" rather than the "noble" gases, and they were assigned the group number zero, presumably to convey that they had no valence electrons. A very influential textbook published in 1962 stated categorically that the noble gases "form no chemically bound compounds" [1]. A very useful model, i.e. the electron octet model, had become an impediment to scientific progress in this particular area. Yet, in the very year when this textbook was published, three research groups reported the successful preparation of three different xenon compounds: the solid, ionic compound $Xe^+[PtF_6]^-$ (not yet completely characterized) could be synthesized by direct combination of gaseous PtF_6 and Xe [2]; XeF_2 could be synthesized by electric discharge through a mixture of gaseous Xe and F_2; and XeF_4 by heating a mixture of Xe and F_2 to 400°C. The synthesis of krypton diflouride was published in the following year [3], while the first argon compound, HArF, was characterized in a low temperature matrix in year 2000 [4,5].

19.1 Hypervalent compounds of the noble gases

The molecular structures of the five homoleptic noble gas compounds that are stable enough to be studied in the gas phase are shown in Fig. 19.1. If the noble gas atoms are assumed to form single electron pair bonds to the fluorine atoms, and if the Xe–O bonds in XeO_4 are assumed to be double (four electron bonds), then the K and Xe atoms in the trifluorides are surrounded by 10 electrons in the valence shell, the Xe atom in XeF_4 by 12, the Xe atom in XeF_6 by 14, and in XeO_4 by 16 electrons. All these compounds may thus be described as containing hypervalent noble gas atoms.

Problem 19.1 Discuss the structure and bonding in KrF_2, XeF_2 and XeF_4 in terms of the VSEPR and MO models.

The mean Xe–F bond distances decrease linearly with the number of fluorine atoms from 198 in the difluoride to 189 pm in the hexafluoride. This decrease of the bond distances

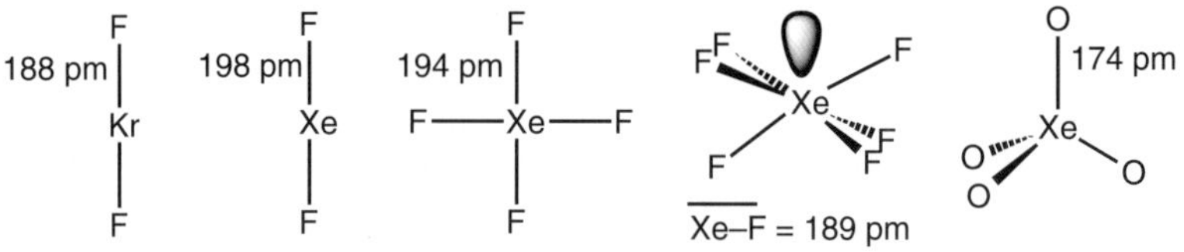

Fig. 19.1. Gas phase structures KrF_2 and XeF_2 (both linear), XeF_4 (square planar) and XeF_6 (distorted octahedral?), and XeO_4 (tetrahedral).

might be interpreted as a sign that the bonds become stronger, but both experimental and calculated mean bond energies indicate that they remain essentially constant or become slightly weaker with increasing number of fluorine substituents: The MBEs of XeF_2, XeF_4 and XeF_6, obtained by high level quantum chemical calculation of the atomization energies, are 126, 122 and 116 kJ mol^{-1} respectively. The experimental counterparts are 130, 130 and 125 kJ mol^{-1} [6].

The Lewis formula of xenon hexafluoride suggests that the central atom is surrounded by a total of seven electron pairs, viz. six bond pairs and one lone pair. The molecule thus has two more valence electrons than TeF_6. Several investigations published in the 1960s concluded that electron diffraction data were inconsistent with reasonably rigid octahedral models. These observations may be interpreted by assuming an equilibrium structure of C_{3v} symmetry where a stereochemically active electron lone pair caps one of the triangular surfaces of a distorted F_6 octahedron. See Fig. 19.1. This C_{3v} structure contains two sets of symmetry inequivalent F atoms: three atoms define the triangle capped by electron lone pair and three the triangle opposite the lone pair. Gas phase ^{19}F NMR spectra, however, show that all F atoms are NMR equivalent [7]. So, if the equilibrium structure has C_{3v} symmetry, the molecule must undergo rapid geometric rearrangements as the lone pair moves from on triangular face of a distorted F_6 octahedron to another.

Alternatively the observed gas electron diffraction patterns may be interpreted in terms of an extremely flexible octahedral equilibrium structure in which the energy required to deform the molecule under C_{3v} symmetry is so low compared to the thermal energy that a significant fraction of the molecules have instantaneous structures far from octahedral symmetry during the brief ($<10^{-16}$ s) electron scattering process.

Very recent, state of the art quantum chemical calculations led to the conclusion that the energy difference between octahedral and capped octahedral structures is less than 1 kJ mol^{-1} [6]. This difference is too small to allow us to draw a conclusion as to which structure is the lower. The calculations and experimental evidence thus agree that the energy difference between octahedral and capped octahedral structures is so small that XeF_6 gas at a given instant will contain a significant number of molecules with either geometry, but we still do not know which of them represents the equilibrium structure.

The Xe–O bond distance in the tetraoxide, 174 pm, or 15 pm shorter than the mean Xe–F bond distance in XeF_6, is consistent with a description in terms of double Xe=O bonds. So is the tetrahedral coordination geometry. In the next and final chapter we shall describe and discuss the structures of oxides of some other main group elements, viz. carbon, nitrogen, phosphorus, sulfur and chlorine.

19.2 Two compounds of Xe^+

$XeCH_3$ has never been observed as a neutral molecule, but the cation has been detected in mass spectra. The Xe–C bond dissociation energy, defined as the energy of the reaction

$$XeCH_3^+(g) = Xe(g) + CH_3^+(g)$$

is 228 ± 10 kJ mol^{-1} [8]. Should we be surprised? Perhaps not: Xe^+ is isoelectronic with iodine, and the I-C bond dissociation energy of ICH_3 is 232 kJ mol^{-1}.

Similarly, the Xe–C bond distance in the pentafluorophenyl xenon cation, $[F_5C_6Xe]^+$, is 208 pm [9] as compared to a I–C bond distance of 213 pm in ICH_3.

References

[1] F. A. Cotton and G. Wilkinson, *Advanced Inorganic Chemistry, Interscience*, New York, 1962.
[2] N. Bartlett, *Proc. Chem. Soc.*, 1962, 218.
[3] J. J. Turner, and G. C. Pimentel, *Science*, 140 (1963) 974 and D. R. MacKenzie, *Science*, 141 (1963) 1171.
[4] L. Khriachtchev, M. Peterson, N. Runeberg, J. Lundell, and M. Räsänen, *Nature*, 406 (2000) 874.
[5] For a fuller account of noble gas chemistry up to and including 1962 see P. Laszlo and G. J. Schrobilgen, *Angew. Chem. Int. Ed. Engl.*, 27 (1988) 479.
[6] D. A. Dixon, W. A. de Jong, K. A. Peterson, K. O. Criste, and G. J. Schroblingen, *J. Am. Chem. Soc.*, 127 (2005) 8627.
[7] M. Gerken, P. Hazendonk, J. Nieboer, and G. J. Schrobilgen, *J. Fluorine Chem.*, 125 (2004) 1163.
[8] J. K. Hovey and T. B. McMahon, *J. Am. Chem. Soc.*, 108 (1986) 528.
[9] H. J. Frohn, A. Klose, T. Schroer, G. Henkel, V. Buss, D. Opitz, and R. Vahrenhorst, *Inorg. Chem.*, 37 (1998) 4884.

Chapter 20

Structure and bonding in oxides and oxoacids of carbon, sulfur, nitrogen, phosphorus, and chlorine

Introduction

While the properties and structures of hydrogen or halogen derivatives of main group elements generally vary in a regular manner as each group is descended, this is not the case for oxides of the elements in Groups 14–17. In particular the composition and structures of the oxides of the first member of each group (the second period element) often differ markedly from the composition and structures of the oxides of the third and later period elements. Some examples:

(i) While carbon dioxide is a gaseous molecule at normal temperatures and pressures, silicon dioxide is a nonvolatile solid. In the CO_2 molecule the carbon atom forms two double bonds to the oxygen atoms, in solid silicon dioxide each Si atom forms single bonds to the four nearest O atoms.
(ii) Dinitrogen trioxide is a planar molecule, $ONNO_2$, with three NO double bonds and a very weak NN single bond. Diphosphorus trioxide, on the other hand, forms dimeric species, P_4O_6, where each P is joined to three O atoms through single bonds.
(iii) Nitrogen forms the oxoacid HNO_3. The analogous oxoacid of phosphorus is unknown. Phosphorus, on the other hand forms H_3PO_4 for which no N analog is known.
(iv) Sulfur forms a dioxide and a trioxide. The oxygen analog of the dioxide is known (ozone), but no oxygen analog of SO_3 is known.
(v) Sulfur forms the oxoacid H_2SO_4, no analogous O compound is known.
(vi) Chlorine forms the dioxide ClO_2, the heptoxide, Cl_2O_7, and the oxoacid $HClO_4$. No analogous compounds of F are known.

The composition and properties of the oxides and oxoacids of the heavier elements of the group are generally similar to those formed by the third-period element.

20.1 Carbon oxides

Carbon forms three gaseous oxides, carbon monoxide (CO), carbon dioxide (CO_2), and carbon suboxide (C_3O_2).

The three atoms in carbon dioxide are connected in the sequence OCO and the molecule is linear. The C–O bond distance is 116 pm as compared to the single C–O bond distance in dimethylether 142 pm, and the mean C–O bond energy is 805 kJ mol^{-1} as compared to a mean C–O bond dissociation energy of 358 kJ mol^{-1} in $(CH_3)_2O$.

The only reasonable Lewis structure for CO_2, viz. ::O::C::O::, indicates the formation of double bonds between C and O and the presence of two non-bonding electron pairs at

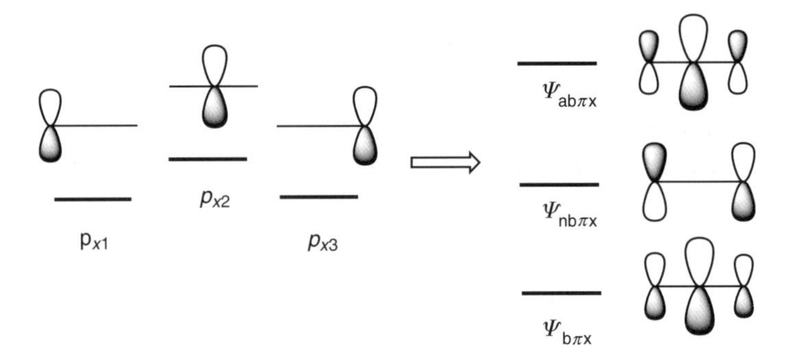

Fig. 20.1. Formation of three-center π molecular orbitals in CO_2 by linear combination of the $2p_x$ AOs of the three atoms. Since the electronegativity of the O atoms is greater than that of C, their $2p_x$ orbital energies are lower.

each O atom. This Lewis structure is consistent not only with the short and strong bonds, but also with the linear structure.

Molecular orbital calculations suggest the following (somewhat simplified) description of the bonding: two electron pairs occupy the essentially non-bonding $2s$ orbitals on the O atoms. The C atom is sp^1 hybridized. Each sp^1 hybrid combines with a $2p_z$ ($2p_\sigma$) orbital on one of the O atoms to form a $2c$ σ-bonding orbital which is filled by another electron pair. We have now accounted for eight valence electrons, the remaining eight enter π orbitals formed by combination of the $2p_x$ or $2p_y$orbitals of the three atoms. See Fig. 20.1. The three $2p_x$ orbitals combine to form one $3c$ π-bonding orbital, Ψ_{bpx}, one nonbonding π-orbital Ψ_{nbpx}, and one $3c$ π-antibonding orbital Ψ_{abpx}. The bonding and the non-bonding MOs are filled by two electrons each. The last four electrons enter the analogous bonding and non-bonding π MOs formed from the three $2p_y$ AOs. This MO description is in agreement with the Lewis structure in so far as four electron pairs are assigned to bonding molecular orbitals while four pairs are assigned to non-bonding orbitals.

The carbon monoxide molecule is isoelectronic with N_2, and one possible Lewis structure is **:C:::O:** corresponding to a triple bond between the atoms. This Lewis structure suggests that the CO bond in this molecule should be significantly stronger and shorter than in CO_2. If the six bonding electrons are assumed to be equally shared between the atoms, the formal charge of the C atom is -1 and on the O atom $+1$. These formal charges suggest that the molecule may have a large electric dipole moment.

An alternative Lewis structure, **:C::O::**, yields formal charges equal to zero for the two atoms and suggest that the CO bond should be of the same length and strength as in CO_2. The bond dissociation energy of CO is in fact 1077 kJ mol^{-1}, 34% larger than the MBE of 805 kJ mol^{-1} in CO_2. The bond distance is 113 pm as compared to 116 pm in CO_2. The electric dipole moment is very small ($\mu_{el} = 0.11$ Debye).

The large difference between the dissociation energy of CO and the mean bond energy in CO_2 is by itself sufficient to justify a description of the CO bond as triple in accordance with the Lewis structure **:C:::O:**.

The valence electron configuration of the N_2 molecule is $2s\sigma_b^2 2s\sigma_a^2 2p\pi_b^4 2p\sigma_b^2$ where each of the molecular orbitals is formed by combination of one valence shell atomic orbital from each atom. See Section 9.5. Molecular orbital calculations on the CO molecule yield very similar results. Four valence electrons are found to occupy the bonding and antibonding MOs formed by linear combination of the valence shell s atomic orbitals of the two atoms, and may be regarded as effectively nonbonding. The net bonding is due to one electron pair occupying a bonding orbital $2p\sigma_b$ formed by combination of the two valence shell p_σ orbitals, and to two electron pairs occupying the two bonding MOs formed by combination of the four valence shell p_π atomic orbitals. This description is consistent with the Lewis structure :C:::O:.

Finally we turn our attention to the suboxide. Experiments show that the atoms are connected in the sequence OCCCO. The Lewis structure of is presumably ::O::C::C::C::O::. The observed C–C and C–O bond distances are indeed consistent with a double bond description O=C=C=C=O. The VSEPR model then leads to the prediction that the molecule is linear. This is not strictly true: the molecule is bent at the central atom with $\angle CCC = 156°$! The potential energy maximum at $\angle CCC = 180°$ (see Fig. 9.1) is, however, less than 0.25 kJ mol^{-1} above the minimum, and the molecule is therefore often described as *pseudolinear* [1]. Why the small energy hump at $\angle CCC = 180°$ is there, nobody seems to know.

Problem 20.1 The coordination geometry of the carbon atom in H_2CO is trigonal planar. Write down a reasonable Lewis structure and suggest a description of the bonding in terms of $2c, 2e$ bonds.

20.2 Some sulfur oxides

Sulfur forms at least 13 binary oxides of composition S_mO_n. We shall, however, confine our discussion to the three simplest, viz. SO, SO_2 and SO_3. The monoxide is very unstable, but may be generated in the gas phase, and is sufficiently long-lived to be characterized structurally before disintegrating. The SO bond distance is 148 pm, about 10% shorter than the single bond S–O distance in HSOH, 166 pm [2]. The dissociation energy 519 kJ mol^{-1}. The SO bond is thus stronger than the double bonds in O_2 ($D = 496$ kJ mol^{-1}) or S_2 ($D = 423$ kJ mol^{-1}).

The SO molecule has the same number of valence electrons as the O_2 molecule. The valence electron configuration of the latter is $2s\sigma_b^2 2s\sigma_a^2 2p\pi_b^4 2p\sigma_b^2 2p\pi_a^2$ where each of the molecular orbitals is formed by combination of one valence shell atomic orbital from each atom. See Section 9.6. A similar set of molecular orbitals for SO is obtained by combination of the valence shell orbitals of sulfur and oxygen, and the electron configuration may be written as $s\sigma_b^2 s\sigma_a^2 p\pi_b^4 p\sigma_b^2 p\pi_a^2$. This means that SO, like O_2, is a paramagnetic molecule with two unpaired electrons, and that the net bonding is due to two electrons in a σ-bonding and two electrons in a π-bonding orbital. The bond distance, the bond dissociation energy, and the electron configuration all suggest a description in terms of a double S=O bond and a Lewis structure in which each of the two atoms carry two non-bonding electron pairs: ::S::O::.

The molecular structure of sulfur dioxide is angular, while the structure of the trioxide is trigonal planar. See Fig. 20.2. It is noteworthy that the bond distances in SO_2 and SO_3 are 5 or 6 pm *shorter* than in the monoxide, while the mean bond energies in SO_2 and SO_3 are respectively 3% *greater* and 9% *smaller* than the dissociation energy of the monoxide. These observations indicate that the SO bonds in these molecules are best described as double.

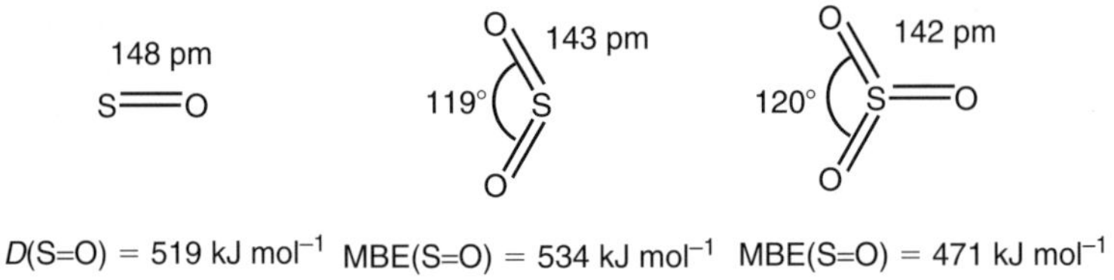

148 pm 143 pm 142 pm

S=O 119° 120°

D(S=O) = 519 kJ mol^{-1} MBE(S=O) = 534 kJ mol^{-1} MBE(S=O) = 471 kJ mol^{-1}

Fig. 20.2. The gas phase molecular structures and S=O bond energies of sulfur oxide, dioxide and trioxide.

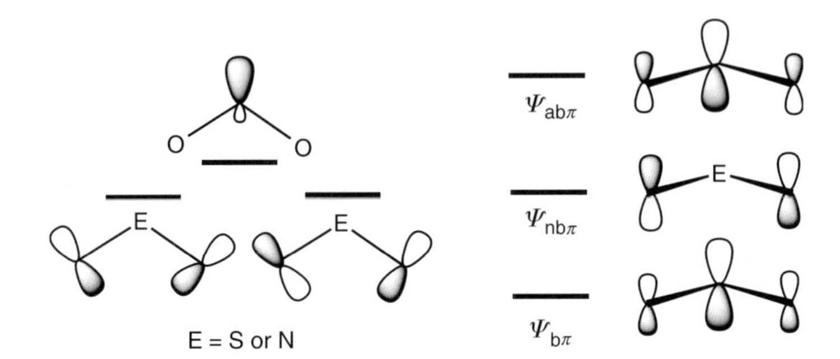

Fig. 20.3. Schematic representation of some non-bonding and π-bonding molecular orbitals in SO_2 or NO_2. Left: the non-bonding sp^2 atomic orbital on the central atom and two delocalized lone pair orbitals formed by linear combinations of oxygen p orbitals in the molecular plane. Right: the π molecular orbitals formed by linear combination of the valence shell p_π AOs of the three atoms. Comparison with the π orbitals of CO_2 (Fig. 20.1) shows that the change from a linear to an angular structure has little effect on the π orbitals perpendicular to the molecular plane.

This in turn suggests that the sulfur atom in the dioxide should be described as tetravalent (as in SF_4) and the sulfur atom in SO_3 as hexavalent (as in SF_6).

The simplest Lewis formula for SO_2 would place two electron pairs between the S and each of the two O atoms, one electron lone pair at the S and two electron lone pairs at each O atom: :S(::O::)$_2$. This Lewis formula implies that the sulfur atom in SO_2 accomodates five electron pairs in the valence shell (as in SF_4). Similarly the simplest Lewis formula of the trioxide would place two electron pairs between the S and each O and two non-bonding electron pairs at each of the latter, S(::O::)$_3$, indicating that the S atom in SO_3 accomodates six electron pairs in its valence shell (as in SF_6) [3]. Note that both the angular structure of SO_2 and the trigonal planar structure of SO_3 are in agreement with the VSEPR model.

Molecular orbital calculations on SO_2 suggest the following description of the bonding. Two valence electron pairs occupy the essentially nonbonding 2s orbitals on the oxygen atoms. The S atom is approximately sp^2 hybridized. Two sp^2 hybrids combine with p orbitals on the two O atoms to form two 2c σ-bonding orbitals which are filled by two electron pairs. We have now accounted for eight valence electrons. Six of the 10 remaining electrons occupy the three nonbonding orbitals shown to the left in Fig. 20.3, while four occupy the 3c π-bonding orbital $\Psi_{b\pi}$, and the nonbonding π-orbital $\Psi_{nb\pi}$ shown on the right. The rest of

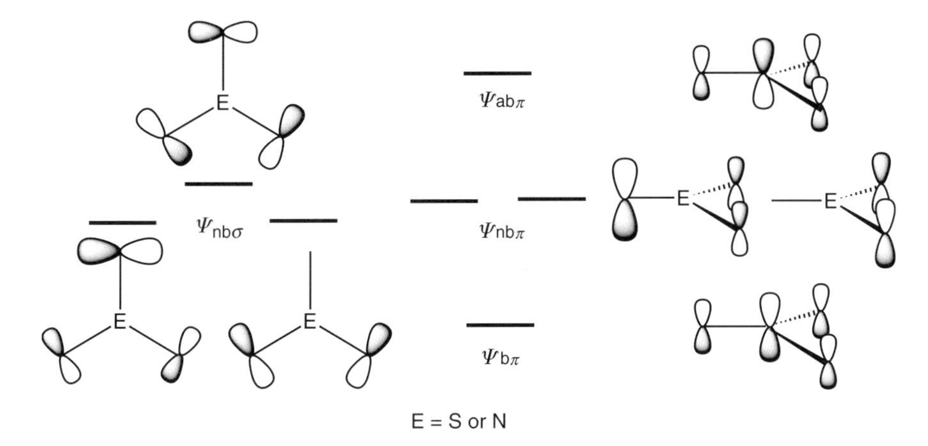

Fig. 20.4. A schematic representation of some non-bonding and π-bonding molecular orbitals in SO_3 or NO_3. Left: the three essentially non-bonding σ molecular orbitals formed by linear combinations of oxygen p orbitals in the molecular plane. In SO_3 these orbital are completely filled by six electrons, in NO_3 the highest, non-degenerate orbital contains a single electron. Right: the four π molecular orbitals formed by linear combination of the valence shell p_π AOs of the four atoms.

the valence electrons in the molecule occupy non-bonding orbitals confined to the oxygen atoms. Even though the Lewis formula implies that the sulfur atom in SO_2 accommodates five electron pairs in the valence shell, the molecular orbital description indicates that that it accommodates only four. While the SF_4 molecule was found to be held together by two $2c, 2e$ and one $3c, 4e$ σ-bond, the OSO molecule is held together by two $2c, 2e$ σ-bonds and one $3c, 4e$ π-bond.

The SO_2 molecule contains two more valence electrons than CO_2. If the structure had been linear, the two additional electrons would have had to enter a $3c$ *antibonding* π orbital. When the molecule bends and the S atom is rehybridized, the three π orbitals in the molecular plane are converted into the three non-bonding σ orbitals indicated at the left of Fig. 20.3.

A simple MO description of SO_3 is obtained by assuming the sulfur atom to be sp^2 hybridized. Three electron pairs occupy $2c$ SO σ-bonding orbitals Three electron pairs occupy the $2s$ atomic orbitals on the O atoms and three more pairs occupy three essentially non-bonding molecular orbitals formed by linear combination of the in-plane O p_σ orbitals perpendicular to the O–S bonds. See Fig. 20.4 left. Finally, the three remaining electron pairs occupy one four-center bonding and two degenerate, non-bonding π-orbitals shown to the right in Fig. 20.4.

20.3 Sulfur oxofluorides

Sulfur forms one oxofluoride in which the S atom is tetravalent and two oxofluorides in which it is hexavalent. See Fig. 20.5. The S=O bond distance in OSF_2 is about 3 pm shorter than in SO_2, the S–F bond distances 1 or 2 pm shorter than the mean S–F bond distance in SF_4. The S=O bond distances in O_2SF_2 or OSF_4 are 1 or 2 pm shorter than in SO_3, the S–F bond distances in the former are about 3 pm shorter than, the S–F bond distances in the

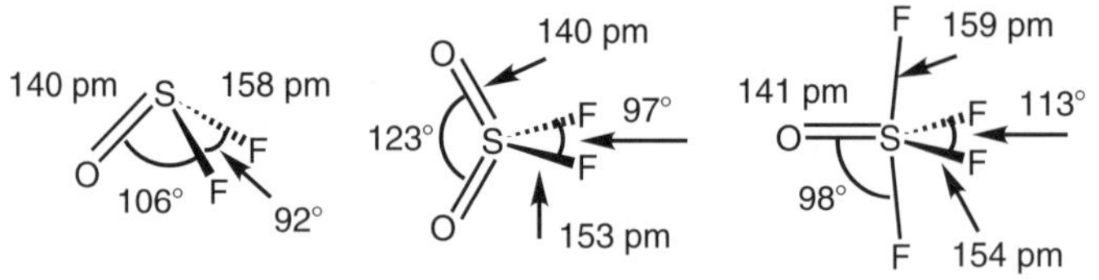

Fig. 20.5. The gas phase molecular structures of the oxofluorides OSF_2, O_2SF_2 and OSF_4.

latter are indistinguishable from those in SF_6. The bond distances in the oxofluorides are thus consistent with the assumption that the SO bonds are double and the SF bonds single.

The energies of atomization of the three molecules calculated from their enthalpies of formation are [4]

$$\Delta U_{\text{atom}}(OSF_2) = 1222 \text{ kJ mol}^{-1}$$

$$\Delta U_{\text{atom}}(O_2SF_2) = 1684 \text{ kJ mol}^{-1}$$

$$\Delta U_{\text{atom}}(OSF_4) = 1973 \text{ kJ mol}^{-1}$$

If we assume the SO bond energy in OSF_2 to be equal to the MBE in SO_2 and the mean S–F bond energy to be equal to the mean bond energy in SF_4, we obtain an estimated atomization energy of OSF_2 (1206 kJ mol^{-1}) which reproduces the experimental value with a deviation of just 1%. Similarly transfer of bond energies from SO_3 and SF_6 leads to an estimated atomization energy of atomization O_2SF_2 equal to 1594 kJ mol^{-1}, about 5% lower than the experimental value, and to an estimated atomization energy of OSF_4 which is about 10% lower than the experimental value, indicating that the S=O and S–F bonds in the latter molecule are – if anything – stronger than in our reference molecules. Like the bond distances, the energies of atomization of the oxofluorides indicate that the SO bonds are double and the S–F bonds single.

Problem 20.2 Draw Lewis structures of the three sulfur oxofluorides in Fig. 20.5. Are the structures in agreement with the VSEPR model?

20.4 Sulfuric acid and related compounds

The gas phase molecular structure of sulfuric acid is shown in Fig. 20.6. The molecule has a twofold symmetry axis bisecting the O–S–O and O=S=O valence angles. Comparison of the S=O bond distance and O=S=O valence angle with the corresponding parameters in O_2SF_2 as well as comparison of the valence angles spanned by the single S–O or S–F bonds, indicates that bonding in the two molecules must be very similar. In the crystalline phase each sulfur acid molecule is linked to four neighboring molecules through four S=O$\cdots$H–O hydrogen bonds [5]. The four O$\cdots$H distances are about 200 pm, i.e. very similar to that found in the water dimer.

Gaseous H_2SO_4 reacts with one water molecule to form a complex in which the two molecules are joined by two hydrogen bonds, one of them very short [6]. See Fig. 20.6. As expected, the shorter hydrogen bond is the one characterized by the larger O$\cdots$H–O angle.

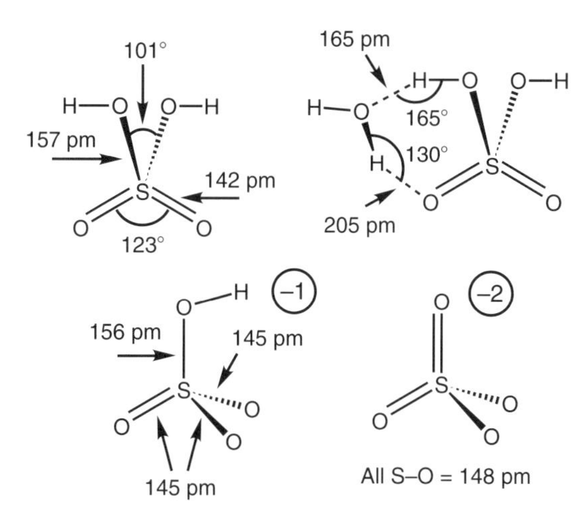

Fig. 20.6. Above: the gas phase structures of sulfuric acid and sulfuric acid monohydrate. Below: the structure of the hydrogen sulfate ion in crystalline $(H_3O)(HSO_4)$ and of the sulfate ion in crystalline $(H_3O)_2(SO_4)$ $2H_2O$.

When this complex enters the solid state, the proton in the stronger bond is completely transferred from the acid to the water molecule: X-ray diffraction shows that the crystal consists of oxonium hydrogensulphate, $(H_3O)(HSO_3)$. The O–H bond distance and the HOH valence angle of the pyramidal oxonium ion are similar to those listed in Table 17.4. The structure of the hydrogen sulfate anion is shown in Fig. 20.6 [5]. Three SO bond distances are about 3 pm longer than the S=O double bond distance in H_2SO_4, the single S–OH bond distance is essentially unchanged. As in crystalline H_2SO_4, all H atoms are involved in hydrogen bonding. As a consequence, cations and anions are connected through an intricate three-dimensional network of hydrogen bonds.

Sulfuric acid dihydrate, $H_2SO_4 \cdot 2H_2O$, crystallizes as dioxonium sulphate. The sulfate ions are tetrahedral with S=O bond distances of 148 pm, six pm longer than the S=O double bond distance in gaseous H_2SO_4, but indistinguishable from the bond distance in gaseous sulfur monoxide [7]. All the oxonium hydrogen atoms are engaged in O–H$\cdots$O=S hydrogen bonding.

20.5 Mono-nitrogen oxides and their N–N bonded dimers

Nitrogen forms no less than eight binary oxides that are sufficiently stable have their molecular structures determined in the gas phase, viz. the three mono-nitrogen oxides $NO_k, k = 1$, 2 or 3, and the five di-nitrogen oxides $N_2O_k, k = 1-5$.

Before proceeding to a discussion of their structures, we pause to establish a set of reference N–N, N–O single and N=O double bond distances and bond energies. See Fig. 20.7. The reference N–N single bond dissociation energy is calculated as the energy of the reaction

$$N_2H_4(g) = 2NH_2(g)$$

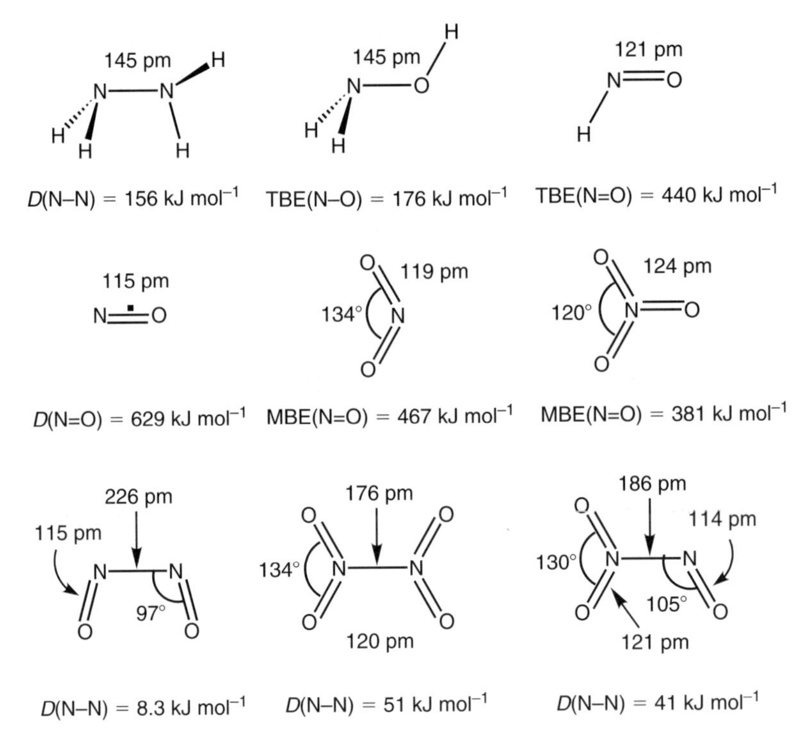

Fig. 20.7. First row: bond distances and bond energies in prototype molecules with N–N and N–O single and N=O double bonds. Second row: bond distances and NO bond energies in the three gaseous mononitrogen oxides. Third row: bond distances and N–N bond dissociation energies of three gaseous dinitrogen oxides.

while reference thermochemical N–O and N=O bond energies are calculated from the energies of atomization of hydroxylamine (H_2NOH) and nitrosyl hydride (HNO) respectively by transferring mean N–H and O–H bond energies from ammonia and water [8,9].

The three dinitrogen oxides N_2O_2, N_2O_3 and N_2O_4 form molecular crystals at low temperatures. Evaporation of N_2O_2 leads to reversible dissociation according to

$$N_2O_2(g) = 2NO(g)$$

with the mole fraction of the monomer (NO) increasing with increasing temperature and decreasing pressure. Similarly evaporation of N_2O_4 yields a gaseous mixture of monomer (NO_2) and dimer (N_2O_4), while evaporation of solid dinitrogen trioxide yields a gaseous mixture of undissociated N_2O_3, NO, and NO_2. It has, however, been possible to determine the structures of both monomers and dimers by spectroscopic techniques. See Fig. 20.7. We begin by discussing the structures of monomeric NO and NO_2.

As noted by Lewis in 1917, the NO, and NO_2 molecules are among the handful of main group element molecules with an odd number of electrons that are stable in the gas phase at normal temperatures and pressures. The bond distance in the monoxide, 115 pm, is about 6 pm shorter than our reference value for a N=O double bond distance, the dissociation energy, 629 kJ mol^{-1}, is about 40% higher than the reference value for a double NO bond [10]. The NO molecule has one electron more than N_2. The additional electron enters an antibonding π-orbital to yield the valence shell electron configuration $2s\sigma_b^2 2s\sigma_a^2 2p\pi_b^4 2p\sigma_b^2 2p\pi_a^1$. The additional electron is expected to cancel the bonding effect of one of the four electrons in π-bonding orbitals, and the net bonding to be due to *five* electrons in bonding MOs. This molecular orbital description indicates a Lewis structure in which the nitrogen atom carries one electron lone pair, the oxygen atom two, and five electrons – corresponding to a bond order of 2.5 – are placed between the two nuclei: **:N.::O::**. This description is consistent with both the high dissociation energy and the short bond.

The structure of NO_2 is angular, the NO bond distance and the mean bond energy are both reasonably close to our reference values for double bonds. The bond distance and bond energy thus suggest a Lewis structure where each oxygen atom carries two electron lone pairs and the nitrogen a single, unpaired electron, while two electrons pairs are placed between the nitrogen and each of the oxygen atoms, **·N(::O::)**$_2$. Note that this Lewis formula yields formal atomic charges equal to zero. The NO_2 molecule has a total of 17 valence electrons, while SO_2 has 18. The Lewis formulae of the two compounds differ only in the number of non-bonding electrons assigned to the central atom, in NO_2 there is one, in SO_2 two. The molecular orbital description of NO_2 is similar to that of SO_2, the only difference being that the non-bonding p_σ orbital on N contains one rather than two electrons. See Fig 20.3.

Problem 20.3 Compare the O=E=O valence angles in NO_2 and SO_2. Can the difference be explained in terms of the VSEPR model?

No nitrogen oxide of composition NO_3 has ever been isolated. The NO_3 molecule may, however, be generated by gas phase reactions like

$$NO_2(g) + O_3(g) = NO_3(g) + O_2(g)$$

and is sufficiently long-lived to be characterized structurally before it disintegrates. The molecule is trigonal planar. The MBE is 35% larger than the reference value for an N–O single bond and 15% lower than the reference value for an N=O double bond. The bond distance is 21 pm shorter than the reference value for a single bond and 3 pm longer than the reference value for an N=O double bond. We choose, therefore, to describe the three NO bonds in the molecule as double like the SO bonds in SO_3. Since NO_3 has one valence electron less than SO_3, this description would imply that one of the three oxygen atoms carries three, rather than four non-bonding electrons.

Molecular orbital calculations on NO_3 indicate that one electron pair occupies a symmetric four-center π-bonding orbital, while two electron pairs occupy non-bonding π orbitals localized on the three oxygen atoms. See Fig. 20.4 right. The unpaired electron is found to occupy a three-center non-bonding p_σ orbital localized on the oxygen atoms. See Fig. 20.4 left since this orbital is non-degenerate, no Jahn–Teller distortion of the molecule is expected [12].

Problem 20.4 Is the Lewis structure **.:O::N(::O::)**$_2$ consistent with the bond distances, bond energies and molecular orbital description of NO_3? Can you find a better Lewis structure?

Problem 20.5 Suggest reasonable Lewis structures for the molecules HNO, N_2O_2, N_2O_3 and N_2O_4 under the assumption that all NO bonds are double.

We now turn our attention to the structure of the dimers, N_2O_2, N_2O_3, and N_2O_4. See Fig. 20.7. The three molecules are all planar, N–O bond distances and ONO valence angles are very close to those found in the mononitrogen oxides. As should be clear from Problem 20.5, reasonable Lewis structures can be drawn for these molecules under the assumption that all N–O bonds are double and all N–N bonds single. On this background the observed N–N bond distances are surprisingly long: The N–N distance in N_2O_4 is 176 pm, 31 pm longer than our reference value, in N_2O_5 it is 41 pm longer,[13] and in N_2O_2 it is 81 pm or 56% longer than the reference value!

The long N–N bond distances are matched by extraordinarily small dissociation energies according to

$$O_mN{-}NO_n(g) = NO_m(g) + NO_n(g).$$

The dissociation energy of ONNO is just 8 kJ mol^{-1} [10], the dissociation energy of $ONNO_2$ 36 kJ mol^{-1} [14], and that of O_2NNO_2 53 kJ mol^{-1} [15]. Thus the dissociation energy of the strongest of the three bonds amounts to no more than one third of our reference value for a single N–N bond!

20.6 Hartree–Fock and configuration interaction calculations on ONNO and O_2NNO_2

Hartree–Fock (HF) molecular orbital calculations on ONNO, $ONNO_2$ and O_2NNO_2 yield N–N bond distances of about 160 pm [16]. Since the calculations fail to reproduce the observed bond distances, examination of the wavefunctions obtained can hardly be expected to yield useful information about the nature of the bonding. Indeed, the experimental bond distances can only be reproduced if the wavefunctions are modified to include electron correlation effects. This may be done in several ways. The best method is probably to write the wavefunction of each molecule as a linear combination of Slater determinants. The first determinant corresponds to the ground state HF wavefunction, the others to excited electron configurations where one or more electrons have been promoted to higher energy molecular orbitals. Calculations based on wavefunctions formed by linear combinations of such Slater determinants are referred to as Configuration Interaction (CI) calculations.

CI calculations on N_2O_4 with a wavefunction consisting of 976 determinants does indeed reproduce both the experimental bond distance as well as the dissociation energy [17]. As expected the largest coefficient (0.90) is that of the determinant corresponding to the ground state electron configuration, the second largest coefficient (0.18) is that of the determinant in which two electrons have been promoted from the N–N σ bonding to the N–N antibonding σ orbital. Such promotion is expected to increase the N–N bond distance and decrease the dissociation energy.

The close agreement between the CI calculations and experiment has, however, been obtained at a prize: The assumption – firmly imbedded in the minds of many chemists – that a molecule may be adequately described in terms of a single, unique electron configuration has been abandoned.

Similar CI calculations on N_2O_2 have been equally successful [18].

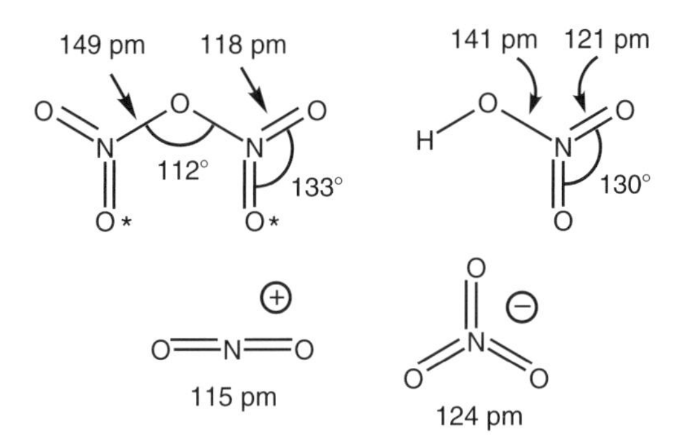

Fig. 20.8. Above: the molecular structures of dinitrogen pentoxide and nitric acid in the gas phase. Below: the structures of the nitryl cation and the nitrate anion found in crystalline N_2O_5.

20.7 Nitric acid and nitric acid anhydride (N_2O_5)

Gaseous N_2O_5 molecules consist of two planar NO_3 fragments joined at an O atom. See Fig. 20.8. The NO_2 units are, however, rotated some 30° out of the plane of the paper, presumably to relieve steric repulsion between the two oxygen atoms indicated by asterisks. The N–O single bond distance is 5 pm longer than our reference value. The mean N–O single bond dissociation energy, defined as half the energy of the reaction

$$N_2O_5(g) = O(g) + 2NO_2(g),$$

is 151 kJ mol^{-1}, about 15% smaller than our reference value for an NO single bond.

The nitric acid molecule is perfectly planar. The N–O single bond distance is identical to our reference value. The NO double bond distance is equal to our reference NO double bond distance in HNO.

The structure of nitric acid is retained in the crystalline phase. The anhydride, however, crystallizes as an ionic compound, nitryl nitrate or $(NO_2)^+(NO_3)^-$. The nitryl cation, which is isoelectronic with carbon dioxide, is linear. The nitrate ion which, is valence shell isoelectronic with sulfur trioxide, is trigonal planar.

Problem 20.6 The NO bond distance in the nitryl cation is 4 pm shorter than in the neutral NO_2 molecule, while the bond distance in the nitrate anion is equal to that in the neutral molecule. Can these differences be explained in terms of the molecular orbital model?

20.8 Dinitrogen monoxide

The three atoms in N_2O are connected in the sequence NNO and the molecule is linear. See Fig. 20.9. The NN bond distance is about 3 pm longer than in the N_2 molecule and about 12 pm shorter than the prototype NN double bond in HN=NH. The NO bond distance is 2 pm shorter than our reference value for a double NO bond distance.

What about the bond energies? The energy of atomization of N_2O is 1108 kJ mol^{-1}. The sum of the dissociation energy of the N_2 molecule and our reference value for an N=O double bond energy is 1385 kJ mol^{-1}. It is clear, then, that one or both bonds in N_2O are

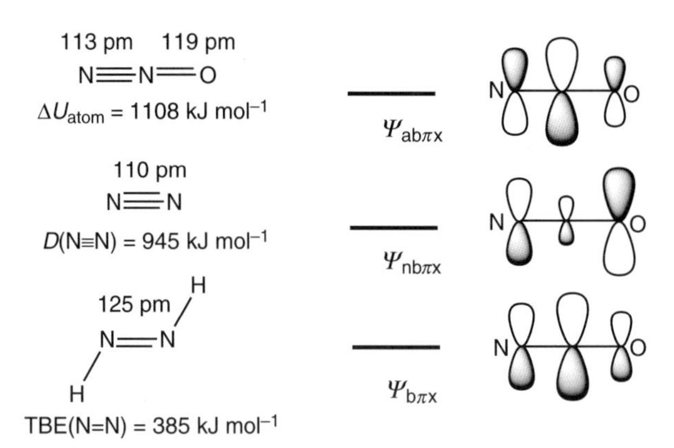

Fig. 20.9. Left: the molecular structures of N_2O, N_2 and HNNH in the gas phase; the energy of atomization of N_2O; the dissociation energy of N_2, and the thermochemical NN bond energy of HNNH. Right: the three-center π molecular orbitals in NNO formed by linear combination of the p_x AOs of the three atoms. Compare with the corresponding orbitals in CO_2, Fig. 20.1.

weaker than in the reference molecules. Combination of the energy of atomization with the reference value for an N=O double bond energy yields a thermochemical NN bond energy of 668 kJ mol^{-1}. This estimate is nearly 30% lower than the dissociation energy of N_2, but nearly 75% higher than the N=N double bond energy calculated from the energy of atomization of HN=NH by transfer of the mean N–H bond energy in NH_3. We conclude, therefore, that both the bond distances and the energy of atomization of N_2O are consistent with a triple NN bond and a double NO bond as indicated in the figure.

Note that N_2O molecule is isoelectronic with carbon dioxide. As in the case of CO_2, molecular orbital calculations suggest that two electron pairs occupy essentially non-bonding 2s orbitals at the terminal atoms, and that the central atom is sp^1 hybridized. Each sp^1 hybrid combines with a $2p_\sigma$ AO on one of the terminal atoms to form a 2c bonding orbital which accommodates an electron pair. The three $2p_{\pi x}$ orbitals combine to form a three-center π orbital that is strongly bonding between the central N atom and both terminal atoms. This orbital is occupied by an electron pair while another pair enters the intermediate-energy π-orbital indicated in the figure. The non-bonding character of the intermediate-energy π orbital of CO_2 is imposed by the molecular symmetry, which requires that the coefficient of the C atom $2p_\pi$ orbital is zero. No such requirement is imposed in the less symmetric N_2O molecule. The coefficient of the $2p_x$ orbital on the central N atom is small, but not zero, and the resulting orbital is weakly bonding between the N atoms and weakly antibonding between the N and O atoms. See Fig. 20.9. Finally four electrons enter the two lowest π MOs formed by linear combination of the three $2p_y$ AOs.

20.9 Phosphorus oxides

Phosphorus forms five oxides of composition P_4O_n, $n = 6, 7, 8, 9$ or 10. The molecular structures of the first and the last member of the series have been determined in the gas phase. See Fig. 20.10. The four phosphorus atoms in tetraphosphorus hexoxide, P_4O_6, are

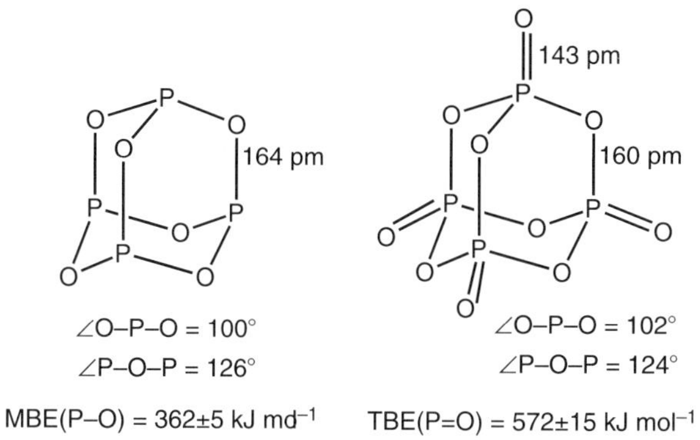

164 pm

∠O–P–O = 100°

∠P–O–P = 126°

MBE(P–O) = 362±5 kJ md^{-1}

143 pm

160 pm

∠O–P–O = 102°

∠P–O–P = 124°

TBE(P=O) = 572±15 kJ mol^{-1}

Fig. 20.10. The gas phase molecular structures and PO bond energies of P_4O_6 and P_4O_{10}.

situated at the corners of a perfect tetrahedron and connected through symmetric P–O–P bridges along the edges. The O atoms are clearly divalent and the P atoms trivalent.

The 1998 edition of the standard English language compilation of thermodynamic data, the *NIST-JANAF Thermochemical Tables*, gives the standard enthalpy of formation of gaseous P_4O_6 at 298 K as -2214 ± 34 kJ mol^{-1} [19]. This value is based on an experimental determination of the enthalpy of formation for the crystalline compound published in 1952. This determination was 11 years later shown to be in error by about 600 kJ mol^{-1} [20]! (The reader has now been warned: Even the most authoritative compilation of experimental data may contain entries that are in serious error!) The new value for the enthalpy of formation of the solid leads to a gas phase enthalpy of formation of $\Delta_f H° = -1606 \pm 40$ kJ mol^{-1} [21]. Calculation of the energy of atomization and division by 12, yields a P–O mean bond energy of MBE(P–O) = 362 kJ mol^{-1}.

The structure of the tetraphosphorus decoxide, P_4O_{10}, is closely related to that of the hexoxide: the four additional oxygen atoms have entered terminal positions at the four phosphorus atoms. The terminal PO bonds are 17 pm shorter than the bridging. The enthalpy of formation of the decoxide is well known. Calculation of the energy of atomization and transfer of the P–O single bond energies from P_4O_6 yields a thermochemical P=O double bond energy of 572 kJ mol^{-1}. Thus both the bond distance and the bond energy indicate that the terminal bonds are double.

The solid face structures of the intermediate members of the series P_4O_n show that they differ only in the number of terminal, doubly bonded O atoms.

In Fig. 20.11 we compare the structures of PF_3 and PCl_3 with the structures of the corresponding phosphoryl trihalides, OPX_3. Both the PO bond distances and the thermochemical bond energies calculated by transfer of the P–X bond energies from the trihalides, indicate that the PO bonds are best described as double.

We end our discussion of oxides, oxohalides, and oxoacids of C, N, P, S, and Cl by noting that in all the compounds that we have considered, two-coordinate oxygen atoms are best described as forming two single bonds, while (with the possible exception of the O atoms in ClO_2) all mono-coordinate oxygen atoms are best described as doubly bonded.

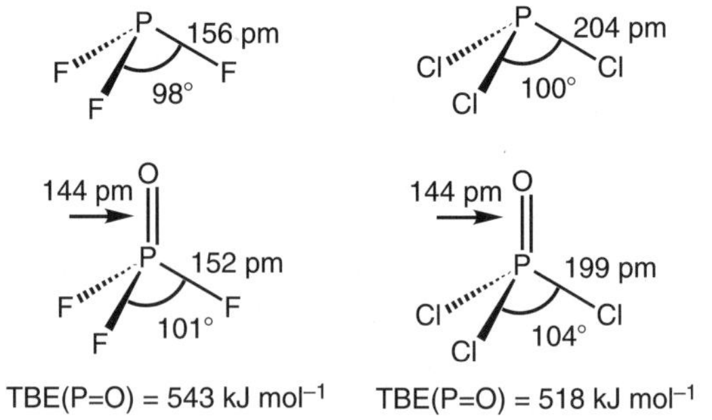

Fig. 20.11. The molecular structures and bond energies in phosphorus trihalides and phosphoryl trihalides.

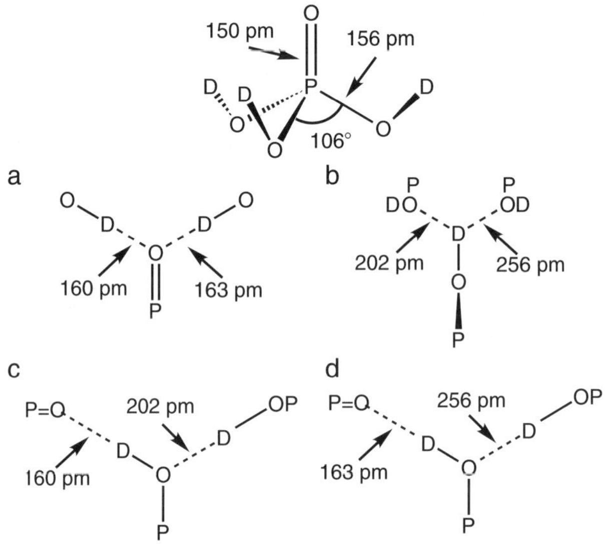

Fig. 20.12. Above: the crystal structure of deuterated orthophosphoric acid, D_3PO_4. The molecule has an approximate threefold symmetry axis through the P=O bond. Below: the intermolecular hydrogen bonds. (a) The phosphoryl group in each molecule forms two very short hydrogen bonds to hydroxyl groups in neighboring molecules; (b) The D atom in one of the three hydroxyl groups forms *two* relatively long hydrogen bonds to hydroxyl groups in neighboring molecules; (c) and (d) Each of the two remaining OD groups participates in one very short hydrogen bond to a phosphoryl group and one relatively long hydrogen bond to the hydroxyl group depicted in (b).

20.10 The crystal structure of orthophosphoric acid

The gas phase structure of H_3PO_4 is unknown, but the crystal structure of deuterated orthophosphoric acid, D_3PO_4, has been determined by neutron diffraction [22]. See Fig. 20.12. Each D_3PO_4 molecule forms eight hydrogen bonds to neighboring molecules. The four O–D$\cdots$O $=$ P interactions with D$\cdots$O distances of about 162 pm may stabilize the crystal with as much as 50 kJ mol^{-1} each. Two O–D$\cdots$OD interactions with D$\cdots$O distances are about 200 pm, i.e. similar to those found in the water dimer, may stabilize the crystal by about 15 kJ mol^{-1} each. The last two hydrogen bonds with O$\cdots$D distances equal to 256 pm are presumably very weak.

The most surprising aspect of the structure is the short PO single bond distance of 155 pm, 5 pm shorter than the corresponding distance in P_4O_{10}, and the long PO double bond distance, 150 pm or 7 pm longer than in P_4O_{10}. In the decoxide the single and double bond distances differ by 17 pm, in the crystalline acid the difference has been reduced to about 5 pm. Quantum chemical calculations indicate that the single and double PO bond distances in an isolated H_3PO_4 molecule should differ by 12 or 13 pm.[23] It seems likely, therefore, that the PO double bond in the crystalline phase is significantly lengthened, and the P–O single bond significantly shortened, by the hydrogen bonds.

20.11 Chlorine oxides and perchloric acid

The molecular structures of the three chlorine oxides that are sufficiently stable to be investigated in the gas phase are shown in Fig. 20.13. In Chapter 17 we described the OCl bond in oxygen dichloride as a single bond. The ClO bond in ClO_2 is seen to be 23 pm shorter and 25% stronger than in OCl_2. We choose therefore to follow the majority of chemists and describe the ClO bonds in ClO_2 as double.

ClO_2 has 19 valence electrons, one more than SO_2. Molecular orbital calculations show that the additional electron has entered the antibonding π orbital, $\Psi_{ab\pi}$. See Fig. 20.3. The molecule is thus held together by four electrons in two $2c$ σ-bonding orbitals plus *one* electron in a $3c$ π-orbital. In this book we have used bond strength and bond length as criteria for distinguishing between single and double bonds. Other chemists would use Lewis structures or the number of electrons in bonding orbitals obtained by MO calculations. In most cases the two sets of criteria give the same results, but OCl_2 represents a dilemma: If you define the multiplicity of a bond in terms of the number of electrons in bonding orbitals, a net number of five bonding electrons for two OCl bonds would indicate that the bonds are best described as single. If you define a double bond as one that is significantly

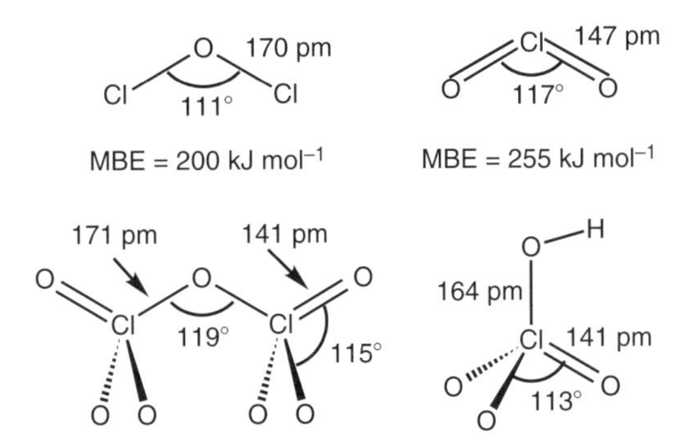

Fig. 20.13. The molecular structures of dichlorine oxide (or oxygen dichloride), chlorine dioxide, dichlorine heptoxide and perchloric acid in the gas phase.

shorter and stronger than a single bond, then the bonds are best described as double. There is no common agreement at this point, so the reader may choose for herself!

The bridging O atom in the heptoxide, Cl_2O_7, is bonded to two Cl atom with bond distances close to those in OCl_2. See Fig. 20.13. In addition each Cl atom is bonded to three terminal chlorines with bond distances about 6 pm shorter than those in ClO_2. The bridging O–Cl bonds are best described as single, the terminal as double. Similarly the bond distances in perchloric acid suggest that the Cl–O(H) bond is single while the three terminal ClO bonds are double.

We end our discussion of oxides, oxohalides, and oxoacids of C, N, P, S, and Cl by noting that in all the compounds that we have considered, two-coordinate oxygen atoms are best described as forming two single bonds, while (with the possible exception of the O atoms in ClO_2) all mono-coordinate oxygen atoms are best described as doubly bonded.

20.12 The shape of triatomic molecules formed from elements in Groups 14 to 18

In Table 20.1 we compare the valence angles of some representative triatomic molecules or ions formed from elements in Groups 14 to 18. It is seen that the shapes of the molecules are determined by the number of valence electrons: carbon dioxide and all other 16 valence electron species are linear, and species with 17 to 20 valence electrons are angular. When the number of valence electrons is increased to 22, the shape reverts to linearity. The author is not aware of any exception to this trend.

CO_2 and other 16-electron species are held together by two $2c, 2e$ σ-bonds and two $3c, 4e$ π-bonding systems. Retention of a linear structure for the 17-electron molecule NO_2 or an 18-electron molecule like SO_2 would require the 17th and 18th valence electrons to enter antibonding π orbitals. One of the three-center π-bonding systems is therefore broken by distortion to an angular structure, while the other π system remains intact. The additional electron in the 19-electron molecule ClO_2 resides in the remaining $3c$ antibonding π orbital. Retention of the π-bonding system in a 20-electron molecule like OCl_2 or SF_2 would require *two* electrons to occupy the $3c$ antibonding π orbital. In these molecules the π interactions cannot be broken by transition from a linear to an angular structure, but the antibonding is reduced by reduction of the valence angle and elongation of the bond distances. As we have seen, the OCl bond in OCl_2 is 23 pm longer than in ClO_2, and the SF bond in SF_2 is 16 pm longer than the SO bond in SO_2 (despite the fact that F has a smaller bonding radius than O and that the electronegativity difference between S and F is greater than between S and O). The structures of the 20-electron molecules OCl_2 or SF_2 are therefore described in terms of single O–Cl or S–F σ bonds.

Table 20.1. Some triatomic molecules ABC formed from elements in Groups 14 to 18: the total number of valence electrons and ABC valence angles.

ABC	No. val. el.	∠ABC	ABC	No. val. el.	∠ABC
OCO	16	180°	OSO	18	119°
NNO	16	180°	FNO	18	110°
FCN	16	180°	ClSnCl	18	99°
ONO^+	16	180°	OClO	19	117°
ONO	17	134°	ClOCl	20	111°
ONO^-	18	118°	FXeF	22	180°

The number of valence shell atomic orbitals in OCl_2 or SF_2 (12) is two larger than the total number of valence electron pairs (10). This is just enough to allow the formation of two $2c$ σ-bonding orbitals, while eight electron pairs are lodged in non-bonding AOs. In the 22-electron molecule XeF_2 the number of valence shell AOs exceeds the number of electron pairs by *one*. This is only enough to allow the formation of a $3c, 4e$ σ-bonding system as described in Section 15.12.

References

[1] J. Koput, *Chem. Phys. Lett.*, 320 (2000) 237.
[2] O. Baum, S. Esser, N. Gierse, S. Bruenken, F. Lewen, J. Hahn, J. Gauss, S. Schlemmer, and T. F. Giesen, *J. Mol. Struct.*, 795 (2006) 256.
[3] For a discussion of alternative Lewis formulae see D. K. Straub, *J. Chem. Ed.*, 72 (1995) 889.
[4] The enthalpy of formation of OSF_4(g) was taken from S. T. Arnold, T. M. Miller, and A. A. Viggiano, *J. Phys. Chem. A*, 106 (2002) 9900.
[5] E. Kemnitz, C. Werner, and S. Trojanov, *Acta Cryst*, C52 (1996) 2665.
[6] D. L. Fiacco, S. W. Hunt, and K. R. Leopold, *J. Am. Chem. Soc.*, 124 (2002) 4504.
[7] I. Taesler and I. Olovsson, *J. Chem. Phys.*, 51 (1969) 4213.
[8] The enthalpy of formation of hydroxylamine was taken from R. S. Saraf, W. J. Rogers, M. S. Mannan, M. B. Hall, and L. M. Thomson, *J. Phys. Chem. A*, 107 (2003) 1077.
[9] The enthalpy of formation of nitrosyl hydride was taken from W. R. Anderson, *Combustion and Flame*, 117 (1999) 394.
[10] E. Wade, J. I. Cline, K. T. Lorenz, C. Hayden, and D. W. Chandler, *J. Chem. Phys.*, 116 (2002) 4755.
[11] The enthalpy of formation of nitrogen trioxide was taken from H. F. Davis, B. Kim, H. S. Johnston, and Y. T. Lee, *J. Phys. Chem.*, 97 (1993) 2172.
[12] W. Eisfeld and K. Morokuma, *J. Chem. Phys.*, 113 (2000) 5587.
[13] J. L. Domenech, A. M. Andrews, C. P. Belov, G. T. Fraser, and W. J. Lafferty, *J. Chem. Phys.*, 100 (1994) 6993.
[14] The enthalpy of formation of $ONNO_2$ was taken from L. V. Gurvich, I. T. Veyts, and C. B. Alcock, *Thermodynamic Properties of Individual Substances*, Hemisphere, New York, 1991.
[15] F. R. Ornellas, S. M. Resende, F. B. C. Machado, and O. Roberto-Neto, *J. Chem. Phys.*, 118 (2003) 4060.
[16] SCF calculations with a 6-311++G** basis yields an N–N bond distance equal to 161 pm in ONNO, 159 pm in O_2NNO_2 and 160 pm in O_2NNO, A. Haaland and N. Strenalyuk, unpublished results.
[17] C. W. Bauschlicher, A. Kormicki, and B. Roos, *J. Am. Chem. Soc.*, 105 (1983) 745. For the results of more recent calculations on N_2O_4 see S. S. Wesolowski, J. T. Fermann, T. D. Crawford, and H. F. Schaefer, *J. Chem. Phys.*, 1997 (106) 7178 and references therein.
[18] M. Tobita, S. A. Perera, M. Musial, R. J. Bartlett, M. Nooijen, and J. S. Lee, *J. Chem. Phys.*, 119 (2003) 10713 and references therein.
[19] *NIST-JANAF Thermochemical Tables*, Fourth Edition, M. W. Chase, Jr., Ed., *J. Phys. Chem.* Ref. Data, Monograph No. 9, 1998.
[20] S. B. Hartley and J. C. McCoubrey, *Nature*, 198 (1963) 476.
[21] *Thermodynamic Properties of Specific Substances*, V. P. Glushko Ed., NAUK Publishers, Moscow, 1978.
[22] R. H. Blessing, *Acta Cryst.*, B44 (1988) 334.
[23] DFT calculations at the B3LYP/6-311++G** level, N. Strenalyuk and A. Haaland, unpublished result.

Index